Global Conference on Recent Developments in Computer and Communication Technologies (GC-RDCT 2021)

IOP Conference Series: Materials Science and Engineering
Volume 1187

Mangalore, India
29 – 30 July 2021

ISBN: 978-1-7138-4647-5
ISSN: 1757-8981

Printed from e-media with permission by:

Curran Associates, Inc.
57 Morehouse Lane
Red Hook, NY 12571

Some format issues inherent in the e-media version may also appear in this print version.

Institute of Physics
Dirac House, Temple Back
Bristol BS1 6BE UK

Phone: 44 1 17 929 7481
Fax: 44 1 17 920 0979

techtracking@iop.org

Additional copies of this publication are available from:

Curran Associates, Inc.
57 Morehouse Lane
Red Hook, NY 12571 USA
Phone: 845-758-0400
Fax: 845-758-2633
Email: curran@proceedings.com
Web: www.proceedings.com

TABLE OF CONTENTS

Preface

Peer review declaration

Group Leverage Centrality and its Applications in Brain Networks 1
Amogh Gupta, Harish Balaji, R. Sundareswaran, V. Mahesh, B. Geetanjali

Simulation of Different Third Harmonic Injected PWM Strategies for 5-Level Diode-Clamped Multilevel Inverter 10
Lakshmi Swarupa Malladi, Shravani Chapala, Sarada Veluri

High Performance Wallace Tree Multiplier Using Majority Gate Based Adders 19
M Mummudi Murasu, Sanjana Sujith, A Anita Angeline, P. Sasi Priya, V S Kanchana Bhaaskaran

Improving the Relation Classification Using Convolutional Neural Network 24
S Kamath, K G Karibasappa, Anvitha Reddy, Arati M Kallur, B B Priyanka, B P Bhagya

Application of Natural Betalain Dye from Beetroot for Improved Efficiency in Dye Sensitized Solar Cell 30
K Soni, A Sheikh, V Jain, N Lakshmi

Integrated Multi-Operational Antenna System Design for CR Applications 42
Rajeev K. Parida, Arjuna Muduli, Dhruba C. Panda, Deepak Kumar Nayak

Power-efficient Strategies for Sensing in Autonomous Mobile Robots, a critical requirement of I4.0 standard 50
Jyoti Bali, H Arpitha, N Anushree, Arunkumar Giriyapur

Low Power Memory System Design Using Power Gated SRAM Cell 60
Srijani Pal, Divya S Salimath, Banusha Chandran, A Anita Angeline, V S Kanchana Bhaaskaran

Replacing two As atoms in silicene nanoribbons in the presence of an external electric field............ 70
Hoang Van Ngoc

Efficient Waste Management Models for Smart Cities: A Case Study 77
K. Reji Kumar, E.N. Satheesh

Design and simulation of Sierpmski Triangular Fractal Antenna for S-Band 84
Jahanvi H. Vyas, NB Gohil, Ved Vyas Dwivedi

Deep Neural Network for Electroencephalogram based Emotion Recognition 90
Shruti Garg, Soumyajit Behera, K Rahul Patro, Ashwani Garg

Real-time Flow Visualization using Projection Mapping Technique 102
T Nishanka, D Prajwal, M Nishmitha, Vinayambika S Bhat

Restoration of degraded underwater images with attenuation coefficients as a clue............ 114
C S Dikshith, Sameeksha M Vernekar, Chaitra Desai, Ujwala Patil, Uma Mudenagudi

Design and evaluation of hybrid SHE+STT-MTJ/CMOS full adder based on LIM architecture 126
Prashanth Barla, Vinod Kumar Joshi, Somashekara Bhat

Design & Simulation of S Band Planar Microwave Filter for Wireless Application 135
Bhumikaba Gohil, Khyati Chavda

A Different Shaped Radiating Element Wide Band Multi-Band Massive MIMO Antenna for 5G/WLAN applications with Enhanced Correlation Coefficient 144
Shrenik Suresh Sarade, S. D. Ruikar

Influence of pH and salinity on the photocatalytic dye degradation and heavy metal ion reduction with cobalt ferrite photocatalysts 156
Anna Heba Chakkunny, Bintu Thomas, Libu K Alexander

Mathematical model and analysis of PV Converter - Inverter System 163
K R Pushpa, R S Geetha

Ring-Based Quantum Network with Quantum Key Distribution (QKD) and Quantum Digital Signature (QDS) 175
Osman Ceylan, Ihsan Yilmaz

Analysis of an algorithm used for finding Unipolar Orthogonal Codes 184
J N Vinutha, Hitesh Trivedi, Suraiya Tarannum, M N Harshavardhana

Modelling of partial discharge pulse propagation in transformer windings using multi conductor transmission line modelling scheme 192
Annie Achankunju, Jineeth Joseph, Ancy Sara Varghese, K R M Nair

Simulation Tests of Tokyo Quantum Network 202
Osman Ceylan, Ihsan Yilmaz

Selection of an efficient wind turbine generator at Government science college premises Chitradurga, India 209
G H Smitha patil, R Megha, Y T Ravikiran, M Prashantkumar, K Udaykumar, B G Kumaraswamy

Low-compute facial expression recognition using fiducial feature-sets 215
A.V. Divekar, D.C. Gharpure

Symmetric Key Cryptosystem based on Sequential State Machine 223
Basappa B Kodada, Demian Antony D'Mello

UWB dual notch implementation using folded bi-section stepped impedance resonator 233
Thomas George, B Lethakumary, M A Abdul Hakeem

A Novel approach in Hybrid Median Filtering for Denoising Medical images 238
R. Seetharaman, M. Tharun, K. Anandan

Comparison of word embeddings in text classification based on RNN and CNN 247
Merlin Susan David, Shini Renjith

Automatic Detection of New Vessels in Retinal Photographs using Machine Learning 256
M Moorthi, M. Gopikrishnan, V Pooja, M R Ragavi Priya, D K Vasanthakumar

Fruit Classification Comparison Based on CNN and YOLO 265
Riyanshu Raj, S S Nagaraj, Saurav Ritesh, T A Thushar, V M Aparanji

Development of data driven adaptive edge detectors for image processing 272
H G Likhith, H L Nishanth, E Lohit, M S Rudramurthy, S A Sushma, T G Keerthan Kumar

Assessment of Asphalt Road Pavement Rehabilitation and Maintenance Practices: The Case of West Showa Zone in Oromia, Ethiopia 289
Lensa Merga Kenea, Leevesh Kumar, Sitesh Kumar Singh, Geleta Chala, Jifara Chimdi

Effect of Thermal Barrier Coating in Sustainable Power production of Gas Turbines 297
S Ghosh, M Saha, S Bakshi, S Mondal

Comparison of D-Vine and R-Vine Techniques for Virtual Network Embedding Problem 305
Preksha Satish, Deeksha Lingraj, S Anjan Kumar, T G Keerthan Kumar

Power Allocation Scheme for Non-Orthogonal Multiple Access System 313
R. Seetharaman, M. Tharun, S. Gayathri, K. Anandan

Author Index

Preface

Computer and Communication Technology (CCT) are now involved in governmental, industrial, and business domains more than ever. Thus, it is important for CCT personnel to continue academic research to improve technology and its adoption to modern applications. The up-to-date research and technologies must be distributed to researchers and CCT community continuously to aid future development. The intent of the activity is to let the participants acquaint with transcendental growth, recent trends, innovations and security issues involved in the domain of communication technologies, high performance computing, big data, social media, hardware & software design, advanced software engineering, Internet of Things (IoT), e-governance etc. and their impact on societal applications through various brainstorming sessions. The conference will facilitate the young researchers, industries and research agencies especially, those, who are carrying out their research work in the aforesaid domain of Computer Science, Information Technology, Electronics and Communication Engineering with valuable discussions in order to make the outcomes more realistic. The technical committee of GC-RDCT received 135 papers, out of which 48 are shortlisted for the Conference for oral. Through this Conference, we have tried to bring in researchers from various universities and institutions to interact and deliberate within and get ideas for their future scientific endeavors. Due to the prevailing COVID-19 pandemic, the conference was organized in FULL VIRTUAL MODE. The conference was held in Google Meet online platform. All the speakers and the presenters have been connected through goggle meet platform and the inaugural function of the conference was held physically at AJIET and was live streamed through Google Meet and Youtube.

I am pleased to acknowledge the unstinting support from my colleagues at AJIET, the International Advisory, the Technical Program, and the Local Organization Committee members. I am incredibly grateful to our Management, Principal & Administrative Staffs, for providing required support.

Dr. Nagesh H R
Organizing Chair, GC-RDCT 2021

IOP Conf. Series: Materials Science and Engineering 1187 (2021) 011001 doi:10.1088/1757-899X/1187/1/011001

Committee

Chief Patron
Dr. A. J.Shetty
President, LMET, Mangalore

Patron
Mr. Prashanth Shetty
Vice President, LMET, Mangalore

Conference Chair
Dr. Shantharama Rai C.
Principal, AJIET, Mangalore

Organizing Chair
Dr. Nagesh H R, AJIET, Mangalore

Information Contact
Dr. Govindraj P, AJIET, Mangalore
Dr. Gnane Swarnadh Satapathi, AJIET, Mangalore

Publicity and Promotions
Mr. Basappa K, AJIET, Mangalore
Dr. Sujaya C, AJIET, Mangalore
Mr. Mohan A. R., AJIET, Mangalore
Mrs. Thrapthi Shetty, AJIET, Mangalore

Local Arrangements Committee
Mr. John Prakash Veigas., AJIET, Mangalore
Mr. Manjukiran B., AJIET, Mangalore
Mr. Ragavendra H, AJIET, Mangalore
Mr. Prateeksha N Rai, AJIET, Mangalore

Finance Committee
Dr. Sadananda Kumar N, AJIET, Mangalore
Mrs. Sharon C. D'Souza, AJIET, Mangalore

Sponsorship Committee
Prof. Prakash K Shetty, AJIET, Mangalore
Mr. Kiran Kumar V G, AJIET, Mangalore
Mr. Chanchal Antony, AJIET, Mangalore

Editorial Committee
Dr. Sangeetha D M, AJIET, Mangalore
Dr. Anotony P J, AJIET, Mangalore
Dr. G. Swarnadh Satapathi, AJIET, Mangalore

Dr. Govindaraju , AJIET, Mangalore
Dr. Sankappa Rai, AJIET, Mangalore

Technical Program Committee – Chairs
Prof. K. Chandrasekaran, National Institute of Technology Surathkal, Karnataka
Dr. Varadraj P. Gurupur, University of Central Florida, Orlando, United States

Technical Program Committee
Dr. Das B B, National Institute of Technology, Surathkal, Karnataka
Dr. Neena S P Panandikar, Don Bosco College of Engineering, Goa
Dr. Amjad Anvari Moghaddam, Aalborg University, Denmark
Dr. Rashed Al Mamun, Sylhet Agricultural University, Bangladesh
Dr. Goutham Sarang, VIT University, Chennai
Dr. Anotony P J, AJIET, Mangalore
Dr. Murugappan, Kwait institute of Science and Technology, Kwait
Dr. Manjaiah, National Institute of Technology Warangal, Andhrapradesh
Dr. Shrikanth S Rao, National Institute of Technology, Surathkal, Karnataka
Dr. Adake C V, KLE Dr. M.S. Sheshgiri College of Engineering and Technology, Belgaum
Dr. Rajesh Rai, AJ Institute of Engineering and Technology, Mangalore
Dr. Faridin Naim Tajuddin, Universiti Malaysia Perlis, Malaysia
Dr. Shrikanth Bontha, National Institute of Technology, Surathkal, Karnataka
Dr. Kumar G N, National Institute of Technology, Surathkal, Karnataka
Dr. Sreeramkumar R, King Abdulaziz University Jeddah, Saudi Arabia.
Dr. Patel R N, National Institute of Technology, Raipur
Dr. Joseph Xavier R, Akshaya College of Engineering and Technology, India
Dr. Yasser S Abdalla, Suez University, Suez, Egypt
Dr. Senthilkumar S, SVS College of Engineering, Coimbatore
Dr. Sanjeevikumar Padmanaban, Aalborg University, Denmark
Dr. Vinod T D'Souza, AJIET, Mangalore
Dr. Shobha M E, Manipal Institute of Technology, Manipal
Dr. Moorthi S, S R Engineering College, Warangal, India
Dr. Rajamani R, PSG College of Technology, Coimbatore, India
Dr. Venkataramani N, Indian Institute of Technology Bombay
Dr. Swaminathan K, National Institute of Technology Surathkal, Karnataka
Dr. Bishakh Bhattacharya, Indian Institute of Technology Kanpur
Dr. Kotteswaran, Veltech University, Chennai
Dr. Balasubramanian M, Indian Institute of Technology Madras
Dr. Deshannavar U B, KLE Dr MSSCET, Belagavi
Dr. Srinivasan P, Birla Institute of Technology and Science (BITS), Pilani
Dr. Mohd Fareq Abdul Malek, University of Wollonong, Dubai, UAE
Dr. Shanthi N, Kongu Engineering College, Tamil Nadu
Dr. Sanjay Misra, Covenant University, OTA, Nigeria
Dr. rer. nat. Ravi Kumar, Indian Institute of Technology-Madras
Dr. Baharudin Ismail, Universiti Malaysia Perlis, Malaysia
Dr. Mani V N, C-MET, Government of India, Hyderabad
Dr. Rahul Mitra, Indian Institute of Technology Kharagpur
Dr. Sampath V, Indian Institute of Technology Madras
Dr. Patil S F, KLE Dr. M.S. Sheshgiri College of Engineering and Technology, Belgaum

IOP Conf. Series: Materials Science and Engineering 1187 (2021) 011002 doi:10.1088/1757-899X/1187/1/011002

Peer review declaration

All papers published in this volume of *IOP Conference Series: Materials Science and Engineering* have been peer reviewed through processes administered by the Editors. Reviews were conducted by expert referees to the professional and scientific standards expected of a proceedings journal published by IOP Publishing.

- **Type of peer review:** Single-blind
- **Conference submission management system:** gc-rdct@ajiet.edu.in
- **Number of submissions received:** 135
- **Number of submissions sent for review:** 83
- **Number of submissions accepted:** 62
- **Acceptance Rate (Number of Submissions Accepted / Number of Submissions Received X 100):** 45.92
- **Average number of reviews per paper: Two**
- **Total number of reviewers involved:** 21
- **Any additional info on review process:** All the papers have been screened for the scope and technical quality as soon as they have been received. Only the papers passed the screening were assigned for technical review. Every paper has been reviewer by two independent reviewers. Reviewers have been informed to check the paper's quality based on the novelty, technical contributions proposed and the discussion on the results presented. Review criteria are: Scope, novelty, originality, organization, clarity of the abstract, quality of figures and tables, results and discussion and referencing/citation.
- **Contact person for queries (please include: name, affiliation, institutional email address)**

Dr. Sangeetha D M
Professor
A.J. Institute of Engineering and Technology,
Mangaluru, Karnataka, India.
Email: dr.sangeetha@ajiet.edu.in

IOP Conf. Series: Materials Science and Engineering 1187 (2021) 012001 doi:10.1088/1757-899X/1187/1/012001

Group Leverage Centrality and its Applications in Brain Networks

Amogh Gupta[1], Harish Balaji[2], R. Sundareswaran[3] , V. Mahesh[4], B. Geetanjali[5]

[1,2]Department of Computer Science and Engineering,
Sri Sivasubramaniya Nadar College of Engineering, Chennai, India
[3]Department of Mathematics
Sri Sivasubramaniya Nadar College of Engineering, Chennai, India
[4,5]Department of Biomedical Engineering
Sri Sivasubramaniya Nadar College of Engineering, Chennai, India
[1]amogh17009@cse.ssn.edu.in, [2]harish17050@cse.ssn.edu.in,
[3]sundareswaranR@ssn.edu.in, [4]maheshv@ssn.edu.in, [5]geethanjalib@ssn.edu.in

Abstract. The concept of leverage centrality is the relationship between the degree of a node relative to its neighbours and operates under the principle that a node in a network is central if its immediate neighbours rely on it for information. It was specifically formulated for brain networks. In this paper, we have built upon leverage centrality and introduced group leverage centrality, which is a measure of how important a subset of nodes is in the network. Then, the effects of meditation on different lobes of the brain were quantified using this new centrality measure

1. Introduction

Graph theory is particularly useful in the study of the brain. The human brain can be modelled as a graph in which neurons are taken as nodes or vertices, and the physical links between the neurons are taken as edges. As the brain network is highly complex and vast, it can be clustered into working regions, or lobes, to effectively model it. These clusters could be considered as the regions of interest (Korhonen et al., 2017) [1]. It has been shown that meditation plays a vital role in improving brain performance (Fox et al., 2014) [2], and helps to improve grey matter concentration, focus, attention, and reduce work stress. Different lobes of the brain have different functions. With the electroencephalography results of the brain (before and after meditation) we can analyse the importance of the lobes, based on the electrical activity. To quantify which lobe is more active, we employ centrality measures. Centrality measures were introduced to study and identify the key nodes in the network. Over the years, many centrality measures have been developed and used in different real life applications (Das et al., 2018) [9]. In 2010, another centrality measure- leverage centrality- was introduced to study and identify the key nodes in the context of brain networks (Joyce et al., 2010) [4]. Since every lobe of the brain is a set of nodes, we have introduced a new centrality measure for a subset of nodes- group leverage centrality, an extension of leverage centrality.

2. Leverage Centrality

Leverage centrality considers the degree of a node relative to its neighbours and operates under the principle that a node in a network is central if its immediate neighbours rely on that node for information. Multiple examples and theorems have been studied and derived by Vargas et al. (2017) [3], which are given below.

2.1 Definition

Leverage centrality is a measure of the relationship between the degree of a given node v, and the degree of each of its neighbours, averaged over all neighbours, and is defined as shown below:

$$l(v) = \frac{1}{k_v} \sum_{j \in N_v} \left(\frac{k_v - k_j}{k_v + k_j}\right) \quad (1)$$

Here, k_v stands for the degree of node v and N_v stands for the set of nodes adjacent to the node v.

2.2 Example

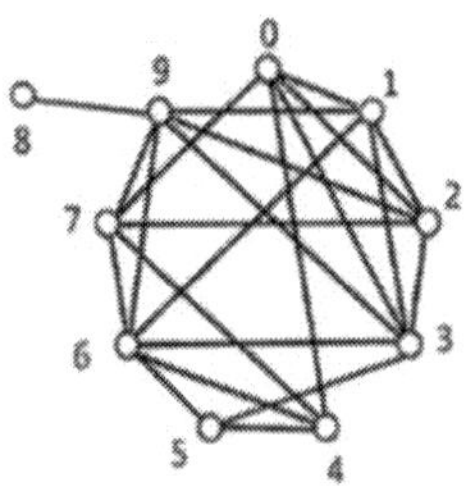

0	[illegible]
1	[illegible]
2	0
3	[illegible]
4	[illegible]
5	[illegible]
6	[illegible]
7	[illegible]
8	[illegible]
9	[illegible]

Figure 1: A Graph with different Leverage Centralities

2.3 Some Propositions of Leverage Centrality

- For any graph $G, \sum_{v \in G} l(v) \leq 0$. The leverage centrality $l(v)$ is zero in any regular graph. The sum of leverage centralities of all nodes in G will be less than 0 in any non-regular graph.
- The number of vertices with positive leverage centrality is $n - 1$.
- A vertex of lowest degree(highest degree) cannot have a positive (negative) leverage centrality. In any graph G, all the vertices except one to have negative leverage centrality and one have positive leverage centrality. There are $n - 1$ vertices with negative leverage centrality in a star graph $K_{1,n-1}$.

The concept of group centrality was introduced by Everett and Borgatti in 2010 [10]. Also, they also defined degree, closeness and betweenness centralities in terms of groups and classes as well as individuals.

2.4. Group Leverage Centrality

In literature , researchers have studied the Leverage Centrality and its different properties. In this work , we define a new metric of Leverage Centrality called the *Group Leverage Centrality,* which is leverage centrality not for a single node but for a group of nodes. We are also introducing two different ways in which Group Leverage Centrality could possibly be defined.

2.4.1 Total Group Leverage Centrality: Total Group Leverage Centrality (TG) is defined as the average of leverage centralities of all the nodes in a given set of nodes S, where S is the subset of V, the vertices of the graph in consideration.

$$TG(S) = \frac{1}{|S|}\sum_{v\in S} l(v) \qquad S \subseteq V \tag{2}$$

2.4.2 Complement Leverage Centrality: Complement Group Leverage Centrality (CG) is defined as the average of leverage centralities of all the nodes in a given set of nodes S, where S is the subset of V (the vertices of the graph in consideration) and only the vertices in the set V\S are considered for computing the leverage centralities. Here, k_v and N_v represent the degree and set of neighbours of node v respectively.

$$CG(S) = \frac{1}{|S|}\sum_{v\in S}\left(\frac{1}{k_v}\sum_{i\in S'\cap N_v}\left(\frac{k_v-k_i}{k_v+k_i}\right)\right) \qquad where \qquad S' = V - S \tag{3}$$

3. Algorithm

3.1 Total Group Leverage Centrality

Total Group Leverage Centrality is found by finding the average of all the leverage centralities of the vertices (all neighbours) present in the specified subset.

Input: Graph $G = (V, E)$, subset S containing n nodes

Output: $TG(S)$

TSum← 0

for node v in S do {

TgSum ← TgSum + l(v) //l(v) gives the leverage centrality of node v

}

TG ← TgSum/n

return TG

3.2 Complement Group Leverage Centrality

Complement Group Leverage Centrality is found by finding the average of all the leverage centralities of the vertices while considering only the neighbours outside the specified subset. Here, N(v) is the set of nodes which are neighbours to node v.

Input: Graph G = (V,E), subset S containing n nodes

Output: CG(S)

CgSum ← 0

for node v in S do {

complementNodeCentrality ← 0

for node i in N(v) do {

if $i \in S'$ do {

complementNodeCentraltiy ← complementNodeCentrality + (k(v)-k(i))/(k(v) + k(i))

}

}

CGSum ← CGSum + complementNodeCentrality/degree[v]

}

CG ← CGSum/n

return CG

4. Group Leverage Centrality in Brain Networks

The concept of Group Leverage Centrality can be used to quantify the effect of meditation on different parts of the brain network and their relation to other parts.

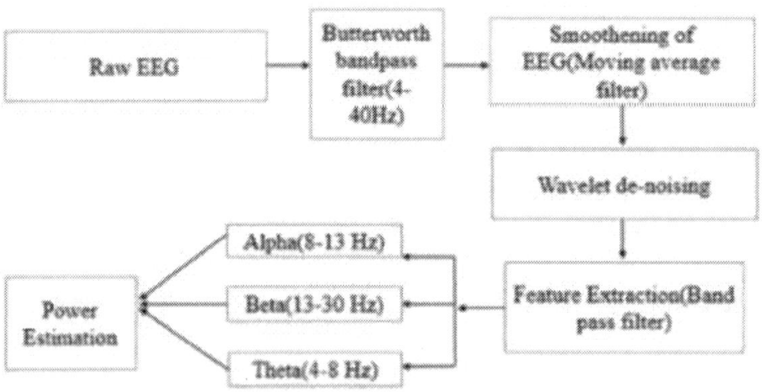

Figure 2: Flowchart for signal processing

4.1 EEG Data Acquisition

The tests were conducted in the Biomedical Engineering Department of Sri Sivasubramaniya Nadar College of Engineering, Kalavakkam. The modality of study used was electroencephalography (EEG). RMS EMG (SuperSpec) software was used to acquire the EEG. The 10-20 Electrode placement system was used to record the signals. The scalp electrodes were placed based on the different lobes in the brain, C3, C4, CZ, central electrodes; F3, F4, F7, F8, FZ, frontal lobe electrodes and FP1, FP2 were placed on the fronto planar area; P3, P4 and P, parietal lobe electrodes and O1, O2 and Oz were occipital lobe electrodes. For reference on the right and left earlobes, Electrodes A1 and A2 were placed. A ground electrode was placed on the mastoid process behind the ear. The odd nmbers depict that they are located on the right side of the brain and even numbers depict that they are

IOP Conf. Series: Materials Science and Engineering 1187 (2021) 012001 doi:10.1088/1757-899X/1187/1/012001

located on the left side of the brain. The electrodes used were silver/silver chloride (Ag/AgCl) electrodes placed on the scalp using a 10 20 EEG paste.

4.2 EEG Data Extraction

In this stage, the raw EEG was averaged by using a smoothing filter and separated into different frequency bands namely alpha, beta and theta (4 - 8 Hz). We removed the delta band. Their individual power spectral densities and their total and relative powers were calculated as shown in Fig. 2 (above) and the values were stored for later analysis.

4.3 Statistical Analysis

All the values that were calculated were tested for skewness and they were found not to be normally distributed. Hence the parameters that were measured were tested non-parametrically. The three independent samples are subjects classified as normal (DASS-normal). All the analysis was pre-meditation results. The task performances were tested. Using the Friedman test, the relative powers of all the four frequency bands were tested for significant differences. The significant level was taken as $p = 0.05$. A related sample test was done between task and rest for all 19 electrode locations to measure the individual significance between each of the variable pairs. The entire analysis was done in IBM SPSS Statistics, Version 24. For all the three groups, the mean relative alpha, beta, theta and delta component energies were found for all the 19 electrode locations during both rest and while performing the various linguistic and non-linguistic activities. The Friedman test was done for all the groups separately and a statistically significant difference was noted in the measured parameters (relative alpha, beta, theta and delta component energies).

4.4 Functional Brain Graph Modelling

The working procedure to get data from the EEG values and conversion to graph models has been studied and described in Graph theoretical analysis of structural and functional systems (Bullmore & Sporns, 2009) [5]. The functional brain was analysed by using an EEG graph. The correlation matrix was then derived from the time series data. This matrix can be converted to adjacency matrix- here we used a text file to store the edged directly without the need of adjacency matrix. Based on the given data, the graphs were modelled by fixing threshold values.

Following gives the brain graph model for alpha, beta and theta bands before and after meditation.

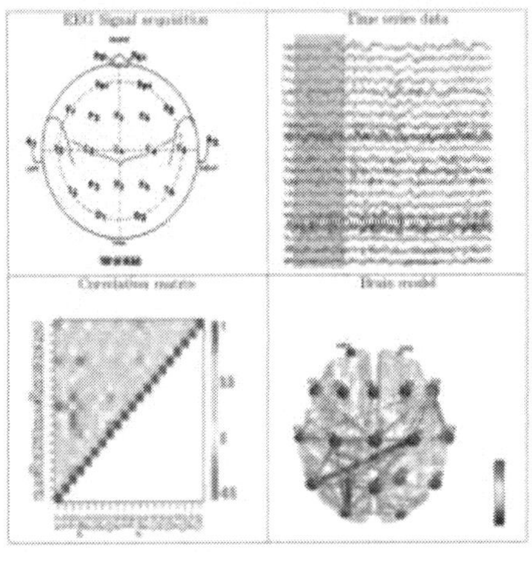

Figure 3: Process of Converting EEG Data to Graph Model

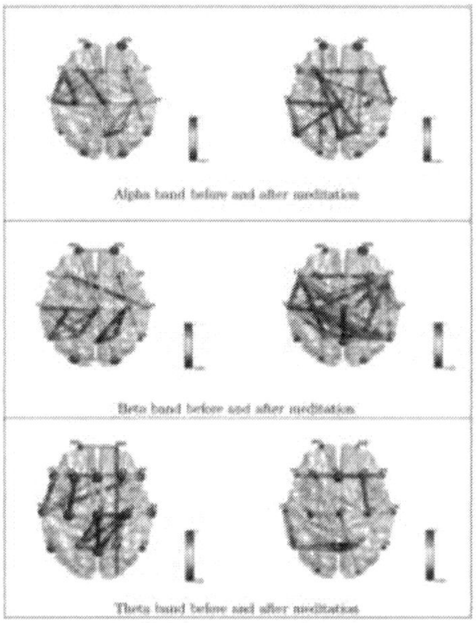

Figure 4: Alpha, Beta, and Theta Bands Before and After Meditation

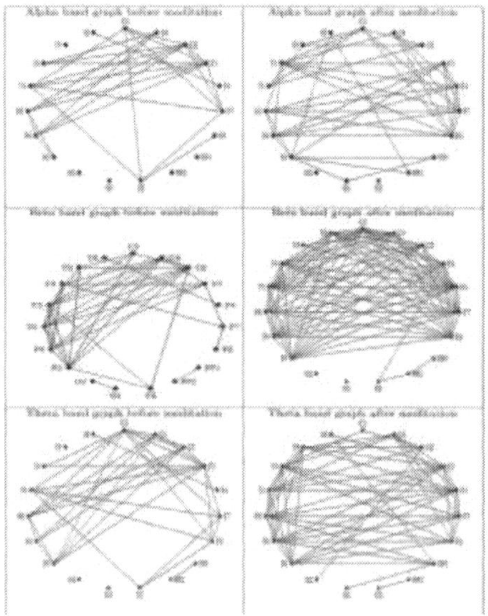

Figure 5: Graphs Obtained After Modelling

After obtaining the graph, the **complement group leverage centrality** was computed for subsets based on the lobes for all the graphs in Fig. 5.

5. Results

The following table contains the computed values of the group leverage centrality. A detailed description of the nodes and the subsets in the table is presented in the discussions section. A few readings in it are left blank because complement group leverage centrality cannot be defined for a subset which contains isolated nodes.

Table 1: Complement Group Leverage Centralities of the Lobes (Subsets) Before and After Meditation

Subsets	**Alpha**		**Beta**		**Theta**	
Complement Group Centrality Meditation	**Before**	**After**	**Before**	**After**	**Before**	**After**
C3,C4,CZ	0.0132	- 0.0526	- 0.0076	0.0225	0.0457	- 0.2809
F3,F4,F7,F8,FZ	0.0146	0.0476	- 0.1368	0.0197	0.0152	0.0057
O1,O2	-	- 0.2777	0.0	-	-	- 0.6333
FP1,FP2	-	- 0.2777	0.0	0.0	0.0	0.0095
P3,P4,PZ	- 0.0111	0.0643	0.0781	- 0.01	- 0.1253	0.0660
T3,T4,T5,T6	- 0.1528	- 0.1093	- 0.0276	- 0.1908	- 0.2735	- 0.0927

6. Discussions

In this section, the group complement leverage centrality values before and after meditation graphs of alpha, beta and theta bands are compared. Alpha band is concerned about the relaxation of the body. Beta band involves high cortical activation of the brain areas and they are activated during stress and tension. Theta band is related to the body's overall Deep relaxation. We have discussed the results of the impact of meditation with respect to the following groups/subsets of nodes.

1. **P3, P4, PZ:** These 3 nodes represent the central parietal lobe electrode locations. They are associated with cognitive processing and contribute to the activity of how things are perceived and segregated. From the results tabulated below, we can very well see that meditation has profound effects in the functioning of these three bands. The value of Alpha has increased to a positive CG; Beta band, which involves increased cortical activity has decreased to a negative value and Theta band which induces deeper relaxation to the body has also increased significantly to a positive value. Thus meditation has an enormous amount of positive effect on the perception capability of the brain and the connectivity between the different lobes.

2. **F3, F4, F7, F8, FZ:** The Dorsolateral Prefrontal cortex is related to short term memory that has been created based on the recently acquired information. It has effects on the logical performance, working memory, intentional and motivation focus. From the table, we can see that the Alpha band value has increased slightly, a dip in the value of Theta band and an increase in Beta. All these results essentially tell that meditation improves the connectivity and information encoding power of the brain.

3. **C3, C4, CZ:** These nodes refer to the central electrodes that control the sensory motor cortex, for example, these neurons are put to use and activated during hand movements. It is related to somatosensory information, thus the influence of meditation doesn't really matter.

4. **O1, O2:** Occipital lobes represent the primary visual areas of the brain. As far as Alpha and Beta

bands are concerned, the isolated nodes O1, O2 which did not communicate with the other lobes showed some range of connectivity (CG is still negative therefore they are not very influential) after the mediation period. In Beta band, after meditation the only connectivity between the 2 nodes has disappeared.

5. **FP1, FP2:** The Prefrontal cortex electrode locations (FP1,FP2) are influential in the arousal, motivation and attention of the person. Before meditation, no active connections were noted. However after meditation the brain activation is visible in Alpha (from None to -0.2777) and in Theta band (to a positive value). No improvement during the Beta band.

6. **T3, T4, T5, T6:** The anterior temporal lobe electrode locations (T3,T4 limbic- amygdala) is associated with emotions while posterior temporal lobe electrode locations (T5,T6- hippocampus) are related to some memory related functions. From the table, we do not see any significant changes in the CG values after meditation in all the three bands.

7. Conclusion

The concept of Group Leverage Centrality to define the centrality of a group of nodes with respect to a brain network has been introduced. Furthermore, two different definitions for the same have been established of which the Complement Group Leverage Centrality has been used to find the effects of meditation on different lobes of the brain. We notice that meditation has profound effects on the human mind, and can cause increased cognitive processing and perception, decreased stress, and general well-being.

8. Acknowledgements

The authors also would like to thank SSN Management for their support and encouragement.

9. References

1. Korhonen,O., H. Saarimaki, E. Glerean, M. Sams and J. Saramaki. Consistency of regions of interest as nodes of fMRI functional brain networks, Network Neuroscience, 1:3, 254 - 274, 2017.

2. Fox, K.C., S. Nijeboer, ML. Dixon, JL. Floman, M. Ellamil, SP. Rumak, P. Sedlmeier, K. Christoff. Is meditation associated with altered brain structure? A systematic review and meta-analysis of morphometric neuroimaging in meditation practitioners, Neuroscience and Biobehavioral Reviews, 43,.48-73, 2014.

3. Vargas, R., A. Waldron, A. Sharma, R. Flórez and D.A Narayan. A graph theoretic analysis of leverage centrality. *AKCE Int J Graphs Comb.* 14(3): 295- 306, 2017.

4. Joyce, K.E, P.J Laurienti, J.H. Burdette and S. Hayasaka. A new measure of centrality for brain networks, *PLoS One*, 5,1-13, 2010,.

5. Bullmore, E., O. Sporns. Complex brain networks: Graph theoretical analysis of structural and functional systems, Nature Reviews Neuroscience, 10, 186-198, 2009.

6. Garey, L. Cortex: statistics and geometry of neuronal connectivity, 2nd edition, Journal of Anatomy, 194, 153-157, 1999.

7. Mahesh, V., G. Bacak-Turan, G. Balaraman, R. Sundareswaran and R. Sujatha. Group closeness centrality of graphs Springer- Conference Proceedings: Recent Advances in Computational and Engineering Mathematics, 2019.

8. Das, K., S. Samanta and M. Pal. Study on centrality measures in social networks: a survey. Soc Netw Anal Min. 8(1):13, 2018.

9. Lee, S.H., C.E. Kim, I.S. Lee, W.M Jung, H.G Kim, H. Jang, et al. Network analysis of acupuncture points used in the treatment of low back pain. Evid Based Complement Alternat Med 2013:402180.

10. M. G. Everett and S. P. Borgatti, The Centrality of Groups and Classes, Journal of Mathematical Sociology 1999, Vol. 23(3), pp. 181-201.

IOP Conf. Series: Materials Science and Engineering 1187 (2021) 012002 doi:10.1088/1757-899X/1187/1/012002

Simulation of Different Third Harmonic Injected PWM Strategies for 5-Level Diode-Clamped Multilevel Inverter

Lakshmi Swarupa Malladi[1], Shravani Chapala[2], Sarada Veluri[3]

[1]Professor, EEE Department, CVR College of Engineering, Telangana, India
E-mail: swarupamalladi@gmail.com
[2]Asst.,Professor, EEE Department, CVR College of Engineering, Telangana, India
E-mail: shravanic2@gmail.com
[3]Asst.,Professor, EEE Department, CVR College of Engineering, Telangana, India
E-mail: saradav001@gmail.com

Abstract: Multilevel Inverters (MLI) have widespread use in industrial drive applications, HVDC transmission with Uninterruptable Power Supply systems. They also have applications in the stand-alone and the grid connected renewable energy systems. Multi-level Inverters also find optimum application in Stand-alone and Grid- connected Systems. Majority of losses i.e. switching losses are considered with constrained device rating, the traditional two-level inverters face limits especially when operating at high frequency. The solution which is optimum and appropriate by many power quality researchers is the use of MLIs. Among various MLI topologies the Diode-Clamped Multilevel Inverter (DCMLI) give a promising solution as they can give better voltage performance for different switching frequencies and an improved total harmonic distortion (THD) profile without the use of filter. To meet the voltage and frequency requirement various modulation techniques are applied to MLIs. The additional switching states created by these topologies provide greater flexibility to control more circuit breakers. This paper presents the comparison of a 5-level DCMLIs with several modulation schemes. The simulation results are discussed in detail and the comparison table is presented in terms of Output voltage and THD.

Index Terms: MLI (Multi-Level Inverter), THD (Total Harmonic Distortion), DCMLI (Diode-Clamped Multi Level Inverter).

1. Introduction

Various topologies of inverters i.e. DC-AC converters are in the literature [1]. The main aim of the inverter is the conversion along with compliance of the international standards so that it could be adopted for commercial applications. The concept of multilevel converters either for rectification or inverter operation is required for gain significance over past two decades due to the improved power quality, capability of voltage, switching losses to be minimum and better electromagnetic compatibility. They are mainly applicable for applications where voltage is high, power is high due their advantages with synthetization of better waveform with less harmonics to achieve high voltages with less maximum power ratings of the switches. They also reduce the dv/dt stress and hence protect the motor winding insulation.

As per literature survey, the MLIs has been introduced in 1975 as another solution of medium voltage and high-power applications. With the latest technology, many industrial applications demand high power applications almost megawatts range along with medium or low power utilities [2][3]. Use

IOP Conf. Series: Materials Science and Engineering 1187 (2021) 012002 doi:10.1088/1757-899X/1187/1/012002

of a high-power source in industrial loads can be required for high-power machines but it may damage the other loads. In small and medium scale industries require medium voltage viable speed drive applications. The inverter is the integral part in many applications such as air conditioning, Solar PV (photovoltaic) power systems etc.

Specific to the application of distributed power generation like PV, with the basic function of converter for inverting operation like DC-AC conversion, it should also be assured by output power quality. It should have flexible protection scheme as well as system controls. The requirements of such inverters are high efficiency and reliability, good tolerance on a wide range of input variations and low cost. This has driven more research in the inverters leading to solutions such as simpler topologies, lower switching components, better modular designs while fulfilling the primary quality requirements. It has become obvious that instead of connecting a dingle power semiconductor switch, use of these modular inverters has gained much more importance. This helps in easy maintenance and at the same time meets the power demand at high voltages as discussed earlier [3].

From the different multi-level inverters, three-level inverter is considered in [4]. Output voltage has more steps or levels in the increasing voltage as a stepwise type for minimizing total harmonic distortion than the conventional two-level inverter. However, as the number of levels increase there arise problems such as voltage imbalances and increase the control complexity in terms of switching.

By considering Single-phase w.r.t. no. of levels used or adopted per phase as shown in FIGURE 1 with no. of switches. For better understanding each power semiconductor switch is shown by an ideal multi position switch. It can be seen from FIGURE 1(a) that multi-level inverter has step by step levels (values) by considering neutral terminal by the passive element, whereas step by step levels which has an intermediate level to create levels as shown in FIGURE 1(b) and further levels as shown in FIGURE 1(c).

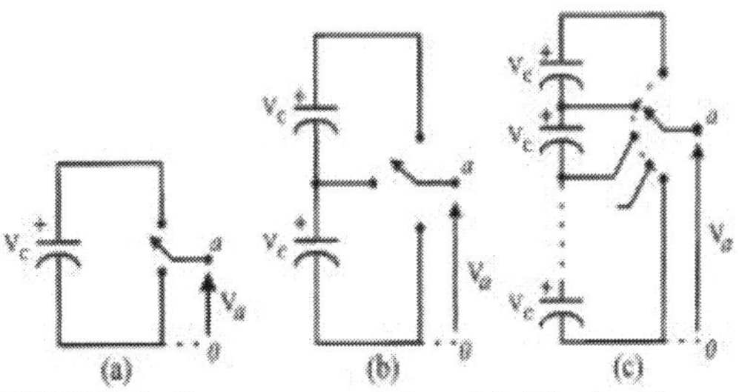

FIGURE 1: Inverter module with Single-Phase two levels (b) three levels and (c) m levels.

The no. of levels of the output voltage are considered with respect to negative or neutral terminal of the converter when acting as an inverter. The no. of steps considered as 'K' and is assumed between the lines and connected in shunt when load is connected. Where m is the no.of steps used in MLI.

K=2m+1 eq. (1)

By considering the steps count as 'P' across line and neutral when load is connected as

P=2K-1 eq. (2)

2. Types of Multi Level Inverters

From literature survey, among various topologies that have been proposed, the main three different topologies that has large impact identified as follows [4]-[9].

- Clamped MLI: Neutral Point or Diode
- Flying Capacitor or Clamped with Capacitor MLI

IOP Conf. Series: Materials Science and Engineering 1187 (2021) 012002 doi:10.1088/1757-899X/1187/1/012002

- Interconnected with many cells with isolated inputs which are based on direct current or voltage (Cascaded H-Bride) MLI

These are the classic topologies shown in FIGURE 2 with several new configurations with some modifications with these topologies.

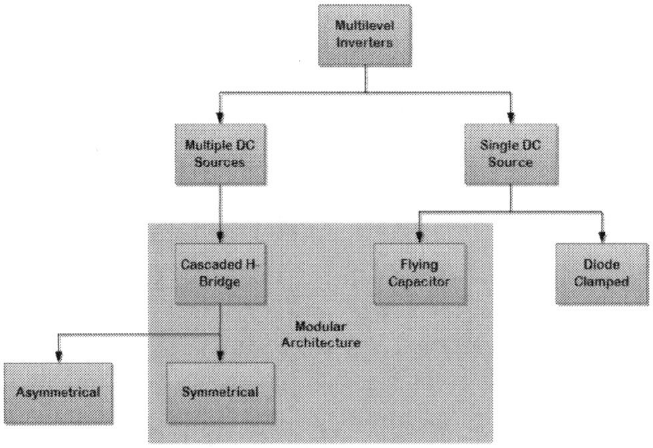

FIGURE 2: Multilevel Inverters Classification

Based on the application of the inverter and its topology and switching frequency, numerous variants of the above modulation techniques have been adopted. Each of them has its own advantages and disadvantages. FIGURE 3 gives the classification of such modulation schemes based on switching frequency for MLIs operation. The switching frequency with respect to high power applications are normally referred in the order of one to many kHz.

Also, for various switching requirements analysis has been done in different modulation techniques and with strategies that can be adopted for these MLIs. Among all of them are listed below.

- SHE (Harmonic Elimination with Selective frequencies)
- SPWM (Pulse Width Modulation with Sinusoidal)
- SVM (Space Vector Modulation)

Pulse Width Modulation with Sinusoidal wave (SPWM): Considering carrier-based method which involves comparing a reference signal(sinusoidal) with a carrier signal(triangular) which has signals that are horizontally and vertically changing to minimize the harmonic content.

Harmonic Elimination with Selective frequencies (SHE): In this type the lower or minimal order frequencies are removed or eliminated from the total frequencies and Fourier analysis is adopted. This eliminates the undesired lower order harmonics which indirectly minimizes the cost of filtering equipment.

Space Vector Modulation (SVM): This method is also a PWM method, but the time is calculated for the switch non-operating region which is represented in the form of space vectors for three-phase from a reference point. The advantage of this method is the inverter is switched off rather than provide a switching that goes to neutral reference temporarily for creating neutral voltage.

One more method, with different series, parallel and series-parallel connections in modulation with the one operating with minimum and maximum switching frequencies. Generally adopted in Cascaded H Bridge inverters with asymmetric configuration which minimize breakdown losses obtained in high-level switches with upper units of power.

IOP Conf. Series: Materials Science and Engineering 1187 (2021) 012002 doi:10.1088/1757-899X/1187/1/012002

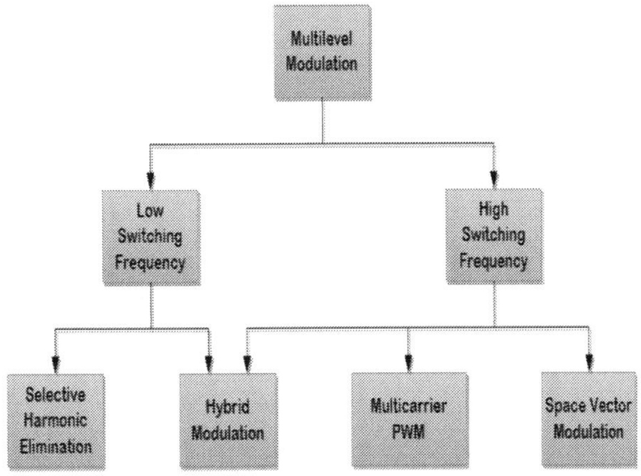

FIGURE 3: Modulation methods classification for MLIs.

3. Diode-Clamped Multi Level Inverters

With respect to the practical applicability, the DCMLI or NPC inverter is first MLI. This topology uses diode as a switching of the switches for increasing or decreasing of DC voltage to obtain desired output voltage levels. They are utilized to control the power semiconductor voltage. V_{dc} is the voltage across switch or capacitor. With the stepwise size 'k' is considered as no. of levels of inverter has 'k-1' capacitors or DC sources, '2(k-1)' switches (IGBT/MOSFET) and '(k-1) *(k-2)' blocking diodes. Considering the number of levels with the increase in the number of levels of output voltage with sinusoidal input thereby reducing the harmonic content.

Principle of Operation and Switching sequence of 5-level Multi Level Inverter - Diode Clamped

DC MLI with five levels has 4 capacitors, each of value Vdc/4. There are eight Power semiconductor switches. For the multi-level inverter with five levels of output voltage across phase will have multi levels, ±vdc/2, ±vdc/4 and 0. One phase/leg of a three phase five-level DCMLI is shown in FIGURE 4. Switching sequence of the switches is included in the Table 1. An entry High in the row denotes that the switch associated with levels include ON/OFF or 0/1 for that output voltage level. An entry Low in the row of the table denotes the Switch is OFF/LOW/0 for that voltage level.
For any output voltage level, we have only four switches conducing at the same time and remaining four switches are OFF.

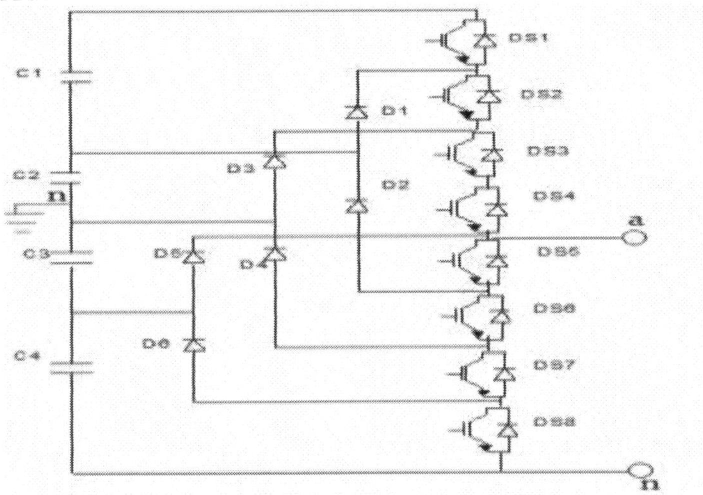

FIGURE 4: Single-phase 5-level DCMLI

TABLE 1. Switching Sequence of 5-level DCMLI

Switch	Van= +Vdc/2	Van= +Vdc/4	Van=0	Van= -Vdc/4	Van= -Vdc/2
DS1	High	Low	Low	Low	Low
DS2	High	High	Low	Low	Low
DS3	High	High	High	Low	Low
DS4	High	High	High	High	Low
DS5	Low	High	High	High	High
DS6	Low	Low	High	High	High
DS7	Low	Low	Low	High	High
DS8	Low	Low	Low	Low	High

The DCMLI provides several voltage levels by connecting the phases to several passive elements. Based on the no. of levels, the passive elements like capacitors are added either in series or parallel. As discussed in the earlier section, there are two capacitors connected via a DC bus which gives three levels in MLI topology. This extra level was because of DC bus neutral point. Even though a neutral point includes a pair with levels of output voltage which is never available, sometimes Multiple Point Clamped (MPC) is used. Capacitor voltage with balance problems are considered with diode inverter is limited in industrial applications. Multi-level Inverter with Diode Clamped Applications are:

- Drives with flexible speed control.
- Compensation with static var
- System interconnections with High voltage
- Transmission lines with High voltage AC and DC

4. Modelling Schemes and Simulation Results

The 5-level DCMLI was modeled, and its performance is studied using the MATLAB_SIMULINK environment. Different PWM methods such as VAPDPWM (Variable Amplitude Phase Disposition PWM), PDPWM (Phase Disposition PWM), POPWM (Phase Opposition Disposition PWM), VFPWM (Variable frequency PWM), Carrier Overlapping PWM-A scheme(COPWM-A) and Carrier Overlapping PWM-B scheme(COPWM-B) schemes are simulated and analyzed. The system is tested for different modulations varies within 0.6 to 1 with their supporting harmonic level values seen by considering Fast Fourier analysis. The measurements of the FFT block show the output voltage in line to line of the inverter in terms of RMS voltage of the inverter in terms of phase value with 'M' as modulation index. Simulation analysis can be concluded, the PDPWM strategy produces energies of the 65th, 70th, 76th, 80th and 84th harmonics. VAPDPWM produces the energies of the 10th, 15th, 75th, 80th, 85th, 100th, 150th, 160th and 180th harmonics.

The PODPWM methods generates the energy of 10th, 80th and 160th harmonics. The VFPWM generates the energies of the 11th, 13th and 16th harmonics. COPWM-A produces 3rd, 78th and 84th harmonics. COPWM-B produces the 3rd, 65th, 77th and 169th harmonics. The modulation index is 0.8 and number of samples is 40. The output line voltage waveform, THD and output phase voltage and its frequency spectrum for 5-level inverter for various PWM schemes with a frequency with a cut-off value 1 kHz as mentioned in Figure 5 to Figure 21.

The harmonic content and disconnection losses estimated with carrier frequency varies by the range of 1500 Hz to 10000 Hz. It is noted that with the increase of no. of levels in MLI, the performance of the system is controlled and minimized by switching losses and overall harmonic distortion. Line starting time is 0.0109 secs. The secondary voltage across phase is 0.02 V with 50 Hz as frequency reference.

IOP Conf. Series: Materials Science and Engineering 1187 (2021) 012002 doi:10.1088/1757-899X/1187/1/012002

1) Phase Disposition PWM Scheme (PDPWM):

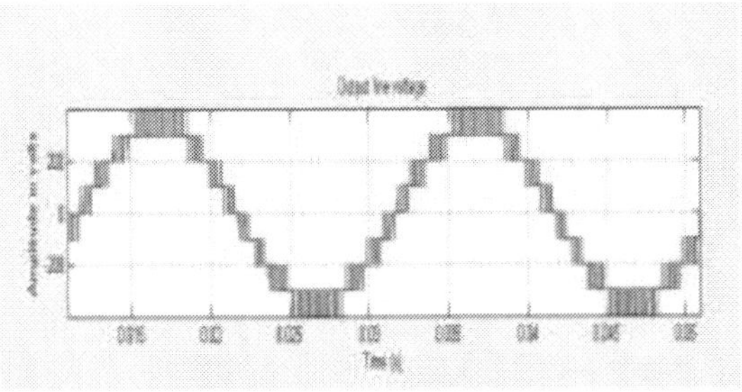

Figure 5: Line voltage Versus Time (secs)

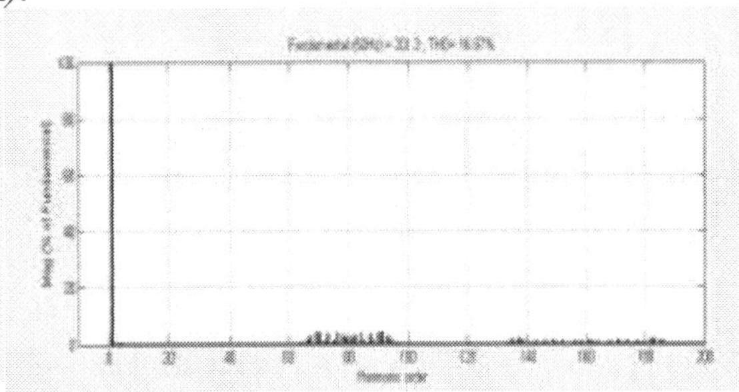

Figure 6: Total harmonic distortion versus Harmonic order

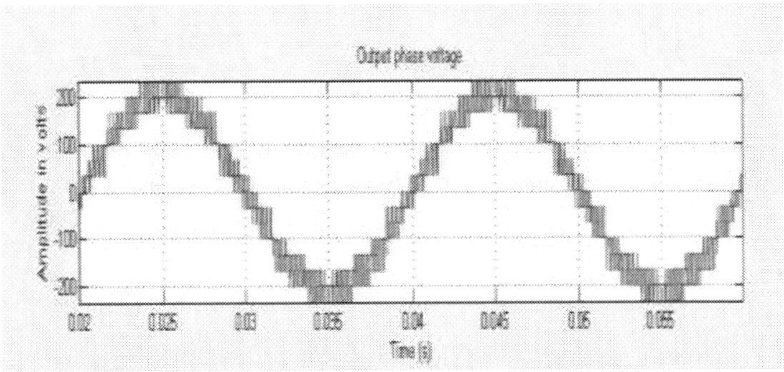

FIGURE 7: Output phase voltage generated by PDPWM schemes

2) Variable Amplitude Phase Disposition PWM Scheme (VAPDPWM):

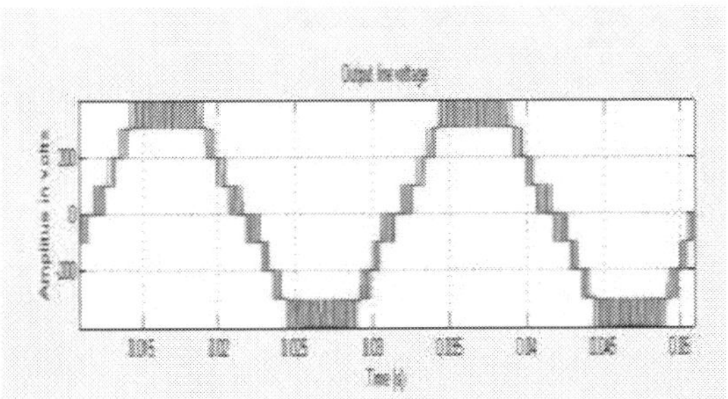

Figure 8: Output Line Generated by VAPDPWM Schemes Schemes

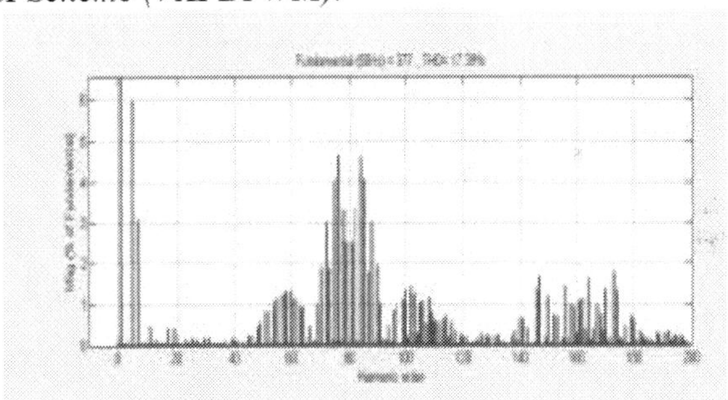

Figure 9: Output Phase Voltage Generated by VAPDPWM

3) Phase Opposition Disposition Scheme (PODPWM)

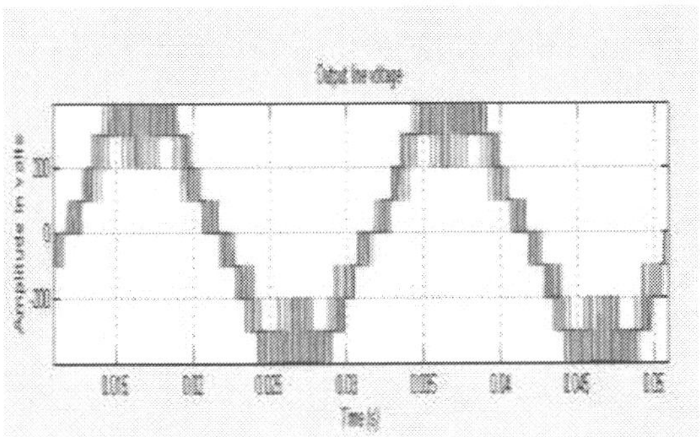

Figure 10. Output line voltage generated by PODPWM schemes

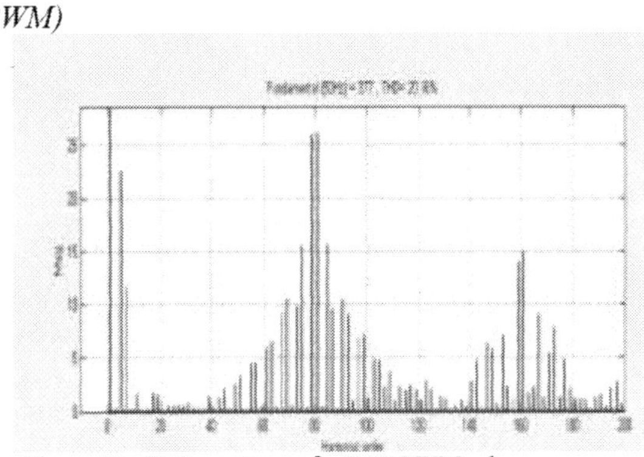

Figure 11: FFT spectrums for PODPWM schemes.

IOP Conf. Series: Materials Science and Engineering 1187 (2021) 012002 doi:10.1088/1757-899X/1187/1/012002

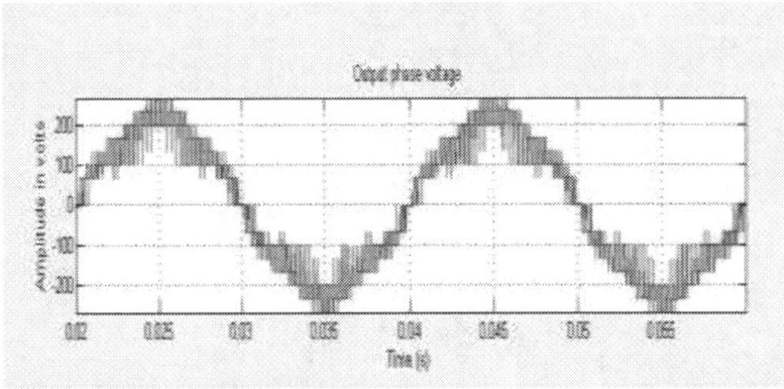

FIGURE 12: Output phase voltage generated by PODPWM schemes

4) Variable Frequency PWM Scheme (VFPWM)

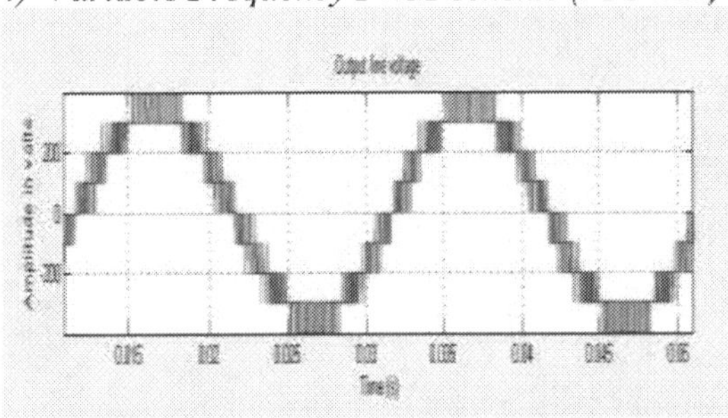

FIGURE 13: Output line voltage generated by VFPWM

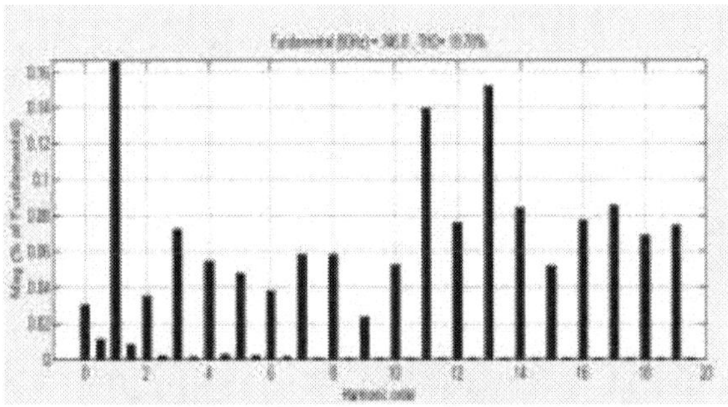

FIGURE 14: FFT spectrums for VFPWM schemes

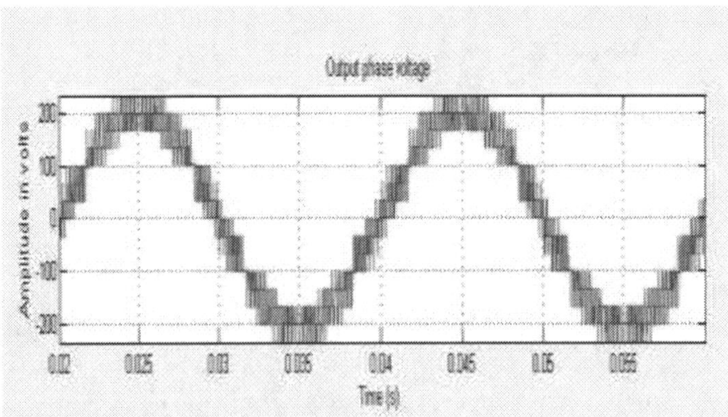

FIGURE 15: Output phase voltage by VFPWM schemes

5) CARRIER OVERLAPPING PWM-A SCHEME

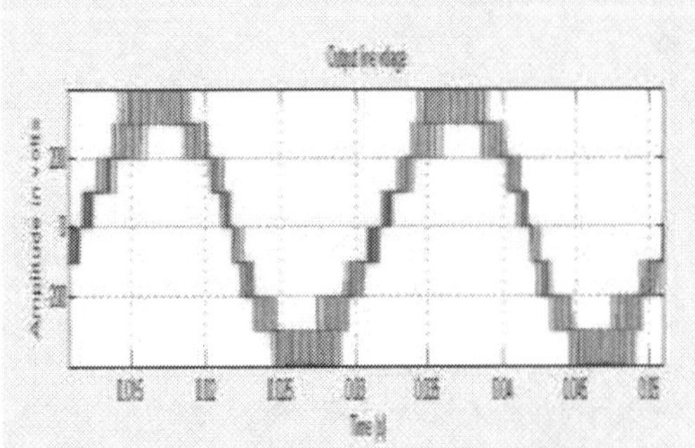

FIGURE 16: Output line voltage generated by COPWM-A

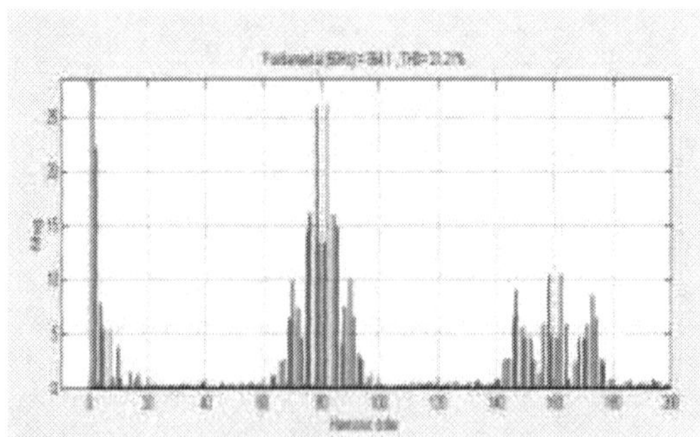

FIGURE 17: FFT spectrums for COPWM-A schemes

IOP Conf. Series: Materials Science and Engineering 1187 (2021) 012002 doi:10.1088/1757-899X/1187/1/012002

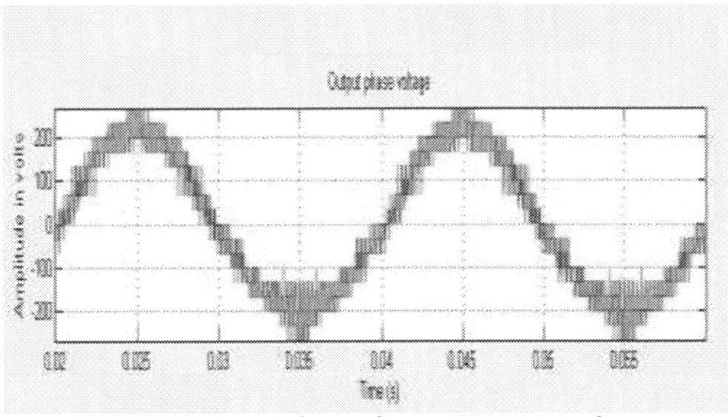

FIGURE 18: Output phase voltage - COPWM-A schemes

6) CARRIER OVERLAPPING SCHEME (COPWM-B)

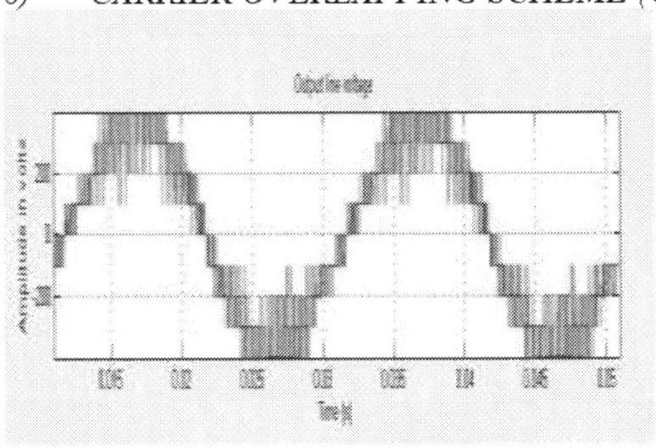

FIGURE 19: Output line voltage - COPWM-B schemes

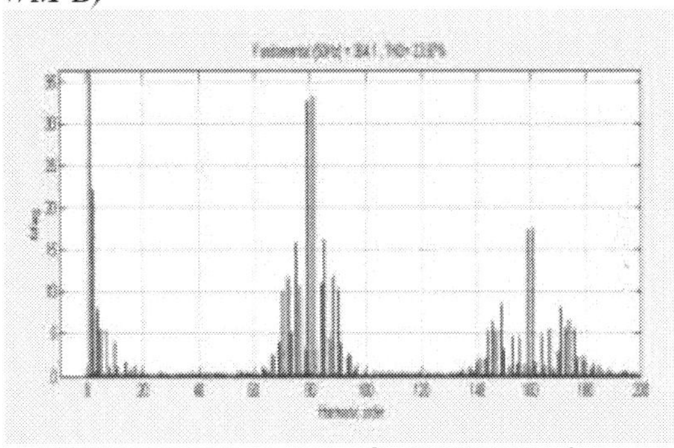
FIGURE 20: FFT spectrums for COPWM-B schemes

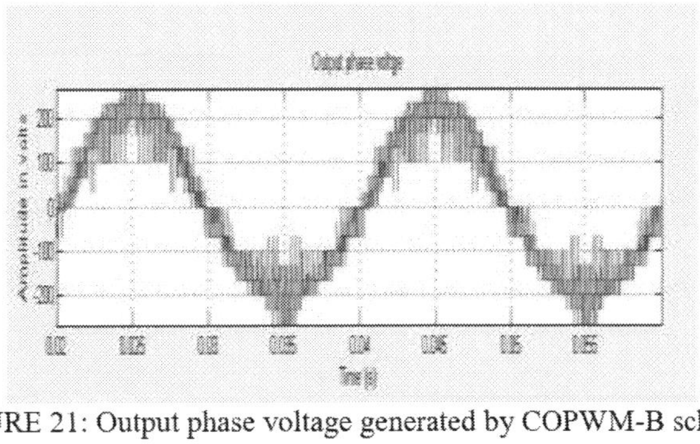

FIGURE 21: Output phase voltage generated by COPWM-B schemes

5. CONCLUSIONS

It concluded from the simulation analysis that the power quality issues related to convention two level inverters can be solved using MLIs. The simulation is done for a three phase 5- level DCMLI with various modulation strategies such as PDPWM, VAPDPWM, PODPWM, VFPWM, CO PWM-A, COPWM-B. The THD content in these various schemes differ with ±5% for the same modulation index. Based on the control strategy that can be used and the application one of them can be selected appropriately.

REFERENCES

[1] J. S. Lai and F. Z. Peng, "Multilevel converters–A new breed of power converters," IEEE Trans. Ind. Applicat., vol. 32, pp. 509–517, May/June 1996.

[2] L. Tolbert, F.-Z. Peng, and T. Habetler, "Multilevel converters for large electric drives," IEEE Trans. Ind. Applicat., vol. 35, pp. 36–44, Jan./Feb. 1999.

[3] R. Teodorescu, F. Beaabjerg, J. K. Pedersen, E. Cengelci, S. Sulistijo, B. Woo, and P. Enjeti, "Multilevel converters — A survey," in Proc. European Power Electronics Conf. (EPE'99), Lausanne, Switzerland, 1999, CD-ROM.
[4] A. Nabae, I. Takahashi, and H. Akagi, "A new neutral-point clamped PWM inverter," IEEE Trans. Ind. Applicat., vol. IA-17, pp. 518–523, Sept./Oct. 1981.
[5] T. A. Meynard and H. Foch, "Multi-level choppers for high voltage applications," Eur. Power Electron. Drives J., vol. 2, no. 1, p. 41, Mar. 1992.
[6] C. Hochgraf, R. Lasseter, D. Divan, and T. A. Lipo, "Comparison of multilevel inverters for static var compensation," in Conf. Rec. IEEE-IAS Annu. Meeting, Oct. 1994, pp. 921–928.
[7] P. Hammond, "A new approach to enhance power quality for medium voltage ac drives," IEEE Trans. Ind. Applicat., vol. 33, pp. 202–208, Jan./Feb. 1997.
[8] E. Cengelci, S. U. Sulistijo, B. O. Woom, P. Enjeti, R. Teodorescu, and F. Blaabjerge, "A new medium voltage PWM inverter topology for adjustable speed drives," in Conf. Rec. IEEE-IAS Annu. Meeting, St. Louis, MO, Oct. 1998, pp. 1416–1423.
[9] R. H. Baker and L. H. Bannister, "Electric power converter," U.S. Patent 3 867 643, Feb. 1975.

High Performance Wallace Tree Multiplier Using Majority Gate Based Adders

Mummudi Murasu M[1], Sanjana Sujith[2], Anita Angeline.A[3], Sasi Priya.P.[4], V S Kanchana Bhaaskaran[5]

[1,2,5]School of Electronics Engineering, VIT Chennai
[3,4]Center for Nano-Electronics & VLSI Design, VIT Chennai

mummudimurasu.m2020@vitstudent.ac.in[1], sanjana.sujith2020@vitstudent.ac.in[2], anitaangeline.a@vit.ac.in[3], sasipriya.p@vit.ac.in[4], kanchana.vs@vit.ac.in[5]

Abstract. An area efficient high performance Wallace Tree Multiplier using Majority Gate based Adders is presented in this paper. The proposed Wallace tree multiplier is designed using 9T majority function based full adder. Design is implemented in Cadence Virtuoso® using a 180nm technology library. The design and analysis of the proposed design offers reduced delay of 33.12% while compared to the conventional wallace tree multiplier design using 16T full adder. The design also offers reduced transistor count of 208 which is minimal compared to that of the conventional design.

Keywords—High speed performance, Wallace Tree Multiplier, Full Adder.

1. Introduction

One of the fundamental building blocks of many VLSI applications are multipliers.To enhance the performance of circuit and system the design of multipliers is very important. Adders are also considered as the major block of many applications, in the proposed design majority gate based adder is used.

A majority gate returns true if and only if more than 50% of its inputs are true[1].Majority gate adders have the advantage of reducing the power consumption and complexity of the design.

High Speed computation especially in digital signal processing applications demands high speed arithmetic circuits. The recent trends in the VLSI circuit regime paves various methods to achieve the same. It also poses lots of difficulties and challenges in optimizing the architectures with respect to all the performance metrics. The demand for circuits which consume less power with reduced area occupancy and increased operating speed are of greater demand in the real time applications. The design of Wallace Tree multiplier caters to achieve the later two metrics to a greater extent [2]. Furthermore, the design using majority gate based adder in multiplier architecture offers multi-fold avantages.

2. Literature Review

Speed of the processors is mainly determined by the speed of the multipliers used in the processor. Hence, multiplication is one of the most important and fundamental operations in various applications. Various researches had been carried out to obtain high speed, low power consumption and reduced area occupancy in the multiplier architectures. The common multiplication operation is performed using addition and shifting operations. Hence, multiplier design incorporates shifter unit and adder unit as major sub-unit. This leads to the necessity in having the adder unit also to offer high speed operation with reduced power consumption. When the performance of the adders improves it enhances the performance of the multiplier circuit.

The design of the full adders using majority gate function reduces the number of transistors when analogized with the full adders designed using conventional methods. In addition, the critical path delay is also found to be minimal. The proposed design is a Wallace tree multiplier by using adders realised with 9T(transistor) majority gate function.

2.1. Array Multiplier

In digital and signal processing methods multipliers play a major role. Conventional multipliers include the logic gate like AND and adders such as half adder and full adder. An array multiplier is realized using an array of full adders and half adders. Partial products computation unit incurs major part of the multiplier area. In the case of having N partial products, it requires N-1, M-bit adders. The shifting of the partial products is performed by simple routing and without the requirement of any logic. It reveals the ease in realising a very efficient transistor level schematic design[1].

2.2. Wallace Tree Multiplier

The key design specification for designing a good chip is to make it low power and high speed. Since multipliers and adders are the prime features for any DSP application, designing a low power and high calibre multiplier will highly augment the throughput of the design. Many Research prove that the wallace tree being a high calibre and area efficient multiplier, it can be highly used as an alternative to conventional array multiplier. The Wallace tree is a variant of long multiplication. The multiplication process in wallace tree multiplier involves three steps, The first step is to multiply each bit of the multiplier with the multiplicand. The weights of the wire relies on the multiplied bits position. In the second step the number of partial products is reduced to two by layers of full and half adders and finally, the wires should be grouped into numbers, and add them using Fulladder.

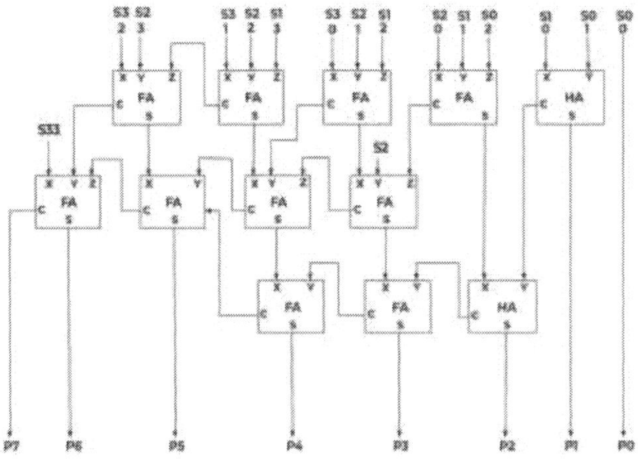

Fig 1. Illustration of Wallace tree multiplier design

2.3. Design of Full Adder using Majority Function

Full adder circuit is basically 2 half adders connected together. They are used to add three 1 bit binary numbers for example A,B,C. They have 2 output states: sum and carry. Sum performs the xor operation of A,B and C. Carry performs the and operation between the inputs A,B and C. In the proposed full adder Sum is generated by majority gate function which reduces the number of transistors. Majority gate is a function that outputs a HIGH when the majority of the inputs are HIGH, else it outputs a LOW. So by majority gate transistor count gets reduced which results in low switch capacitance reduction and also it results in working at ultra low supply voltage. Only 2 transistors are needed from Vdd to ground instead of 6 transistors to make it useful in low power design. Majority (minority) network synthesis utilizes majority (minority) gates as logic elements. A majority (minority) function has got three inputs and one output which produces an output value 1 when a majority (minority) of input values are 1. Thus, the majority functions, M, corresponding to a majority function is given by

$$M(A,B,C) = AB + BC + CA \qquad (1)$$

Sum is calculated with cout compliment

$$SUM = A \oplus B \oplus Cin = Maj(A,B,Cin).(A+B+Cin) + A.B.Cin$$
$$= Cout(A+B+Cin) + ABCin \qquad (2)$$

3. Design and Analysis

Wallace tree multiplier with 9T full adder design and simulations are performed in Cadence® Virtuoso with 180nm technology library. The circuit design is carried out in the Spectre environment and the simulations are performed in ADEL environment at a temperature of 27°C.

Fig. 2 depicts the design of full adder using majority function. It incurs 9 transistors which is much less than the conventional full adder.
Design of the proposed circuit consists of 3 input capacitors, they have been designed using the NMOS transistor available in the 180nm technology by tying the drain and source of the MOSFET.

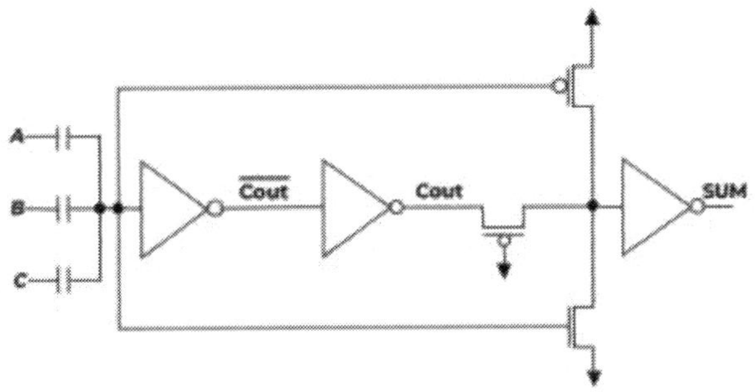

Fig 2. 9T Full adder design based on majority function

The design of Wallace tree multiplier consisting of full adders and half adders is designed with 3 stages of addition. Table 1 depicts the comparative analysis of the performance metrics with the conventional multiplier unit. The depiction of the delay in Fig. 3, demonstrates the delay incurred by the Wallace tree multiplier using majority function to be 1.05ns, which is 0.52 less than the Wallace tree multiplier with 16T adder.

Table 1. Performance analysis of multiplier architecture transistor count and power

Performance metrics	Wallace tree multiplier with 16T full adder [1]	Wallace tree multiplier using 9T majority gate adder [proposed]
Transistor Count	262	208
Power(μW)	136.4	85.1

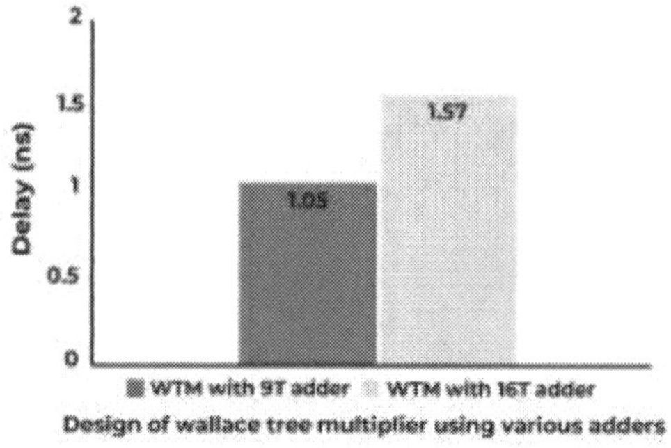

Fig 3. Comparison of delay in Wallace tree multiplier with 9T adder and 16T adder

4. Conclusion

A 4-bit high calibre Wallace tree multiplier using majority function is designed and implemented. The design and analysis using Cadence® Virtuoso with a 180nm technology library offers a reduced transistor count of 208 and power consumption of 85.1μW. It also demonstrates 33.12% reduction in delay while compared to Wallace tree multiplier with 16T full adder.

References

[1] M Naresh, B Suneetha,Assistant Professor, Department of Electronics and Communication Engineering, NallaNarasimhareddy College of Engineering, Hyderabad, Telangana, India. Design of Low Power Full Adder Based Wallace Tree Multiplier Using Cadence 180nm Technology. International Journal of Innovative Research in Science, Engineering and Technology.

[2] Yuvaraj Subramaniam, Srinivasan Alavandar, Department of ECE, Professor, Department of EEE. Low power and high speed computation using hybridized multiplier, IEEE 2013.

[3] Keivan Navi,Mehrad Maeen ,and Omid Hashemipour Faculty of Electrical and computer Engineering of shahid Beheshti University GC,Tehran,Iran.. An energy efficient full adder cell for low voltage, IEICE Electronic Express, Vol6, N0.9, 553-559.

[4] S. Kaviya, D. Kumar PG Scholar, Assistant Professor, Department of ECE, P.A. College of Engineering and Technology Pollachi, Tamil Nadu, India. Design of an Efficient Multiplier Using Transistor Level Modified Adders. Journal of VLSI Design and Signal Processing-ISSN: 2581-8449Volume 5 Issue 2.K. Elissa, "Title of paper if known," unpublished.

[5] Saradindu Panda,A. Banerjee, B. Maji and Dr. A.K. Mukhopadhyay, "Power and Delay Comparison in between Different types of Full Adder Circuits", International Journal of Advanced Research in Electrical, Electronics Engineering, Vol. 1, Issue 3, September 2012.

[6] Divya Bora,U.M. Gokhale. (2015),_Review on Design of High Speed Array Multiplier Using BICMOS Logic‖, International Research Journal of Engineering and Technology (IRJET),Volume02, Issue09, pp. 2395–0072.

[7] Shweta Kumar et al.,(2012), —Performance analysis of gdi based 1-bit full adder circuit for low power and high speed applications‖, International Journal of Science, Engineering and Technology,Volume3, Issue6, pp.37–43.

[8] Resham Singh et al., (2015), —High Performance Low Delay 10T Full Adder‖, International Journal of Innovative Research in Computer and Communication Engineering,Volume3, Issue8.

[9] R. S. Waters, E. E. Swartzlander, "A reduced complexity Wallace multiplier reduction," IEEE Transactions on Computers, vol. 59, no. 8, pp.1134-1137, 2010.

IOP Conf. Series: Materials Science and Engineering 1187 (2021) 012003 doi:10.1088/1757-899X/1187/1/012003

[10] K. G. Krishna, B. Santhosh, and V. Sridhar, Design of Wallace Tree Multiplier using Compressors, International Journal of Engineering Sciences & Research Technology, 2, 2013, 2249-2254.

Improving the Relation Classification Using Convolutional Neural Network

S Kamath, K G Karibasappa, Anvitha Reddy, Arati M Kallur, B B Priyanka, Bhagya B P
KLE Technological University, Hubballi, Karnataka, India

Abstract - Relation extraction has been the emerging research topic in the field of Natural Language Processing. The proposed work classifies the relations among the data considering the semantic relevance of words using word2vec embeddings towards training the convolutional neural network. We intended to use the semantic relevance of the words in the document to enrich the learning of the embeddings for improved classification. We designed a framework to automatically extract the relations between the entities using deep learning techniques. The framework includes pre-processing, extracting the feature vectors using word2vec embedding, and classification using convolutional neural networks. We perform extensive experimentation using benchmark datasets and show improved classification accuracy in comparison with the state-of-the-art methodologies using appropriate methods and also including the additional relations.

Keywords: Relation extraction, NLP, CNN, Word2vec embedding, Deep Learning.

1. Introduction

The task of determining semantic relations from a text is known as relation extraction. Extracted relations are those that exist between two or more entities of the same type and fit into one of the several semantic groups. It is an important part of natural language processing systems that needs to extract explicit facts from text, to ease the task of question answering systems[12] and knowledge base completion. Information extraction, Automatic content extraction [1] document summarization, machine translation, and the creation of thesauri and semantic networks are some of the uses of automatic semantic relations recognition.

The task of assigning specified relation labels to entity pairings that appear in texts is referred to as relationship classification. The following is an analysis of the Relation Extraction (RE) task: We want to figure out the link between e1 and e2 given a sentence S containing two nominals e1 and e2.

For example: "The<e1>burst</e1>has been caused by water hammer<e2>pressure</e2>."

e1 = burst

e2 = pressure

Cause-Effect (e2, e1).

IOP Conf. Series: Materials Science and Engineering 1187 (2021) 012004 doi:10.1088/1757-899X/1187/1/012004

The goal would be to automatically recognize that this sentence expresses Entity-Destination relationship between e1 and e2, for which we use the notation Entity-Destination (e1, e2). Accurate relation classification facilitates precise sentence interpretations, discourse processing, and higher level NLP tasks.

2. Related Work

One of the important factors seen in the Relation extraction is the availability of annotated datasets. Previously distinct supervision paradigm [3] were used instead of depending on the annotated datasets for training also several multilingual datasets [11] were developed for such tasks. With the advancement in the neural network and Deep learning techniques [13] several relation extraction classification models have been proposed considering various features. while syntactic features in the sentences [4] where the distance between the entities play a vital role, several other features include dependency parse features, incorporating knowledge through WordNet, Frame Net and POS [9] are also considered to bring significant improvements in accuracy of extraction. Yet another work considering lexical and sentence level features was proposed in [14] where these approaches depend on additional features.

Relation Extraction has not only seen its advancement in the task of relation classification but also has been used as a pipeline model or pre-processing model for many applications. Extraction task can be used to extract triplets to produce knowledge graphs [5] of relational facts between person location and organization in the text. However, in the wild it is not just the relation between these Named entities but different ways of expressing the common class relations between the entities are used [8]. Another visible change in the progress of Relation Extraction has been in considering the external knowledge [7] through knowledge base to boost the performance of the model. But in real world scenario the prior knowledge resources may not be consistent and may lead to ambiguity between the existing relations. Thus, with the combined features of the syntactic and semantic along with the external knowledge sources several relation extraction models have gained added accuracy.

Until recently the work of Relation Extraction has seen tremendous growth in binary classification where relation between the entities is learnt within the sentences but however when we consider the document as a discourse then the fragments of text [15] are classified as potential relations limiting the total classifications to be very minimal however for the relations across the sentences [6] there are plentiful of possible relations across and within the documents where utmost care must be given to avoid false positives in the relationship across sentences.

3. System model

The block diagram of the proposed system is shown in figure 1 below. The selected dataset from the user is pre-processed using the toolkit followed by the annotation of the dataset, which includes identification of the entities e1, e2,and labels. Further,these annotated sentences are given to the CNN model which uses pre-trained word2vec embeddings, leading to relation extraction between entities in the sentences.

IOP Conf. Series: Materials Science and Engineering 1187 (2021) 012004 doi:10.1088/1757-899X/1187/1/012004

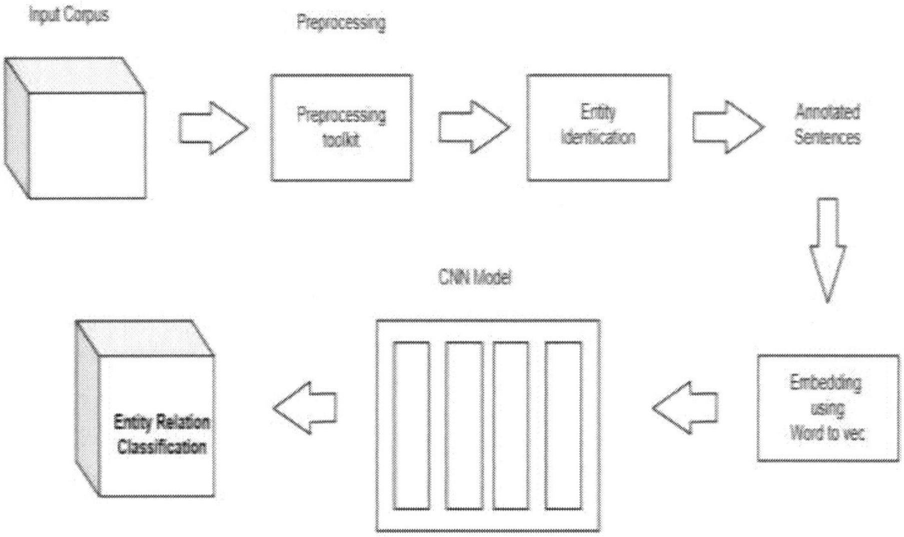

Figure 1. System Design for Relation Extraction

4. Architecture

Many Natural Language Processing tasks have been solved using Deep Neural Networks. Instead of extracting features manually, Deep Neural Network is used to perform optimized feature extraction for effective model construction, and hence it is a well-suited approach used in various domains. In this paper, a convolutional neural network is used to extract relationships using linguistic features and word embedding using a word table to generate feature vectors of the entities in the sentence. Pre-processing is performed to extract the labels from the text and generate tokens for the sentence. As mentioned in figure 2 below it can be observed that the model takes the complete sentence with mentioned entities as an input and output a probability vector corresponding to all possible relation types. Each feature is represented by a vector that is randomly initialized. A pertained word vector is employed for word embedding. The embedding layer maps every feature value with its corresponding feature vector and concatenates them, to get local features from each part of the sentence.

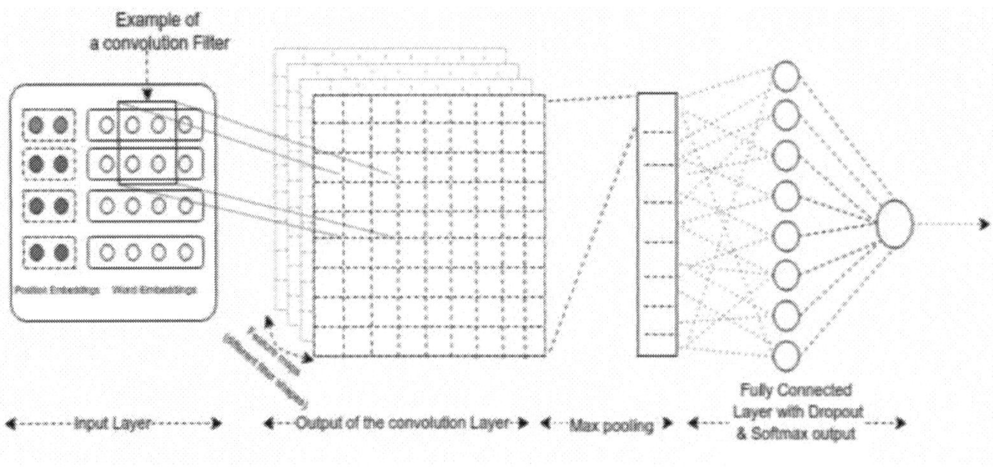

Figure 2. Convolution Neural Network for relation extraction

To get global features through all filters, we apply maximum pooling over time. The idea of max-pooling is to take into account only the most useful features from the full sentence. To make inferences, the pooled features are input into a fully linked feed-forward neural network. In the output layer, we use a softmax classifier with the number of outputs equal to the number of possible relations between entities.

5. Implementation

DATASET: SEMEVAL 2010 task 8 dataset was used for the proposed model with nine common relations mentioned in the state-of-art work as well as many other works. The state-of-the-art was further improvised by introducing 3 new relations into the dataset namely Phase-Of, Participates-In, Located-In. This introduction of new relations was done to increase the efficiency of the model as the same as mentioned in other similar works. Manual annotation was carried out for the new datasets introduced. Datasets with different types of English preposition relations were not considered as they require human annotation including annotating the entities as e1 and e2 and labelling which is a tedious task. The distribution of the dataset among training and testing, for the relations, is mentioned in table 1.

Table 1. Distribution of Dataset

Relation	Train Data	Test Data	Total Data
Cause-Effect	1003	328	1331
Component-Whole	941	312	1253
Content-Container	540	192	732
Entity-Destination	845	292	1137
Entity-Origin	716	258	974
Instrument-Agency	504	156	660
Member-Collection	690	233	923
Message-Topic	634	261	895
Product-Producer	717	231	948
Phase-Of	430	70	500
Participates-In	430	70	500
Located-In	430	70	500
Others	1410	454	1864
Total	9290	2927	12217

The train and test files are loaded to the model to generate labels and sentences text files. Vocabulary files words.txt and labels.txt containing the words and labels in the dataset are generated and stored. The model also saves a dataset_params.json with some extra information. A base model directory is created under the experiments directory which contains a file params.json which sets the hyper-parameters for the experiment. During training, a model is instantiated to train on the training dataset following the hyper-parameters specified in params.json.

Further evaluation is carried out using the given metrics on the development set.

Evaluation on the test set is carried after experimenting and selecting the best model and hyper-parameters based on the performance on the development set, the further performance of the model is evaluated on the test set and overall model performance is analysed after completing many experiments based on the training set's performance. F1- score combines precision and recall, and works also for cases where the datasets are imbalanced as it requires both precision and recall to have a reasonable value. Accuracy is always good to check as one option. F1-score is another. The model achieved Precision: 76.95, Recall:84.30, F1 score: 88.66, and Accuracy: 75.22, after being trained on the training dataset. Summarizing the performance of the model on the test dataset, the model achieved Precision: 78.08, Recall:83.64, F1 score:80.76, and Accuracy:75.44. The performance of the model on the revised train dataset is that the model achieved Precision:78.96, Recall:84.57, F1 score:81.66, and Accuracy:76.69. The performance of the model on the revised test dataset as compared with other datasets is that the model achieved Precision:80.05, Recall:84.73, F1 score:82.32, and Accuracy:77.14.

6. Results and Discussion

A brief description of the comparison of the proposed model with other models and the output of the system is summarized in this section. Table 2 shows the variation of Precision, Recall, F1 score, and Accuracy with different datasets experimented. As stated in [2] increase in the size of data would lead to better results, the same could be observed in the proposed work. The loss is calculated on training and validation and its interpretation is based on how well the model is performing. Loss value implies how well a model behaves after each iteration of optimization. The loss graph was plotted for understanding the behaviour of the model. The graph depicted that for each epoch during training, the loss was decreasing which is ideal for the model to perform in a better way.

Table 2. Comparison of results of data-sets used

Dataset	Precision	Recall	F1 Score	Accuracy
SEM-EVAL 2007 task 4	79.00	79.16	79.08	79.00
SEM-EVAL 2010 task 7	78.08	83.61	80.76	75.44
SEM-EVAL 2010 task 8 with 3 additional relations	80.06	84.73	82.32	77.14

7. Conclusion

In this paper, we have used SEM-EVAL 2010 TASK 8 dataset. The model was trained to learn the relationship among the entities in a sentence using a convolution neural network. To further improve the performance of extracting relations, a pre-trained Word2Vec embedding approach was integrated into the system which improved the f1 score to 80.76. To explore more relations among entities an additional 500 sentences each of 3 new relations was introduced into our dataset. Our results illustrated that the proposed model can keep high-precision over iterations, and adding more data and relations gives improved results. Hence we achieved the improved f1 score i.e, 82.32. Although there are other approaches, which seem to be much more complicated when compared to the proposed model which is simple and significantly gives a better F1 score.

8. References

[1] Stephanie Strassel, Mark Przybocki, Kay Peterson, Zhiyi Song, and Kazuaki Maeda. Linguistic resources and evaluation techniques for evaluation of cross-document automatic content extraction. In *Proceedings of the Sixth International Conference on Language Resources and Evaluation (LREC'08)*, Marrakech, Morocco, May 2008. European Language Resources Association (ELRA).

[2] Iris Hendrickx, Su Nam Kim, Zornitsa Kozareva, Preslav Nakov, Diarmuid Ó Séaghdha, Sebastian Padó, Marco Pennacchiotti, Lorenza Romano, and Stan Szpakowicz. SemEval- 2010 task 8: Multi-way classification of semantic relations between pairs of nominals. In *Proceedings of the 5th International Workshop on Semantic Evaluation*, pages 33–38, Uppsala, Sweden, July 2010. Association for Computational Linguistics.

[3] Sebastian Riedel, Limin Yao, and Andrew McCallum. Modeling relations and their men- tions without labeled text. In José Luis Balcázar, Francesco Bonchi, Aristides Gionis, and Michèle

IOP Conf. Series: Materials Science and Engineering 1187 (2021) 012004 doi:10.1088/1757-899X/1187/1/012004

Sebag, editors, *Machine Learning and Knowledge Discovery in Databases*, pages 148–163, Berlin, Heidelberg, 2010. Springer Berlin Heidelberg.

[4] Kun Xu, Yansong Feng, Songfang Huang, and Dongyan Zhao. Semantic relation classifica- tion via convolutional neural networks with simple negative sampling. In *Proceedings of the 2015 Conference on Empirical Methods in Natural Language Processing*, pages 536–540, Lisbon, Portugal, September 2015. Association for Computational Linguistics.

[5] Yuhao Zhang, Victor Zhong, Danqi Chen, Gabor Angeli, and Christopher D. Manning. Position-aware attention and supervised data improve slot filling. In *Proceedings of the 2017 Conference on Empirical Methods in Natural Language Processing,* pages 35–45, Copenhagen, Denmark, September 2017. Association for Computational Linguistics.

[6] Pankaj Gupta, Subburam Rajaram , Bernt Andrassy , Hinrich Schutze, Thomas Runkler. Neural Relation Extraction Within and Across Sentence Boundaries, 14th Jan 2019.

[7] Bo Li, Wei Ye, Canming Huang, and Shikun Zhang. Multi-view inference for relation extraction with uncertain knowledge, 2021.

[8] Linlin Wang, Zhu Cao, Gerard de Melo, and Zhiyuan Liu. Relation classification via multi- level attention CNNs. In *Proceedings of the 54th Annual Meeting of the Association for Computational Linguistics (Volume 1: Long Papers)*, pages 1298–1307, Berlin, Germany, August 2016. Association for Computational Linguistics.R. Nicole, "Title of paper with only first word capitalized," J. Name Stand. Abbrev., in press.

[9] Kamath S., Shivanagoudar C., Karibasappa K.G. (2021) Part of Speech Tagging Using Bi-LSTM-CRF and Performance Evaluation Based on Tagging Accuracy. In: Thampi S.M., Gelenbe E., Atiquzzaman M., Chaudhary V., Li KC. (eds) Advances in Computing and Network Communications. Lecture Notes in Electrical Engineering, vol 736. Springer, Singapore. https://doi.org/10.1007/978-981-33-6987-0_25

[10] Youngseo Son and H. Andrew Schwartz. Discourse Relation Embeddings: Representing the Relations between Discourse Segments in Social Media, 4th May 2021.

[11] Abhyuday Bhartiya, Kartikeya Badola, Mausam. DiS-ReX: A Multilingual Dataset for Distantly Supervised Relation Extraction, 17th April 2021.

[12] Jiayang Cheng1, Haiyun Jiang , Deqing Yang , and Yanghua Xiao. A Question-answering Based Framework for Relation Extraction Validation, 7th April 2021.

[13] Hailin Wang, Ke Qin, Rufai Yusuf Zakari, Guisong Liu, Guoming Lu. Deep Neural Network Based Relation Extraction: An Overview, 6th Jan 2021.

[14] Daojian Zeng, Kang Liu, Siwei Lai, Guangyou Zhou and Jun Zhao. Relation Classification via Convolutional Deep Neural Network, Proceedings of COLING 2014

[15] Youngseo Son and H. Andrew Schwartz. Discourse Relation Embeddings: Representing the Relations between Discourse Segments in Social Media, 4th May 2021.

Application of Natural Betalain Dye from Beetroot for Improved Efficiency in Dye Sensitized Solar Cell

K Soni[1], A Sheikh[1], V Jain[2] and N Lakshmi*[1]

[1]Department of Physics, Mohanlal Sukhadia University, Udaipur, Rajasthan, 313001, India
[2]Department of Physics, Geetanjali Institute of Technical Studies, Udaipur, 302132, Rajasthan, India

*1) N Lakshmi: nlakshmi@mlsu.ac.in

Abstract. C, N, S co-doped TiO_2 (TU-TiO_2) synthesized through solid state method using high energy ball milling was applied onto fluorine doped tin oxide (FTO) glass substrate and sintered at 450° C for half hour for use as a photoanode of dye-sensitized solar cells (DSSCs). Natural Betalain containing dye extracted from beetroot was used as sensitizer, Iodide triodide was used as electrolyte and graphite as counter electrode to obtain FTO/C, N, S-TiO_2/Dye/Electrolyte/Graphite/FTO DSSC. X-ray Diffraction analysis established the formation of single phased materials retaining the basic TiO_2 structure. Optical properties were enhanced with increase in molar concentration of thiourea in TiO_2. Fourier Transform Infrared spectroscopy confirmed the incorporation of sulphur into TiO_2. The highest of 94% relative increase is observed in solar to electric conversion efficiency for C, N, S co-doped TiO_2 compared to un-doped TiO_2. Along with decrease in the band gap making the cell more sensitive to visible light, the improved performance is also due to the rise in charge injection from betalain dye to the C, N, S co-doped TiO_2.

1. Introduction

Dye-sensitized solar cell (DSSC) is a device for the conversion of visible light into electricity, based on the sensitization of wide band gap semiconductors [1] and have become an attractive alternative to conventional solid state solar cells due to their remarkably low cost, ease in manufacturing and fairly good energy conversion efficiency [2- 5]. For example, the photoelectric conversion efficiency of the DSSC TiO_2/2 organic dye/Co(II/III) tris(phenanthroline)-based redox electrolyte /Pt has reached 14.30% [6]. Some of the popular materials used as photoanodes are TiO_2, ZnO, SnO_2 etc. However, since these materials have large band gaps of the order of 3 eV, they do not respond to visible light and become active only under ultra violet (UV) light [7]. Due to this limitation, solar energy cannot be utilized extensively in these materials and it is necessary that the band gap be decreased so that the visible light, which is majorly emitted by the Sun, can be absorbed. One method of doing this is by doping with anions leading to narrowing of band gap [8].

Among the popular photoanode materials, TiO_2 has a relatively smaller band gap along with fairly good conversion efficiency. Studies showing the shifting of the optical absorption edge of TiO_2 towards lower energy i.e., from UV to visible, by doping it with anions such as C [9-10], N [11-12] and S [13-16] have been reported. Recent studies have also revealed that TiO_2 nanofibers enhanced the optoelectrical properties of DSSCs. For instance, Mahmoud et al [17] have synthesized sulphur doped TiO_2 nanofibers and have reported solar to electrical conversion efficiency of 4.27% as compared to the 1.54% efficiency of untreated TiO_2 nanofibers. The choice of sensitizer in DSSCs also plays an important role in the working efficiency and also their stability, cost and visible light absorption characteristics. Most of the dyes used as sensitizers for DSSCs are Ruthenium (Ru) based and are costly [18]. There is thus a need to explore low cost, Ruthenium metal-free dyes for developing cost effective DSSCs. Recently, organic dyes have attracted interest due to their low cost and environmentally friendly nature [19-21]. As compared to synthetic organic and inorganic dyes, extraction of dyes from natural biological resources is an effective way of reducing the overall cost of solar cells. Studies of natural dyes extracted from various plants and fruits containing anthocyanin, betalain, chlorophyll, indigo etc. [22-25] show that these serve as effective sensitizers and have proved to bind easily with the surface of the photocatalyst [26].

To the best of the authors' knowledge, DSSCs prepared by C, N, S co-doped TiO_2 and betalain containing natural dye together have not been reported yet. In this work, our aim is to combine the enhancement of optoelectronic properties of TiO_2 on doping with anions C, N, S along with the use of a natural sensitizer, namely, betalain, present in the juice of beetroots, to prepare a low cost DSSC. The results show that the photovoltaic efficiency of the resulting DSSC is enhanced and is promising especially considering the ease of fabrication, environmentally friendly nature of the dyes and low cost of materials used.

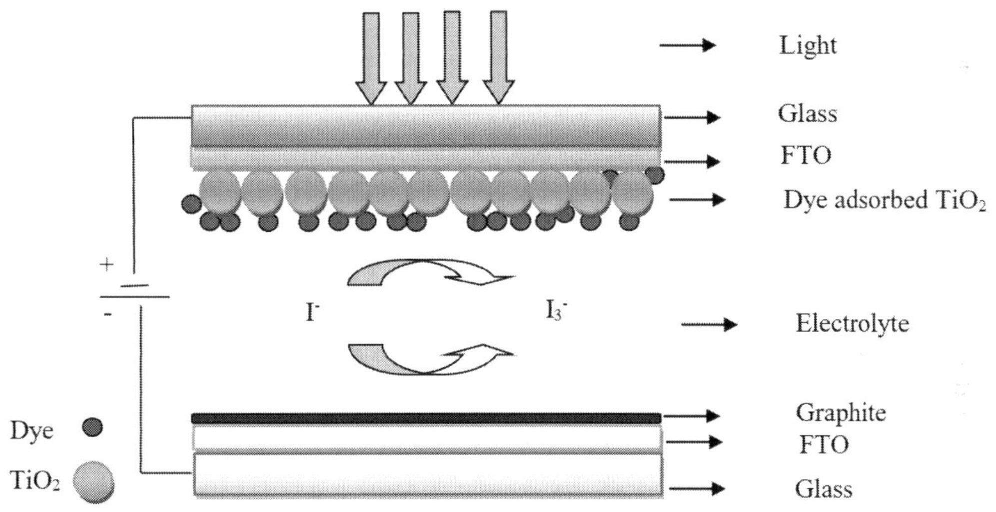

Figure 1. Schematic representation of the DSSC.

2. Experimental

2.1. Main Materials

The following materials have been used without any further purification. TiO_2 particles (Merck India), Thiourea (S.D Fine Chem. Ltd.), I_2 (Aldrich, USA, 99.99% pure), Potassium iodide (S.D Fine Chem. Ltd., 99.8% pure), Ethylene glycol anhydrous (Sigma Aldrich, USA, 99.8% pure), Acetonitrile (Merck, India, 99.5% pure), Ethanol (Jebsen & Jessen GmbH & Co., Germany, 99.9% pure), Acetic acid (Sigma Aldrich, USA, 99% pure), Ethyl Cellulose (Aldrich, USA), Terpinol (Aldrich, USA), Acetone (Merck, India) and FTO glass (USA, Aldrich, 2.2 mm thickness, surface resistivity 8Ω/cm^2).

2.2. Synthesis of C, N, S sulphur-doped TiO_2

Using TiO_2 and thiourea as a precursor, a series of un-doped and C, N, S co-doped TiO_2 samples were prepared by high energy ball milling. Three samples were prepared with a change in molar ratio of 0.25 in thiourea (0.25:4, 0.50:4 and 0.75:4) in order to dope carbon, nitrogen, sulphur into TiO_2. Pure

TiO_2 and thiourea mixed TiO_2 powders were ball milled in a high energy SPEX8000M mixer mill with a milling speed of 1080 rpm for 20 minutes. Tungsten carbide (WC) balls and vials were used fixing the ball to sample weight ratio to be 10:1. For purposes of discussion in the rest of this paper, the samples after milling are designated as TU-x, where, x represents the molar ratio of thiourea.

2.3. Extraction of natural dye
For the extraction of betalain dye, fresh beetroots were peeled, sliced to small pieces and ground to juice using a grinder machine. The juice was sonicated with 75% ethanol acidified with 10 mM HCl for 30 minutes and the solution was then filtered using a filter paper.

2.4. Preparation of DSSCs
Flourine doped Tin Oxide (FTO) slides were successively cleaned in surfactant dissolved distilled water, distilled water and acetone for cycles of 30 minutes each respectively in an ultrasonic bath. The C, N, S co-doped and un-doped TiO_2 powders obtained after ball milling were mixed with acetic acid, ethyl cellulose and terpinol as surfactant and kept suspended for half an hour after which it was sonicated using probe sonicator. The prepared slurries were doctor bladed on the conducting side of the FTO glass, followed by sintering at 450°C for 30 minutes in a furnace and then cooled to room temperature. The films were then kept dipped in 0.5 mM ethanol solution of the natural dye prepared from beetroot for 24 hours for complete dye loading. Graphite, which served as counter electrode, was prepared by rubbing high carbon pencil on FTO glass and sintering at 100 °C for 30 min. The electrodes (dye sensitized TiO_2 photo-anode) and counter electrodes were sandwiched together to make a solar cell with an active cell area of 0.25 cm^2. The two electrodes were held together by binder clips. An electrolyte consisting of 0.5 M I_2 and 0.05 M KI in 10 mL ethylene glycol was injected into the gap of the two electrodes. The schematic diagram for DSSC is shown in Figure 1.

2.5. Characterizations
The X-ray diffraction (XRD) patterns were recorded on a Rigaku Ultima IV X-ray diffractometer at 40 kV and 40 mA over a range of 2θ from 15° to 80° using Cu K_α radiation. Images were obtained using a multimode NT-MDT Solver NEXT Atomic Force Microscope (AFM) by scanning over an area of 10 μm x 10 μm. Field Emission Scanning electron microscope image of the TU-0.75 sample were taken in NOVA NanoSEM 450. The voltage and pressure were maintained at 15kV and 2.44 x 10^{-5} torr, respectively. Optical properties of the C, N, S co-doped TiO_2 were determined using a UV-Vis spectrophotometer (ELICO SL 159) in the range of 250 – 800 nm. Fourier transform infrared (FT-IR) spectra were obtained using Bruker model Alpha IR spectrometer in the range of 4000-420 cm^{-1}. The photovoltaic test of the DSSC was carried out by measuring the I-V characteristics curves using Keithley 2400 source meter under white light irradiation from a 200 W lamp. An irradiation intensity of 100mW/cm^2 was used for measurement of I-V curves.

3. Results and Discussion

3.1. X-ray Diffraction analysis
The best fitted XRD patterns of the un-doped and C, N, S co-doped TiO_2 powders calcined at 400°C for 1 hour are shown in Figure 2(a) which show well resolved diffraction peaks corresponding to the reflections of anatase phase of TiO_2. The diffraction patterns match well with the JCPDS data 894921[ICSD # 076173] and corresponds to a tetragonal structure with spacegroup I41/amd [27]. From the good fitted graphs, it is evident that mixing of thiourea and TiO_2 for the compositions studied results in a single phased material with the basic crystal structure of TiO_2. The average particle size, observed to be about 40 nm for all samples, and strain have been estimated by the Williamson Hall method and are tabulated in Table 1. Also, the refined parameters such as lattice parameters, volume, R_{Bragg}, R_f and χ^2 are given in Table 1. The variation of strain with change in the dopant molar concentration is shown in Figure 2(b) and is observed to range from 11.2 x 10^{-4} to 22 x 10^{-4}. From the figure it is clear that an increase in the thiourea molar concentration leads to an increase in strain with the relative strain of TU-0.75 sample being double than that of pristine TiO_2. This is because the

sulphur atom has a much larger size than oxygen atom and on being incorporated into the TiO_2 lattice without changing the crystal structure, the lattice becomes increasingly strained with an increase in the content of sulphur.

Table 1. Average particle size, strain and refined parameter of undoped and C, N, S co-doped TiO_2.

Sr. No.	Sample Name	Particle Size (nm)	Strain (x 10^{-4})	Lattice parameter a (Å)	c (Å)	Volume (Å^3)	R_{Bragg}	R_f	χ^2
1	TU-0	43	11.5	3.7840	9.5104	136.180	9.73	6.64	3.79
2	TU-0.25	40	9.3	3.7841	9.5099	136.178	8.83	6.57	3.99
3	TU -0.50	44	9.8	3.7843	9.5099	136.197	7.61	6.70	3.90
4	TU -0.75	40	22.4	3.7840	9.5093	136.167	6.55	5.27	3.70

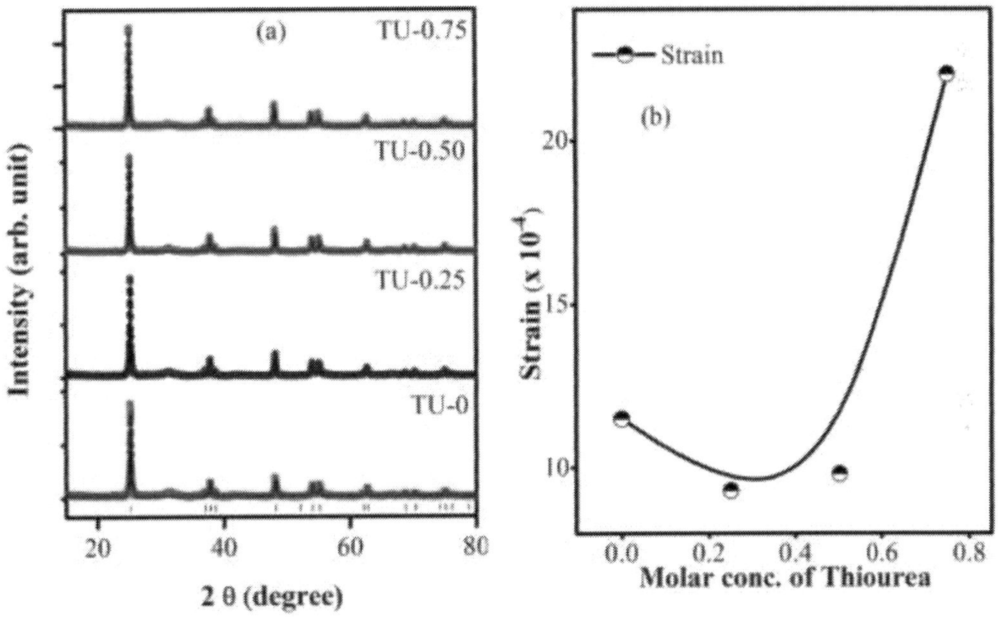

Figure 2. (a) XRD of undoped and C,N,S co-doped TiO_2; (b) variation of strain with change in molar concentration of thiourea in TiO_2.

3.2. Field Emission Scanning Electron Microscopy

Field Emission Scanning Electron Microscopy (FE-SEM) micrograph of C, N, S co-doped TiO_2 (TU-0.75) is reported in Figure 3(a). The surface morphology of TU-0.75 exhibits spherical shape, uniform distribution of grains and slight agglomeration. From the histogram plots in Figure 3(b), the average particle size obtained was 75 nm. Energy dispersive X-ray analysis (EDX) confirms the presence of carbon, nitrogen, sulphur, titanium and oxygen as observed by Figure 3(c). The quantitative results of EDX of C, N, S co-doped TiO_2 sample clearly show that the wt. % of C, N, S, Ti and O are 24.09, 0.45, 0.09, 41.01 and 34.36, respectively.

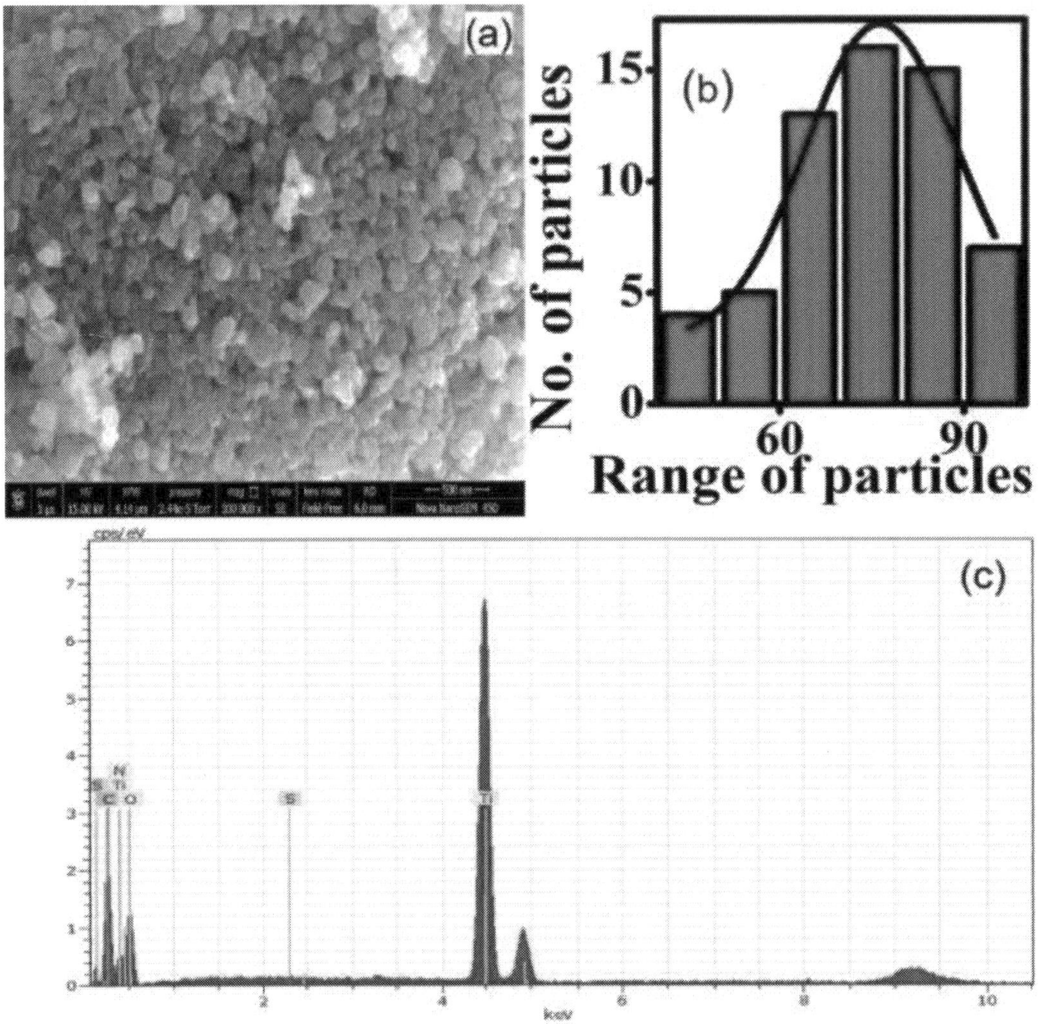

Figure 3. (a) FESEM image of TU-0.75; (b) histogram for particle size determination and (c) EDX of TU-0.75.

3.3. Atomic Force Microscopy

Analysis of the AFM images have been done by 3D and 2D profile analysis and the 3D AFM images are presented in Figure 4. The artefacts obtained in the AFM images were removed using the Gwddyion software and value of parameters such as average roughness, skewness, kurtosis and entropy are given in Table 2. Average roughness is calculated by the arithmetic average of height of the surface above a hypothetical smooth plane. It can be seen from the table that roughness of un-doped TiO_2 is the least and increases with an increase in the content of thiourea doping. This can be related to the increase in strain induced in the lattice due to the larger size of sulphur, as also observed from XRD studies. The Skewness (R_{sk}) shows the degree of symmetry of rough surface profile and can be used as a measure of balance between the peaks and valley of the asperites and describes the profile symmetry about a mean line. Accordingly, R_{sk} would be zero for symmetrical profile, negative for a surface with predominantly planar and valleys morphologies and positive for a surface which has more peaks than valleys [28]. The values obtained for the skew from the curve have positive value indicating that all the samples have more peaks than valleys. Kurtosis (R_{ku}) measures surface sharpness and show the degree of pointedness or bluntness of asperities on surface (R_{ku} is less than 3 for a platykurtic or flat surface). R_{ku} for all samples in the present study have values less than 3 (table 2) implying that the surface is platykurtic. Figure 5 gives the linear fits of skew and kurtosis for varying molar concentration of dopant in TiO_2 from which it is apparent that there is nearly a linear rise in these statistical or functional parameters with an increase in the molar concentration of thiourea. Entropy generally describes the morphology connected to the degree of order or disorder in grain structure distribution and the values obtained, given in table 2, are nearly the same for all four samples.

Table 2. AFM parameters of pure and C, N, S co-doped TiO_2.

Sr. No.	Sample No.	Mean Roughness	Skew	Kurtosis	Entropy
1	TU-0	208 nm	0.677	1.710	-14.58
2	TU -0.25	158 nm	0.305	1.634	-14.66
3	TU -0.50	220 nm	0.772	1.780	-14.62
4	TU -0.75	211 nm	0.786	1.795	-14.80

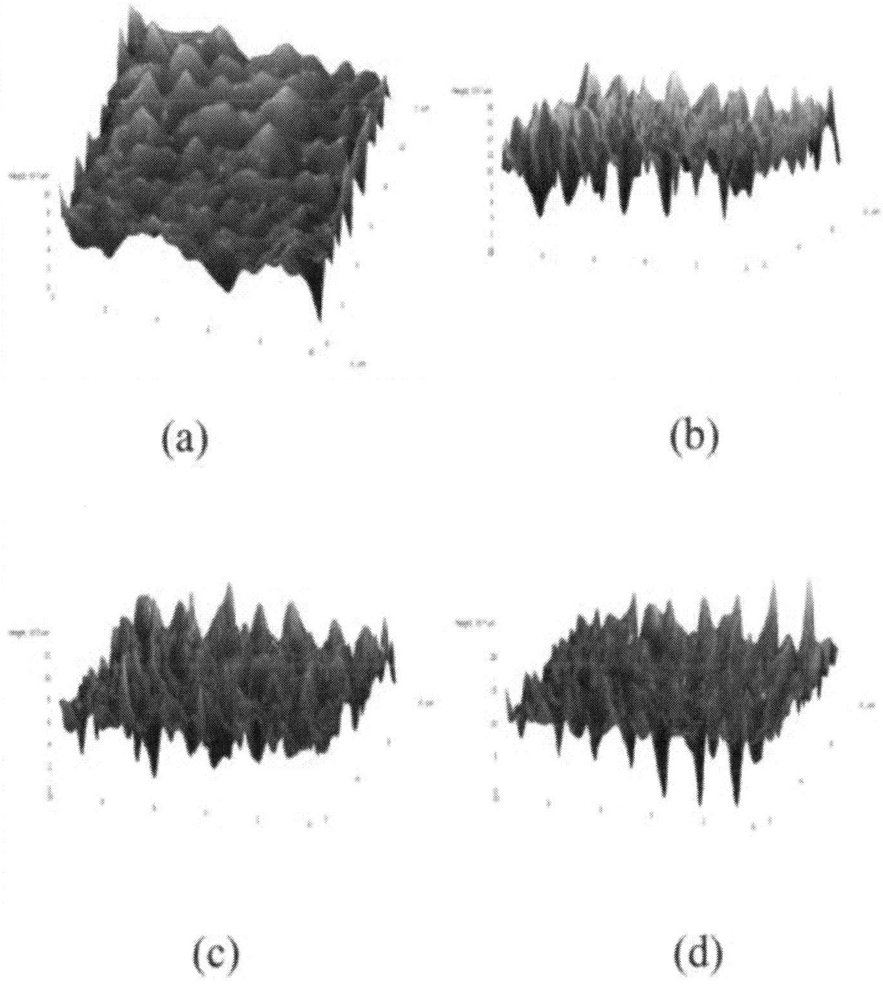

Figure 4. AFM 3D Images of (a) TU-0; (b) TU-0.25 doped TiO_2 ;(c) TU -0.5 doped TiO_2 and (d) TU -0.75 doped TiO_2.

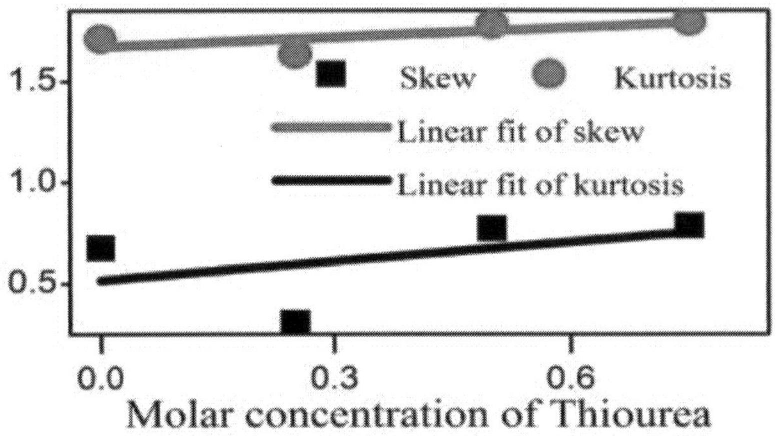

Figure 5. Trends of skew and kurtosis with varying molar concentration of thiourea in TiO_2.

3.4. UV- Visible Spectroscopy

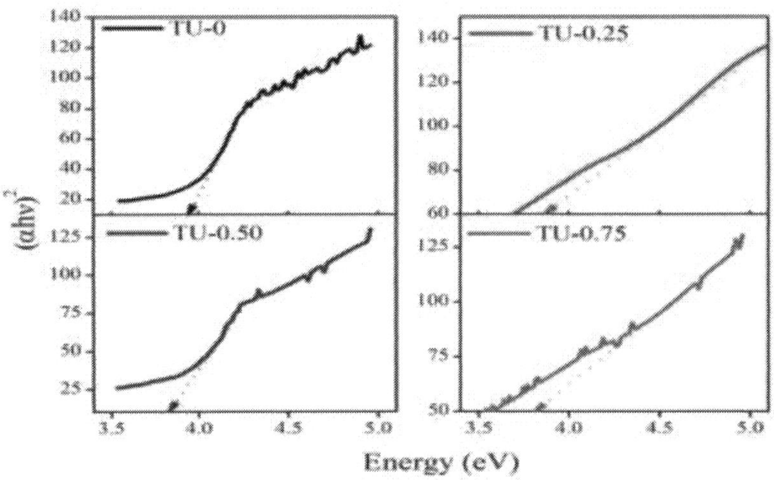

Figure 6. Tauc plot of pristine TiO_2 and C, N, S co-doped TiO_2.

Optical properties of the samples have been measured using UV-visible spectrophotometer and are based on the Beer - Lambert Law which relates the absorption of light with the physical property of the material. The optical absorption measurements of the samples were carried out in the wavelength range of 230-800 nm. In general, although the absorbance was observed to be high in the UV and lower visible regions with decrease at higher wavelengths, the absorption edge of TiO_2 shifts towards higher wavelengths with increase in the molar concentration of thiourea indicating that doping with carbon, nitrogen, sulphur leads to a higher response of TiO_2 to visible radiation. The shifting of the absorption edge towards higher wavelength results in the reduction of band gap energy with an accompanying enhancement in the optoelectronic properties.

The absorption coefficient was calculated using the following equation

$$\alpha = \frac{A(E_g - h\nu)^{1/2}}{h\nu} \quad (1)$$

Where, α represents absorption coefficient, E_g the band gap, h Planck constant and ν the frequency of the radiation. For determining the band gap of the C, N, S co-doped and undoped TiO_2, Tauc plots are plotted as shown in Figure 6. The linear behaviour of the graph indicates direct nature of optical bandgap. Direct bandgap values were evaluated from the intercepts obtained from straight line extrapolation towards the energy axes. The band gap obtained was 3.94 eV, 3.86 eV, 3.84 eV and 3.8 eV for TiO_2, TU-0.25, TU-0.50 and TU-0.75 respectively. A linear variation of the band gap as a function of the molar concentration of thiourea is obtained (Figure 7) from which it is observed that band gap decreases by 0.18 eV with an increase of every 25% in the doping molar concentration of thiourea into TiO_2.

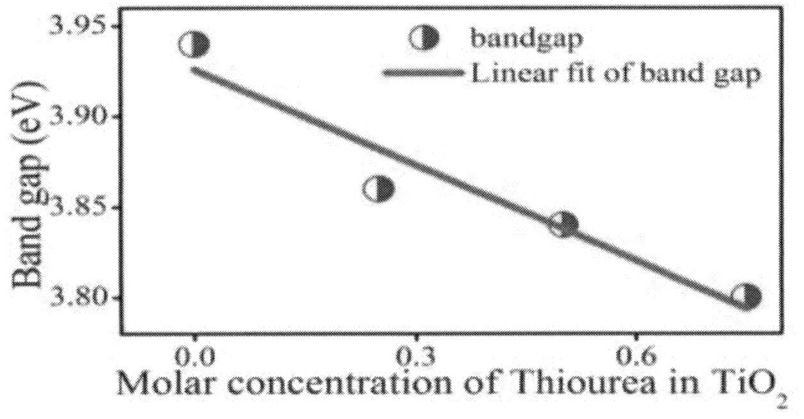

Figure 7. Variation in band gap as a function of change in molar concentration of thiourea in TiO_2.

3.5. Fourier Infrared Spectroscopy

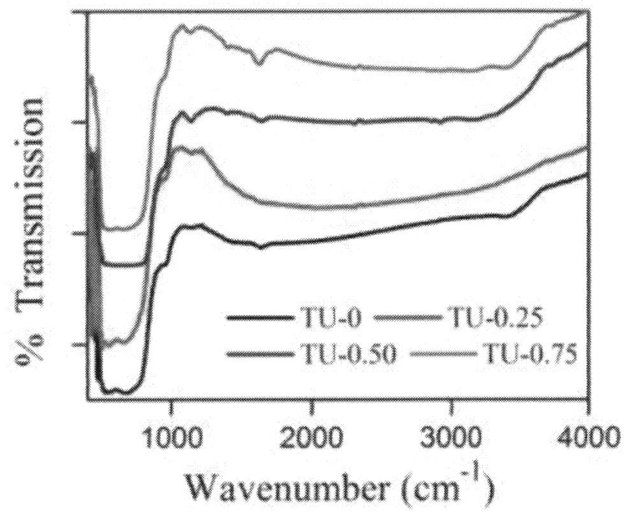

Figure 8. Infrared spectrum of undoped and C, N, S co-doped TiO_2.

Fourier transform infrared (FTIR) spectroscopy is a useful tool in recognizing the chemical bond structure and provides an insight into the essential chemistry of the material such as the existence of interactions present in samples, material phase composition etc. FTIR spectra of pristine TiO_2 and C, N, S co-doped TiO_2 in the range of 420- 4000 cm^{-1} are shown in Figure 8. The peaks in the range of 3400-3100 cm^{-1} and 1700-1600 cm^{-1} are attributed to surface absorbed H_2O and hydroxyl groups respectively [29, 30]. From the figure, a strong band at 1630 cm^{-1} for all the samples is observed which is attributed to the O-H bending vibration of adsorbed water molecule on the surface of material [31]. The intensity of 1630 peak is stronger for TU -0.75 doped TiO_2 as compared to un-doped TiO_2. This is due to the of imbalance of extra positive charge created locally due to the introduction of S^{6+} into the

TiO_2 lattice since Ti is present as Ti^{4+} in the TiO_2 crystal structure resulting in attraction of more hydroxide ions on the crystal surface. Devi et al. [29] have also obtained similar results in their study of S doped TiO_2 for enhancement of photocatlytic activity. For all C, N, S co- doped TiO_2 samples, a band in the range of 1150 – 1130 cm^{-1} is seen, which is indicative of S-O vibration and confirms the incorporation of sulphur in the TiO_2 crystal structure [32]. Various peaks are observed in the range of 800- 420 cm^{-1} which can be attributed to the Ti-O-Ti stretching vibrations [33, 34].

3.6. Device performance

Current density (J) versus voltage (V) plot of DSSCs of undoped and C, N, S co-doped TiO_2 are given in Figure 9 and the short circuit current density (J_{sc}), open circuit voltage (V_{oc}), fill factor and efficiency are summarized in Table 3. The plot comparing the efficiency and fill factor as a function of the doping molar concentration of thiourea in TiO_2 is shown in Figure 10 from which it can be seen that the DSSCs with carbon, nitrogen and sulphur doping show improved performance as compared to the undoped TiO_2. From Table 3, it can be seen that maximum V_{oc} and J_{sc} is observed for TU-0.75 doped TiO_2 due to which the efficiency of TU-0.75 is 94% more than that of undoped TiO_2. Our obtained efficiency is higher than that of reported value by García-Salinas et al. and Sengupta et al. [33, 34]. This is due to the higher obtained maximum point current and maximum point voltage leading to increased fill factor. However, for these solar cells, the observed J_{sc} and V_{oc} are less compared to the DSSCs prepared using commercial dyes (e.g. N719, N3). The lower value of open circuit voltage (V_{oc}) may be due to the molecular structure of the natural dye (betalain) used in these samples. The obtained efficiencies are also lower than the commercial dye based DSSCs and is due to the comparatively weaker binding of natural dyes with metal oxides (e.g. TiO_2, ZnO, SnO_2) along with a lower charge transfer spectra of these dyes under solar spectrum [35].

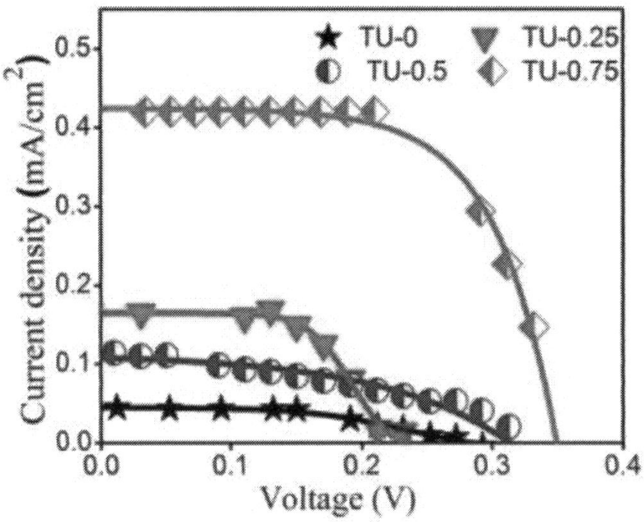

Figure 9. J-V plots of prepared DSSCs (Lines are Sigmoidal fitted).

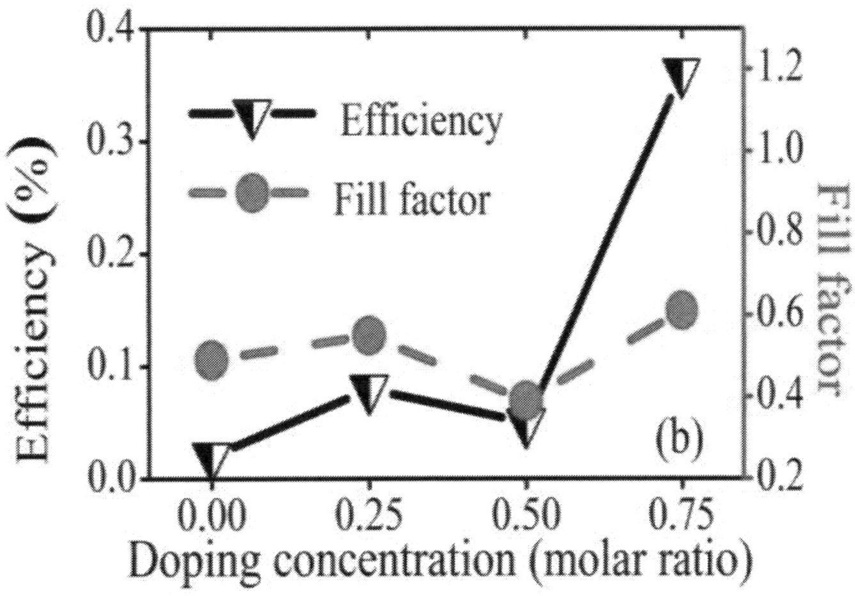

Figure 10. Efficiency and fill factor vs. doping concentration of thiourea of fabricated DSSCs.

Table 3. Photovoltaic parameters of DSSCs sensitized by betalain dyes.

Configuration	Sample name	J_{sc} (mA/cm^2)	V_{oc}(V)	Fill factor	Efficiency (%)
FTO/TU-0/Dye/Electrolyte/Graphite/FTO **(our work)**	TU0	0.04	0.28	0.49	0.02
FTO/TU0.25/Dye/Electrolyte/Graphite/FTO **(our work)**	TU 0.25	0.16	0.24	0.55	0.08
FTO/TU0.50/Dye/Electrolyte/Graphite/FTO **(our work)**	TU 0.5	0.11	0.32	0.38	0.05
FTO/TU0.75/Dye/Electrolyte/Graphite/FTO **(our work)**	TU 0.75	0.42	0.35	0.61	0.36
FTO/ZnO/Dye/Electrolyte/carbon/FTO **(others work)** [36]		0.72	0.46	0.54	0.17
FTO/ZnO/Dye/Electrolyte/Platinum/FTO **(others work)** [35]		0.75	0.45	0.58	0.19

4. Conclusions

Different molar concentrations of thiourea i.e C, N, S co-doped TiO_2 were prepared by high energy ball milling to device a DSSCs. Investigations using X-ray diffraction have ruled out formation of other phases and average particle sizes and strain have also been estimated from the spectra. Dye sensitized solar cells have been fabricated using C, N, S co-doped TiO_2 as photoanode and betalain containing natural dye, extracted from beetroot, as sensitizer. AFM measurements show increased roughness on doping with thiourea. FT-IR results confirm the incorporation of sulphur in TiO_2. Analysis of the optical parameters shows that thiourea doping has enhanced the optical properties. Doping TiO_2 with carbon, nitrogen and sulphur has led to a systematic decrease in the band gap. Results show that C, N, S doping has increased the efficiency of the solar cell with respect to un-doped TiO_2. This study thus shows the promising nature of using the chosen methods/sensitizer with advantages such as the ease of fabrication and cost effectiveness of the overall fabrication process of DSSCs. More importantly, the use of natural dye as sensitizer makes it environmentally friendly and with this simple process a relative increase of highest of 94% in the efficiency is obtained for 75% molar concentration of thiourea doped into TiO_2. Extensive availability of beetroot and ease of extraction of betalain dye from it makes this technique novel and inexpensive one for solar cells applications.

Acknowledgement
We acknowledge support from DST-FIST and UGC-DSA schemes of the Department of Physics, MLSU. Kumavat Soni acknowledges UGC NET-JRF. The authors are grateful to K.D. Patel, Professor, Department of Physics, Sardar Patel University, Gujarat, India for his useful discussion and guidance in electrical properties measurement. FT-IR and UV-Vis optical measurements were performed respectively in the Department of Industrial Chemistry and Department of Botany, MLSU.

References

[1] Grätzel M 2003 *J. Photochem. Photobiol. C: Photochemistry reviews* **4** 145-153
[2] Sharma K, Sharma V and Sharma S S 2018 *Nanoscale Res. Lett.* **13** 381
[3] Hosseinnezhad M, Gharanjig K, Moradian S and Saeb M R 2017 *Energy* **134** 864-870
[4] Umale S, Sudhakar V, Sontakke S M, Krishnamoorthy K and Pandit A B 2019 *Mater. Res. Bull.* **109** 222-226
[5] Kabir F, Sakib S N and Matin N 2019 *Optik* **181** 458-464
[6] Kakiage K, Aoyama Y, Yano T, Oya K, Fujisawa J I and Hanaya M 2015 *Chem. Comm.* **51** 15894-15897
[7] Zheng P, Zhou W, Wang Y, Ren D, Zhao J and Guo S 2020 *Appl. Surf. Sci.* **512** 144549
[8] Phattalung S N, Limpijumnong S and Yu J 2017 *Appl. Catal. B: Environ.* **200** 1-9
[9] Han X, An L, Hu Y, Li Y, Hou C, Wang H and Zhang Q 2020 *Appl. Catal. B: Environ.* **265** 118539
[10] Noorimotlagh Z, Kazeminezhad I, Jaafarzadeh N, Ahmadi M and Ramezani Z 2020 *Catal. Today* **348** 277-289
[11] Huang H, Song Y, Li N, Chen D, Xu Q, Li H and Lu J 2019 *Appl. Catal. B: Environ.* **251** 154-161
[12] Qiu P, Sun X, Lai Y, Gao P, Chen C and Ge L 2019 *J. Electroanal. Chem.* **844** 91-98
[13] Zhou S, Liu S, Su K and Jia K 2019 *J. Alloy. Compd.* **804** 10-17
[14] Jiang L, Luo Z, Li Y, Wang W, Li J, Li J, Ao Y, He J, Sharma V K and Wang J 2020 *Chem. Eng. J.* **402** 125549
[15] Sun Q, Zhang J, Wang P, Zheng J, Zhang X, Cui Y and Zhu Y 2012 *J. of Renew. and Sustain. Energy* **4** 023104
[16] Ma D, Xin Y, Gao M and Wu J 2014 *Appl. Catal. B: Environ.* **147** 49-57
[17] Mahmoud M S, Akhtar M S, Mohamed I M, Hamdan R, Dakka Y A and Barakat N A 2018 *Mat. Lett.* **225** 77-81
[18] Wongcharee K, Meeyoo V and Chavadej S 2007 *Sol. Energy Mater. Sol. Cells* **91** 566-571
[19] Zhang L,Yang X,Wang W, Gurzadyan G G, Li J, Li X, An J, Yu Z, Wang H, Cai B and Hagfeldt A 2019 *ACS Energy Lett.* **4** 943-951
[20] Jia H L, Li S S, Gong B Q, Gu L, Bao Z L and Guan M Y 2020 *Sustain. Energy Fuels* **4** 347-353
[21] Heng P, Mao L, Guo X, Wang L and Zhang J 2020 *J. Mater. Chem. C* **8** 2388-2399
[22] Gokilamani N, Muthukumarasamy N, Thambidurai M, Ranjitha A and Velauthapillai D 2013 *J. Sol-Gel Sci. Technol.* **66** 212-219
[23] Isah K U, Ahmadu U, Idris A, Kimpa M I, Uno U E, Ndamitso M M and Alu N 2015 *Mater. Renew. Sustain. Energy* **4** 39
[24] Syafinar R, Gomesh N, Irwanto M, Fareq M and Irwan Y M 2015 *Energy Procedia* **79** 896-902
[25] Hosseinnezhad M 2016 *Mater. Technol.* **31** 348-351

[26] Hao S, Wu J, Huang Y and Lin J 2006 *Solar energy* **80** 209-214
[27] Soni K, Sheikh A and Lakshmi N 2020 *AIP Conference Proceedings* **2220** 0200741-5
[28] Raposo M, Ferreira Q and Ribeiro P A 2007 *Modern research and educational topics in microscopy* **1** 758-769
[29] Devi L G and Kavitha R 2014 *Mater. Chem. Phys.* **143** 1300-1308
[30] Soni K, Sheikh A and Lakshmi N 2019 *AIP Conference Proceedings* **2162** 0201081-5
[31] Yang G, Yan Z and Xiao T 2012 *Appl. Surf Sci.* **258** 4016-4022
[32] Yang G, Xiao T, Sloan J, Li G and Yan Z 2011 *Chemistry–A European Journal* **17** 1096-1100
[33] Yang G, Jiang Z, Shi H, Xiao T and Yan Z 2010 *J. Mater. Chem.* **20** 5301-5309
[34] Huang D, Liao S, Quan S, Liu L, He Z, Wan J and Zhou W 2007 *J. Mater Sci.* **22** 2389-2397
[35] Sengupta D, Mondal B and Mukherjee K 2015 *Spectrochimica Acta Part A: Molecular and Biomolecular Spectroscopy* **148** 85-92
[36] García-Salinas M J, and Ariza M J 2019 *Appl. Sci.* **9** 2515

Integrated Multi-Operational Antenna System Design for CR Applications

Rajeev K. Parida[1], Arjuna Muduli[2], Dhruba C. Panda[1], Deepak Kumar Nayak[3]

[1]Electronic Science and Technology, Berhampur University, Berhampur, India
[2] Dept. of ECE, Koneru Lakshmaiah Education Foundation, Vaddeswaram, Andhra Pradesh, India
[3]Department of ECE, Budge Budge Institute of Technology, Kolkata, West Bengal, India
rkparida91@gmail.com, arjunamuduli@gmail.com, nayakd990@gmail.com

Abstract. A four-port multi-operational antenna system is designed for Cognitive Radio application. It consists of a sensing ultra-wideband antenna and three communicating narrowband antenna. The CPW fed monopole is a sensing antenna. It covers 2- 11 GHz of the UWB band. The three narrow band antennas are operating at 5.834 GHz, 6.42 GHz, 7.355 GHz, and 8.786 GHz respectively. The mutual coupling is below -15 dB. This integrated UWB/NB antenna system can also be a good candidate for C-band, Maritime radio navigation, and X-band applications.

Keywords *- Cognitive Radio (CR), Ultra-Wideband (UWB), Narrowband (NB)*

1. Introduction

In the last few years, most of the allocated radio spectrum is underutilized for the exponential growth of wireless communication users. For efficient spectrum utilization, Cognitive Radio (CR) emerged as an efficient solution [1, 2]. The antenna is a vital part for CR system [3-12]. This system needs a UWB antenna for continuous spectrum sensing [13]. However, if any channels are available or vacant, the communication is established through a narrowband (NB) antenna. Several communication bands are classified within UWB. To use these bands, more number of NB antennas is needed. However, UWB and the number of NB antennas are responsible for large physical dimensions and coupling issues in a CR antenna system [14-19]. Ebrahim etal. proposed a two-port antenna system of dimension 68 x 54 x 0.79 mm^3 which can operate in quad-band mode and covers 3.1-11 GHz band for sensing operation [20]; In another work [21], the system of dimension 80 x 65 x 1.58 mm^3 operates in single-band mode and used 2-5.5 GHz band for sensing purpose. A various integrated antenna system for cognitive radio is reported in [22, 23]. However, such an antenna system suffers from lower sensing bands, poor gain, and isolation, also larger antenna size. Now a day, researchers are using extra circuitry to improve isolation [24]. Without adding any extra circuitry and components to achieve high isolation is an open challenge. Further, reconfigurable antennas are gaining more popular [25-26]. The reconfiguration can be done by using switching components (i.e p-i-n diodes, varactor diodes, etc.). In [27], the authors highlight the drawbacks of such kind of switching techniques. Again, the number of NB states is limited to the underutilization of available system areas [28].

A compact four-port antenna system is proposed in this article. This CR system is maximally meets the above challenges.

IOP Conf. Series: Materials Science and Engineering 1187 (2021) 012006 doi:10.1088/1757-899X/1187/1/012006

2. Antenna design

The Integrated UWB/NB antenna system is shown in Fig. 1. It consists of one sensing and three communicating antennas integrated in a proper manner.

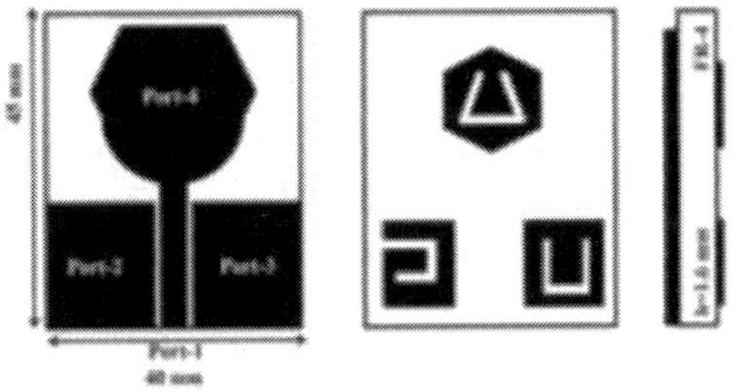

(a) (b) (c)

Figure 1. Integrated four-port antennas structure (a) Top view, (b) Back view, and (c) Side view

A. Ultra-wideband Sensing Antenna

We choose a CPW fed monopole antenna for spectrum sensing (Fig. 2). It helps to easy integration of narrowband communicating antennas within the available area. The geometrical parameters of the CPW-fed monopole antenna are well documented in [13]. The lower cut-edge frequency (UWB antenna) is determined by using Equation -1.

$$f_L = \frac{c}{\lambda} = \frac{7.2}{\{(l+r+g)\times k\}} \text{GHz} \tag{1}$$

where the equivalent cylindrical height, radius, and the gap of the monopole antenna are represented with l, r, and g. Empirically, we fixed the value of k to obtained f_L.

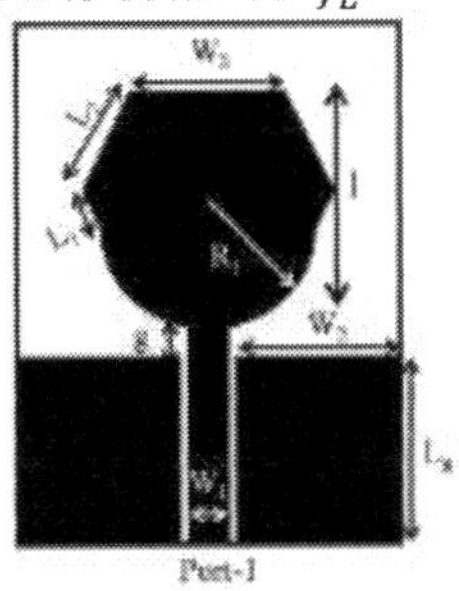

Figure 2. UWB sensing antenna (Antenna-1) [W_1=3 mm, W_2=18.20 mm, W_3= 14.80 mm,L_1=5.83 mm,L_2=14.80 mm, Lg=16 mm, R_1=12 mm, g=0.5 mm]

B. Narrow-band Communicating Antenna

Three NB antennas are chosen for communication purposes are shown in Fig. 3. These are well-known slot antennas reported in [3, 19]. The side edge of the hexagonal patch antenna is 6 mm (Fig. 3 (c)).

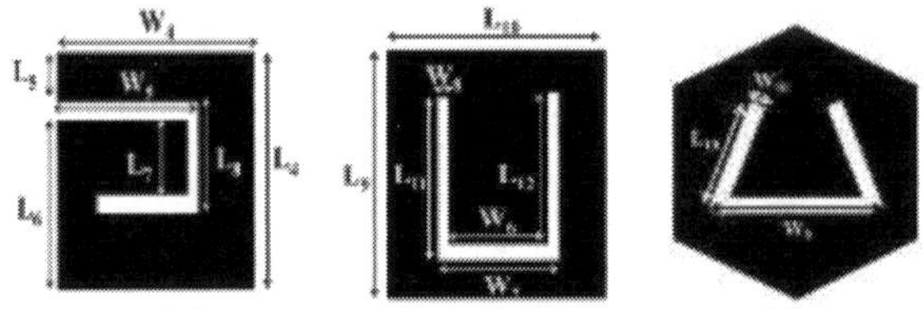

(a) (b) (c)

Figure 3. NB communicating antennas (a) G-shaped (Antenna-2), (b) U-slot (Antenna-3), and (c) Semi-triangular slot (Antenna-4) [W_4=9.50 mm, W_5=7.50 mm, W_6=3 mm, W_7=3.50 mm, W_8=0.25 mm, W_9=6.50 mm, W_{10}= 0.25 mm, L_4=9 mm, L_5=2 mm, L_6=6 mm, L_7=3 mm, L_8=5 mm, L_9=7.50 mm, L_{10}=7 mm, L_{11}=5 mm, L_{12}=4.50 mm,L_{13}=6 mm]

3. Results and discussion

CST 2019 is used for simulating and analyzing the system performances. The system is simulated by using a low-cost FR-4 substrate (ε_r =4.4 and loss tangent=0.02) of size 45 x 40 x 1.6 mm^3.

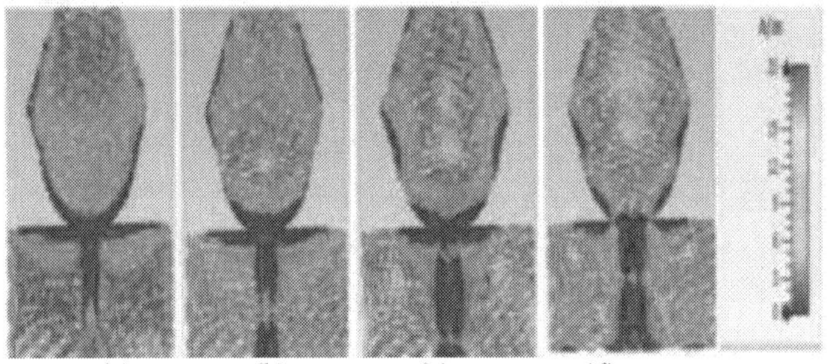

(a) (b) (c) (d)

Figure 4. Simulated surface current densities (Monopole antenna) at (a) 2.35 GHz, (b) 4.52 GHz, (c) 8.1 GHz, and (d) 10.12 GHz

Figure 4 shows the surface current distribution of the UWB monopole antenna at different frequencies. It helps in identifying the areas where the current amplitude is minimal. So we, incorporate the NB communicating antennas in that location (Both center of ground plane and the radiator) where the magnitude of the current is less. Also, they indicate the formation of higher order modes as frequency increases. It helps to increase the NB states by efficiently utilizing the system area.

Figure 5 shows the reflection coefficient vs. frequency plot of the proposed antenna system. It observed that the UWB antenna operates from 2 GHz to 11 GHz, and the NB antennas operating at 5.834 GHz, 6.42 GHz, 7.355 GHz, and 8.786 GHz. The simulated gains of the UWB antenna are 2.07 dBi, 3.36 dBi, 5.51 dBi at 2.35 GHz, 4.52 GHz, and 10.12 GHz.

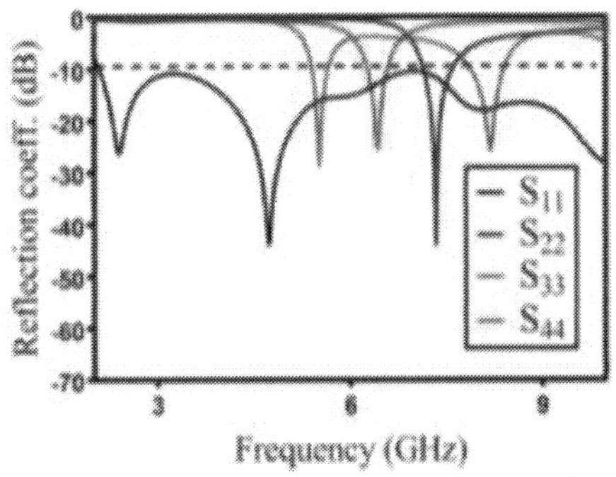

Figure 5. Reflection coefficient of the integrated UWB/NB antennas system

Figure 6 illustrates the isolation plot of the proposed antenna system. The isolation is less than -15 dB throughout the Ultra-wideband range. Figure 7, 8, and 9 shows the simulated radiation patterns for the individual antennas. Table I indicates the performance and applications of NB antennas.

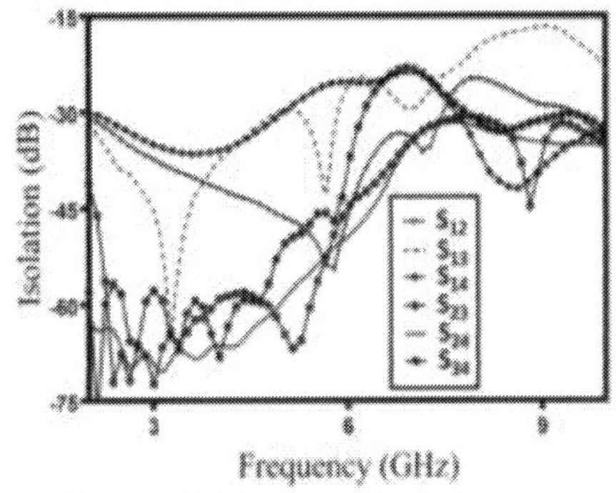

Figure 6. Simulated isolation plot

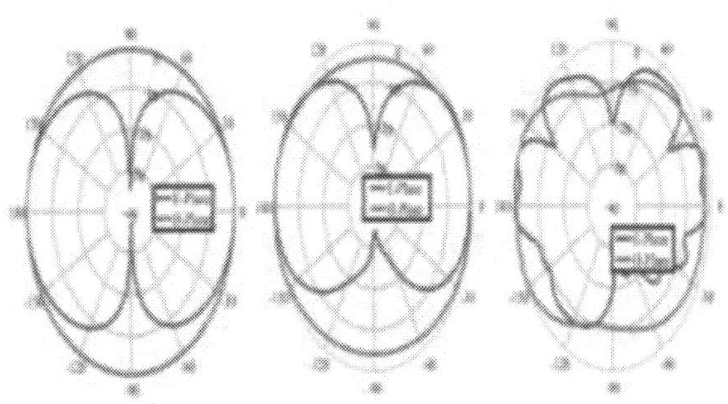

(a) (b) (c)

Figure 7. Radiation patterns of Antenna-1 at (a) 2.35 GHz, (b) 4.52 GHz, and 10.12 GHz

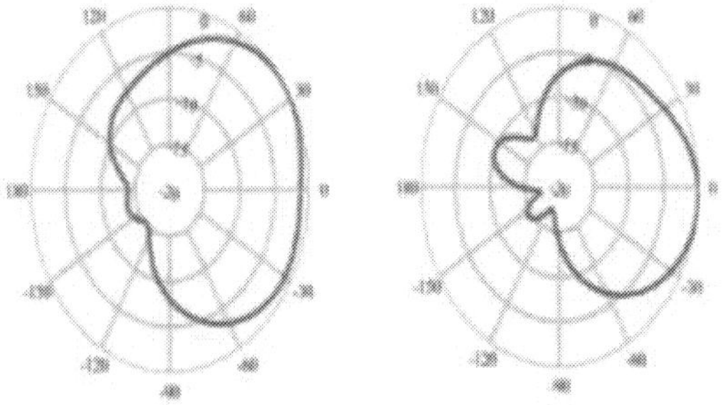

(a) (b)

Figure 8. Radiation patterns of Antenna-2 at (a) 5.834GHz, and (b) 8.786 GHz.

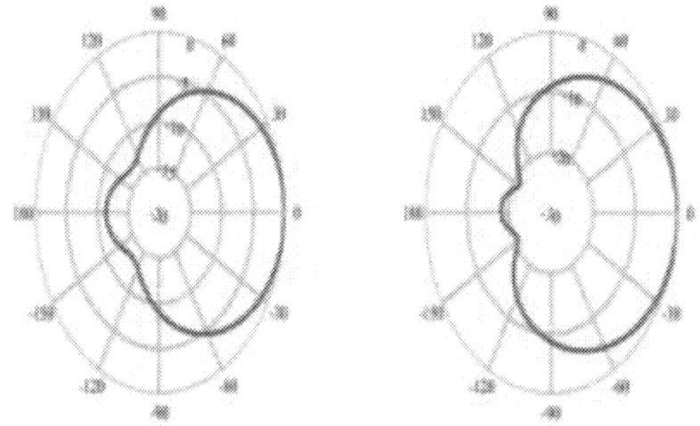

(a) (b)

Figure 9. Radiation pattern (a) 7.355 GHz (Antenna-3), and (b) 6.42 GHz (Antenna-4)

Table 1. Performance and applications of communicating antennas

Parameters	**Port-3 (G-shaped)**	**Port-2 (U-slot)**	**Port-4 (Semi-triangular slot)**
Operating Freq. (GHz)	5.834, 8.786	7.355	6.42
Bandwidth (GHz)	5.70-6, 8.44-9.02	7.1-7.7	6.22-6.55
Gain (dBi)	2.76, 2.69	4.47	3.857
Applications	Maritime Radio Navigation	X-band	C-band

Table II shows comparison between proposed works with previously reported work. It is found that our system has provided maximum number of NBs and compact in size.

Table II. Comparison of the proposed antenna with previously reported antennas

Ref.	UWB Range (GHz)	Number of Narrowbands	Number of Ports	Antenna Size (mm^2)
[20]	3-11	4	2	68 x 54
[21]	2-5.5	1	2	80 x 65
[22]	3.3-12	2	2	58.35 x 75.7
Proposed antenna	2-11	4	4	45 x 40

4. Conclusion

A small four-port antenna system of dimension 45 x 40 x 1.6 mm^3 is proposed. The major contribution is that the number of NB states increases by proper utilization of the available system area. The coupling between the antennas below -15 dB is obtained.

Acknowledgment

R. K. PARIDA (IF160222) thanks the DST INSPIRE Govt. of India for providing financial support in terms of a senior research fellowship (SRF).

References

[1] FCC, "FCC report and order on ultra wideband technology," Federal Commun. Commission, Washington, DC, USA, 2002.
[2] MS.Haykin,"Cognitive radio: brain-empowered wireless communications," IEEE Journal on Selected Areas in Communications, vol. 23, no. 2, pp. 201–220, Feb. 2005
[3] S. Shrestha, S.-K. Noh, and D.-Y. Choi, "Comparative Study of Antenna Designs for RF Energy Harvesting," International Journal of Antennas and Propagation, vol. 2013, pp. 1–10, 2013.
[4] Immadii G., Venkata Narayana M., Anuradha S.Y., Rani M.A., Lokesh K., Meghana B., "Dynamic frequency selection in wireless access systems including the band reallocation and allocation by using cognitive radios," Journal of Advanced Research in Dynamical and Control Systems, 10 (2), pp. 516-522, 2018.
[5] B B Shankar, D Anantha Sai Kumar, G Venkata Kowndinya, Arjuna Muduli, Deepak Kumar Nayak, "An Ultra-Wide Band Octagonal Antenna with Reconfigurable Narrowband Antenna for Cognitive Radio Applications," International Journal of Scientific & Technology Research, ISSN 2277-8616, Vol. 9, pp: 1573-1576, Issue 1 January 2020.
[6] Lakshmi M.V., Pardhasaradhi P., Madhav B.T.P., "Circular monopole reconfigurable antenna with notch band filter characteristics," Journal of Engineering Science and Technology Review, 11 (5), pp. 139-143, 2018.
[7] Teja Babu K., Syam Sundar P., Madhav B.T.P., Prudhvi Nadh B., Kotamraju S.K. "Dual notch UWB monopole antenna with u-shaped slots," ARPN Journal of Engineering and Applied Sciences, 14(11), pp. 2125-2130, 2019.
[8] Rama Krishna T.V., Madhav B.T.P., Geetanjali S., Parnika B., Bhargavi M.L., Tanmai A.S., Anilkumar T. "Design and study of a CPW fed truncated circular patch switchable band-notched UWB antenna," International Journal of Recent Technology and Engineering, 8(1), pp. 3037-3043, 2019.

[9] Prabakaran N., Kalyan S., Murthy K., Sai Baba D., Naveen C., Sai R. (2019), "A coplanar waveguide (CPW) fed circular microstrip antenna for UWB applications," International Journal of Innovative Technology and Exploring Engineering, 8(6), pp. 531-534.

[10] Mishra, S. R., & Lalitha, S. K. "Filtennas for wireless application: A review," International Journal of RF and Microwave Computer-Aided Engineering, 29 (10), 2019.

[11] Murthy K.S.R., Umakantham K., Murthy K.S.N., Madhav B.T.P. (2019), "Design and analysis of a reconfigurable antenna for dual band ISM medical and Wi-Fi applications," International Journal of Innovative Technology and Exploring Engineering, 8(10), pp. 3630-3634.

[12] Gundu R.P., Saradhi P.P., Madhav B.T.P., Nadh B.P. (2019), "Frequency reconfigurable lungs-shaped microstrip antenna using pin diodes and annular ring," International Journal of Innovative Technology and Exploring Engineering, 8(8), pp. 181-184.

[13] R. K. Parida, R. K. Mishra, N. K. Sahoo, A. Muduli, D. C. Panda, and R. K. Mishra, "A Hybrid Multi-Port Antenna System for Cognitive Radio," Progress In Electromagnetics Research, vol. 106, pp. 1–16, 2020.

[14] R. K. Parida, A. K. Sahu, D. K. Naik, N. K. Sahoo, D. C. Panda, and A. Muduli, "A UWB and Multiband Reconfigurable Antenna for Application in Cognitive Radio," in 2019 IEEE Indian Conference on Antennas and Propogation (InCAP), pp. 1–3, Dec. 2019.

[15] A. Muduli and R. K. Mishra, "Modified UWB microstrip monopole antenna for cognitive radio application," in 2015 IEEE Applied Electromagnetics Conference (AEMC), pp. 1–2, Dec. 2015.

[16] A. Nella and A. S. Gandhi, "A Five-Port Integrated UWB and Narrowband Antennas System Design for CR Applications," IEEE Trans. Antennas Propagat., vol. 66, no. 4, pp. 1669–1676, Apr. 2018.

[17] S. Atapattu, C. Tellambura, and H. Jiang, Energy Detection for Spectrum Sensing in Cognitive Radio. New York: Springer-Verlag, 2014.

[18] Y. Tawk, J. Costantine, and C. Christodoulou, *Antenna Design for Cognitive Radio*, Artech House, 289, 2016.

[19] A. Zaidi, W. A. Awan, N. Hussain, and A. Baghdad, "A Wide and Tri-band Flexible Antennas with Independently Controllable Notch Bands for Sub-6-GHz Communication System," Radioengineering, vol. 29, no. 1, pp. 44–51, Apr. 2020.

[20]E. Ebrahimi, J. R. Kelly, and P. S. Hall, "Integrated Wide-Narrowband Antenna for Multi-Standard Radio," IEEE Transactions on Antennas and Propagation, vol. 59, no. 7, pp. 2628–2635, Jul. 2011.

[21] H. Nachouane, A. Najid, A. Tribak, and F. Riouch, "Dual Port Antenna Combining Sensing and Communication Tasks for Cognitive Radio," International Journal of Electronics and Telecommunications, vol. 62, no. 2, pp. 121–127, Jun. 2016.

[22] S. Pahadsingh and S. Sahu, "A two port UWB-dual narrowband antenna for cognitive radios," Microwave and Optical Technology Letters, vol. 58, no. 8, pp. 1973–1978, 2016.

[23] A. Nella and A. S. Gandhi, "A Survey on Planar Antenna Designs for Cognitive Radio Applications," Wireless Pers Commun, vol. 98, no. 1, pp. 541–569, Jan. 2018.

[24] N. Anveshkumar and A. S. Gandhi, "Lumped Equivalent Models of Narrowband Antennas and Isolation Enhancement in a Three Antennas System," Radioengineering, vol. 27, no. 3, pp. 646–653, Sep. 2018.

[25] R. Kumar and R. Vijay, "Frequency agile quadrilateral patch and slot based optimal antenna design for cognitive radio system," International Journal of RF and Microwave Computer-Aided Engineering, vol. 28, no. 2, p. e21176, 2018.

[26] Allam V.K., Madhav B.T.P., Anilkumar T., Maloji S., 'A novel reconfigurable bandpass filtering antenna for IoT communication applications', Progress In Electromagnetics Research C, 96 , pp.13-26, 2019.

[27] N. A. Kumar and A. S. Gandhi, "A Compact Novel Three-Port Integrated Wide and Narrow Band Antennas System for Cognitive Radio Applications," International Journal of Antennas and Propagation, vol. 2016.
[28] S. Sharma and C. C. Tripathi, "An Integrated Frequency Reconfigurable Antenna for Cognitive Radio Application," Radioengineering, vol. 26, pp. 746–754, Sep. 2017.

IOP Conf. Series: Materials Science and Engineering 1187 (2021) 012007 doi:10.1088/1757-899X/1187/1/012007

Power-efficient Strategies for Sensing in Autonomous Mobile Robots, a critical requirement of I4.0 standard

Jyoti Bali[1,5], Arpitha H[2,5] , Anushree N[3,5], Arunkumar Giriyapur[4,5]

[1]jyoti_bali@kletech.ac.in, [2]arpitahuggi16@gmail.com,
[3]nilavaranushree@gmail.com,[4]aaron@kletech.ac.in
[5]KLE Technological University, Hubballi-31, Karnataka, India

Abstract. In a production environment, there are several challenges in meeting the Industry4.0 (I4.0) standard requirements. Energy efficiency is an essential area of focus. In the production setup, the critical and real-time control systems need to be very efficient while implementing functions, namely, accurate sensing, fast processing and precise actuation. Automated Guided vehicles (AGVs) and Automated Guided Vehicles are an integral part of modern and intelligent manufacturing systems. Power consumption in such systems is directly proportional to the performance level achieved. However, there is a need to evolve strategies to reduce power consumption and attain optimal performance. Field Programmable Gate Array(FPGA) based controller solutions can provide competent performance at optimized power consumption. The proposed work discusses the requirements of I4.0 concerning energy efficiency infrastructures for the intelligent manufacturing setup. The need to develop efficient subsystems for sensing, decision-making and actuation based on FPGA is stressed. Thus the focus is on the FPGA based power-efficient models used for sensor fusion technique in Autonomous Mobile Robots. The fundamentals of sensor fusion technique and the need to fuse sensor data for improved decision making and actuation are emphasized.

Keywords: *Industry 4.0, Energy efficiency, Automated Guided Vehicles, Autonomous Mobile Robots, Smart Manufacturing, Field-programmable Gate Array, Sensor Fusion*

1. Introduction

The recent manufacturing applications do range from mobile computing to cloud computing. As a result, intelligent and interconnected manufacturing systems have evolved, hence Industry 4.0. The result is that the systems are specialized as real-time systems enabling the industry to achieve new lean manufacturing levels. The prominent technologies in place are Big Data, Smart Factory concept, Cyber-physical systems, Industrial IOT and Interoperability. Big Data deals with handling massive data collected from various sources within and outside the Manufacturing industry. The analysis of massive data enables intelligent analysis of data and improved decision-making process that makes a significant and remarkable change in how the organizations or the manufacturing setups function. Smart Factory provides the seamless integration of all the processes from planning to execution, enabling improved productivity, maintenance and optimization of each at the individual process level. Cyber-physical systems aim to refine the computation, networking of each of the physical processes and employ feedback control loops to work in real-time. The decision making in the control loops involves artificial intelligence and machine learning ability. Industrial IoT stands for connecting all devices to the internet and each other across the industrial plant. It is based on cloud computing and

IOP Conf. Series: Materials Science and Engineering 1187 (2021) 012007 doi:10.1088/1757-899X/1187/1/012007

intelligent sensor networks. Thus an infrastructure with robust connectivity enables the smart factory to use the gathered data to manufacture, collaborate with other machines, report and learn at highly efficient rates. As a result of seamless integration of organizational data generated from interconnected cyber-physical elements sharing data across the internet for the smart manufacturing setup, intelligent machines and humans' interoperability is achieved. Interoperability enables error-free transmission and compatibility across the factory environment, achieving higher efficiency, accuracy and reliability as per the requirements of Industry 4.0. Automated Guided Vehicles(AGVs), Automatic Storage and Retrieval Systems(ASRS), Autonomous Mobile Robots(AMRs) have emerged as primary solutions for material handling applications in a modern factory environment[1-6].

Autonomous Mobile Robots is an example of the evolution of intelligent technology as per I4.0 needs. AMR senses the surrounding environment and decides its shortest navigation path, avoiding the obstacles along the path traversed. An autonomous mobile robot's objective is to find an efficient navigation route to its final destination, using the support of robust sensor networks and an accurate decision-making facility. The efficiency of the AMR is judged by the saving incurred in reduced power consumption, enhanced battery life, and the distance travelled by it. The onboard facility of intelligence provides Autonomous Mobile Robots with a vital role in a smart factory environment. AMR's have evolved over the years to provide increased flexibility, safety efficiency, higher throughput and assured fast return on investment (ROI). The intelligence level developed on intelligent mobile robots ensures that they can handle the monotonous, hazardous, and laborious, dull and dirty tasks efficiently, replacing humans. AMRs for Logistics for labour-intensive transport applications, E-commerce functions, namely, raw materials transport, sorting of received parcels for delivery and returns, inventory management, and manufacturing applications, as data-centres.
Furthermore, there are research platforms for biotech and health care industry related services, Data Centers, Research and development. There is a widespread use of innovative methodologies employed for healthcare-based solutions for accurate detection of ailments or abnormal conditions in human beings and the systematic analysis of the identified problems followed by accurate diagnosis. Similar facilities are available for monitoring and up-keeping machine health conditions for effective maintenance [7-14].

In a production environment, Automation and Robotics based control systems play an essential role in implementing the objectives set for the attainment of I4.0. Hence the control system solutions need to have high performance and power-efficient characteristics. The adopted control strategy needs to be very efficient, as the accuracy and reliability expected in the robot operation are very high in autonomous applications, replacing human beings. It involves accurate sensing, fast processing and precise actuation. The current era of technological changes has been adopted at an accelerated rate to attain Industry 4.0 standards. The role and need of mobile robots in the present industrial scenario are highly challenging, as the demand for rational approaches to material handling and manipulation is fast growing. Many types of sensors may be used in a mobile robot application to gather information from the surrounding environment. The sensors, in general, possess distinct characteristics, having a unique principle of operation with a wide range of the electromagnetic spectrum, suiting the demands of various applications. A single sensor used in standalone mode provides a limited sensing range in a constrained environment and is associated with operational errors. However, many sensors' synergistic and networked operation provides the broader perspective of the sensed environment in a mobile robot application. The high-speed performance of sensors employed in the critical control systems ensures that the response is always relevant. Since mobile robots operate in a complex, uncertain and highly dynamic environment, the rich information gathered from sensor networks can helping the robot system optimize a strategy for path planning and navigation. It is also essential that how best a controller uses the sensor data tapped from a collection of sensors and fuses them provides the highly reliable and intelligently processed information for robot navigation [1-6].

1.1 Energy requirements of I4.0 standard
Energy efficiency is an important measure considered for I4.0 applications. In countries like Europe and the United States, industrial energy consumption amounts to 25% to 35% of their total energy

production. Intelligent manufacturing systems under I4.0 hence aim for power-efficient and sustainable operations. Similarly, along with the supply of electric power, its constant availability and reliability of performance are critical factors. Hence the energy efficiency and energy management are the critical needs of the I4.0 standard. Hence, innovative strategies and technologies are employed to improve energy efficiency. Management using the IoT enabled monitoring and control services to minimize the disruptions in the intelligent manufacturing setup. The objectives are here to limit energy wastage and to exercise real-time control of power consumption. It is ensured using an array of sensor nodes, monitoring the integration of energy sources across an intelligent network in the company and beyond. It is also an essential factor to improve cost-effectiveness. Thus, the steps initiated in this direction impact even abiding with the environmental regulations and exploiting alternate energy sources. The key drivers for power management functions in line with the demands of the I4.0 standard can be listed as follows:

- Need for meeting requirements in the regulatory, environmental and societal context.
- Need for efficient mechanisms for managing complexities and challenges of industrial ecosystems
- Need for awareness regarding energy management and efficiency
- Need to ensure uninterrupted operations using critical monitoring strategies based on IT-enabled services
- Need for highly reliable, optimized quality of power using intelligent decision-making functions implemented using data analytics and AI-based machine learning ability[1-6].

Intelligent manufacturing setups use intelligent energy management and metering solutions based on Industry 4.0 and IOT to continuously monitor the plant's operations, energy consumption, wastage, or energy leak. All the devices are connected to share information about production and operation status to promote energy efficiency. Wired and wireless sensor networks role is prominent in gathering the relevant data and contextual information across the plant for integrated and intelligent decision-making on energy management. The diagram in Figure 1 shows the prominent components in an environment that even supports collaborative learning [1-6].

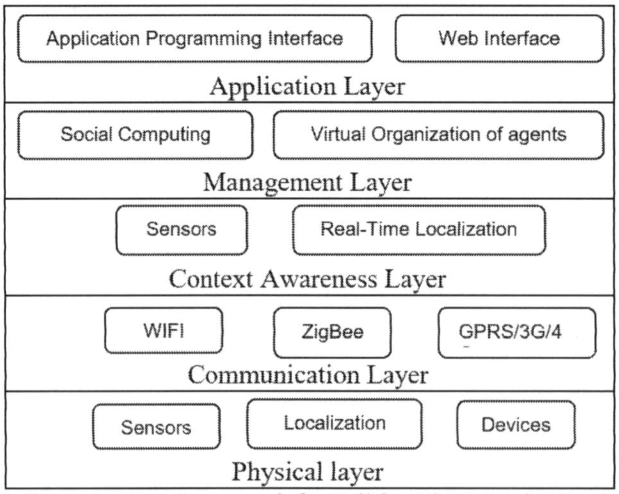

Figure 1. Context-Aware Framework for Collaborative Learning Environment

The Physical layer comprises all the devices and the infrastructure for energy management across the plant. Sensors in standalone and networked form collect the data that help in operations, namely, characterizing the environment, setting machine parameters, monitoring machine status (idle or run), gathering maintenance status parameters, and detecting overriding conditions machine or process parameters. These are essential in estimating the power consumption profile. Localization is a process of identifying, locating, and tracking operators and the machines accurately, using a set of beacons set up throughout the plant is ensured. Devices include all those interfacing equipment needed by the system to generate the localization data acting as the intermediate devices between sensors and the designed localization infrastructure and the internet. Communication Layer allows for accommodating any communication protocol, namely, Modbus, Profibus, Ethernet/IP, Interbus, Control Net, ZigBee, WIFI, and GPRS/3G/4G. The context-aware layer is built using both the wireless sensor network and Real-time localization using the available physical and logical infrastructure. The function, namely, collecting contextual information from sensors and the localization data from beacons, is carried out in the context layer. The management layer provides the services related to communication, logical and intelligent functions correlated with the monitored activities. Decision-making functions are invoked based on the various conditions or events that arise during the operations. Critical decision making is required for cases, namely, machine tool in jammed condition, vibration level attaining the maximum limit of defined thresholds, non-functioning Auto Tool-changer. Turning off of heating process or reducing fuel consumption efficiency can be options for overriding operating temperature conditions in a process setup. Application Layer is concerned with identifying all those events or actions in machine operating conditions, machine status and the process or the tool parameters. Social computing maximizes information functionality to improve relationships between humans and machines [1-6].

Thus the framework provides information on the energy consumption facts based on the data collection and interpretation mechanisms. Recommendation mechanisms are built based on the energy consumption patterns identified and recorded. Based on the requirements and recommendations of the energy-aware framework, the appropriate hardware platforms need to be adopted. FPGA based solutions can provide power-efficient architectures for accurate Data acquisition, decision making and precise actuation in the control loops set up for innovative manufacturing solutions. FPGA based hardware design exploits the inherent features such as reconfigurability, parallelism and competitive cost as compared with processor technology. FPGA's are widely used for real-time applications. FPGA based implementation of algorithms for sensor fusion, image processing and actuation can reduce cost as well as time. It ensures the quick prototyping of complex algorithms, simplifies verification and testing. Hence the feasibility of using FPGA/SOC based solutions for accurate sensing achieved through sensor data fusion techniques, followed by high-speed execution of control strategies, namely, obstacle detection and path planning and finally, the precise actuation for robot navigation. The sensing, processing, and actuation strategies can be implemented using an algorithm sequentially on a sophisticated controller [1-6]. The other alternative is to use multiple processor cores for distributed and parallel computing, ensuring high-speed processing. The case study of the sensor fusion process followed for Autonomous Mobile Robot is discussed further in this direction.

2. Autonomous Mobile Robot and its functionalities

The critical components used in a mobile robot are based on the key functions, namely, Perception, Decision-Making and Actuation. Perception is the built-in capability of a robot system using the appropriate sensing technologies, namely, laser scanning, stereo vision, force-torque sensing, and spectrometry, to see and perceive the environment. Internet data provides enormous data to the robot in either wired or wireless fashion as an endless array of sensors that can support perception. The decision is the action or the outcome planned based on the intelligence models designed and the environment's perceived information. Nowadays, Artificial intelligence models mimic the human brain capabilities restricting to the application demand. As a result, the robots prove to be more innovative and faster in decision making. Actuation is the process of energizing the actuating elements of the robot for motion[15-20]. The critical functionalities associated with any autonomous mobile robot are Navigation, Mapping and Localization, and Obstacle Avoidance. Robot perception of the environment and how decision-making processes are handled are crucial for a mobile robot application. The navigation plan for a mobile robot is implemented based on the data collected from various sensors. Robot perception is enhanced using a robust sensor network. The use of technologies, namely Cloud Computing, Fog Computing, and Edge Computing, enables the mobile robot to network with intelligent systems across the plant and even remotely. Edge computing solutions mainly rely on FPGA based reconfigurable architectures. The algorithm implementation sing FPGA provides significant performance as compared with CPU. Even though building FPGA compatible code, i.e. VHDL or Verilog, is complex and time-consuming, it is worth doing as it provides an actual acceleration. Hence the FPGA implementation of the Central limit theorem for Sensor Fusion is proposed.

2.1 Sensor Fusion Technology in Autonomous Mobile Robot

Among the various tasks performed by a mobile robot, the real-time object detection is a prominent one. It is a computer vision technique that helps the robot to capture, identify and locate the objects. Using the sensor fusion technique, the data is extracted from the information gathered from various sensors and treated suitably as required by the system. The robot then performs the object detection task in order to detect the object and the environment. Object detection task primarily involves capturing and identifying the objects to decide on the navigation path. The popular applications of real-time object detection are, namely, crowd controlling, self–driving cars, face recognition, and surveillance applications. The sensor fusion process yields the data, which is more accurate and dependable an provides better performance as compared to the data collected from a individual sensor data. The process is preferred mainly to obtain noise free and precise measurements required in a mobile robot application to achieve accurate navigation. As each sensor type has its strengths and weaknesses, the algorithm needs to determine the way the data is captured and processed to provide an accurate model. The work proposed here is based on an algorithm that combines IMU positioning and ultrasonic positioning for localization in an Autonomous Mobile Robot. Here the algorithm based on an extended Kalman filter is used to fuse the information collected from IMU and ultrasonic sensors to estimate the target position that can enable in building an accurate and stabilized localization model [15-20].

2.2 FPGA Technologies for Sensor Fusion

The FPGA board for sensor fusion is used to develop the sensor fusing algorithm and compile the program. The mobile robot sensors provide the data (position/detection of an object) from the front view, which is digital, and then sends the data back to FPGA. The interfacing code for the sensors can be written in Arduino IDE kind of platforms, which can be further implemented on FPGA. FPGA board provides the facility to interface different sensors and to implement the logic of sensor fusion process. The fusion process can be implemented either using direct fusion or the indirect fusion process. Direct fusion can be the result of fusing data from sensor types, namely, heterogeneous sensors, homogeneous sensors, soft sensors and also the sensor data history. Indirect fusion process involves data gathering from the environment along with human experiences and knowledge, apart from sensor data. The sensor fusion algorithms use standalone hardware that can be implemented on Field Programmable Gate Array (FPGA) or Application Specific Integrated Circuit (ASIC). FPGAs provide an attractive option for the fusion implementation as compared with that of an ASIC fabrication, which proves to be time consuming and an expensive alternative [15-20].

2.3 Challenges in Sensor Fusion in Real-Time Object Detection

The increasing demand for robust and precise navigation system has given rise to sensor fusion algorithms. However, choosing a suitable algorithm is the biggest challenge to achieve precise output. These algorithms mainly possess two challenges, i.e., these algorithms are computation-intensive and require hybrid but tightly coupled processing resources. Real-time sensor fusion requires high speed performance using the parallel data processing capability of systems of the type FPGA. Similarly, the processing capability needs to be robust that handles a vast amount of information and complex interrelationships in the multisensory information. The prominent requirement is that the real time data processing needs to be achieved with the highest accuracy possible. A rise in the number of sensors used leads to more power usage when used in applications like gesture recognition and indoor navigation. Hence, the right choice of sensor interface can help bring down the power consumption. We know that multi-sensor data fusion helps to improve the system's reliability, but at the same time, additional devices can significantly increase the overall cost of the system. During the monitoring process, similar activities can be interpreted differently by unrelated sensors. The amount of false alarms/output is relatively high. Therefore, it is required to give a higher priority to the more reliable technology. The data obtained from sensors should be merged most efficiently and analyzed for common trends and similarities. The proposed work involves studying sensor fusion techniques using specific methodologies, namely, filtering techniques like Kalman Filter and algorithmic techniques on data fusion. Further study of the real-time sensor fusion technique and applying it to a real-time-based scenario can be challenging. Work can be extended for navigation of mobile robots using sensor fusion in real-time that provides high performance and low-cost implementation [15-20].

3. Proposed Methodologies for Implementation

The discussion of methodologies for sensor fusion, namely, Kalman Filter method, Central limit theorem and Multisensory Laplacian Fusion, are covered under this section.

3.1 Sensor Data Fusion using KALMAN Filter method

The Kalman Filter method employed here is an algorithm used for sensor fusion that uses a series of measurements observed over time and computes estimates of the unknown variables. The algorithm employs basically the two steps , namely, Prediction and Correction. Here the kalman filter is used to process the data collected from the 9-axis IMU sensor to determine the mobile robot orientation. The computation of orientation using the Kalman Filter involves series of complex matrix calculations with trigonometric functions that can further be converted into scalar values. CORDIC algorithm is an hardware efficient algorithm that can be used with functions , namely, trigonometric, hyperbolic and logarithmic functions along with the eigen value estimation.

3.2 Sensor Data fusion using Central limit theorem

Here, the central limit theorem is used in mobile robots for object and range detection to avoid obstacle contact. The sensors used here are infrared and IR sensors to measure the distance from the obstacle. The technique is simple and inexpensive enabling the mobile robot to efficiently detect the obstacle as well compute the distance between the robot and obstacle.

3.3 Real-Time Multisensory Laplacian Fusion

Laplacian fusion technique is implemented using the pyramid decomposition process that utilizes the reusable memory architecture and parallel computing. Hence the fusion algorithm constructs the Laplacian pyramid for every captured image and then extracts local salience information at multiple pyramid levels. Further the extracted and fused pyramidal information is reverse transformed to reconstruct the image. It compares the pixel energy captured locally with that of the most definite and nearest pixel value , as per the choice of salient features from source images.

MATLAB is an excellent tool that enables many features for simulation but using MATLAB algorithm for ASIC or FPGA needs some additional implements and features needed to process the stream of bits. Kalman filter is entirely defined by the State-space model (assumed dynamics of the framework to be filtered) and Measurement model. Suppose a mismatch exists between model (state-space and measurement), the Kalman filter and a "real-life model". In that case, a Kalman Filter with non-optimal characteristics and behaviour is likely got. In order to avoid this problem, other filters such as adaptive Kalman filters and robust Kalman filters can be used.

3.4 FPGA Technologies for Sensor Fusion

There are several hardware platforms based on microcontrollers, Digital Signal processors (DSPs) and FPGA, available to implement the fusion algorithms in an embedded system. The choice of the platform is mainly based on the key features, namely, performance, power consumption and the cost per chip along with the software tool support. As a result of which the system can designed meeting

IOP Conf. Series: Materials Science and Engineering 1187 (2021) 012007 doi:10.1088/1757-899X/1187/1/012007

the constraints in terms of incurring the lower cost of implementation as well meeting of the project deadlines timely. As ASIC devices require special foundry facilities, FPGA proves to be an attractive and natural alternative for such implementation.

FPGA's are mainly used for techniques that need high-speed processing and attain true parallelism and thereby increase the system's performance. The information obtained from the fusion of sensors is complex, and hence using the other microcontroller for prototyping would not be an appropriate choice. The sensor fusion algorithm that captures massive data and strict requirement of real-time response, FPGA's are highly recommended because it possesses high speed, complex logic control ability. FPGA also controls the interface and timing of the sensor signal acquisition, pre-processes data, and transfers the information to DSP in parallel, which may simplify the peripheral circuit and ensure the system's real-time performance.

4. Implementation of Sensor Fusion

The implementation of the Sensor data fusion can be carried out using the two possibilities discussed further using Figure 2 and Figure 3. As per the implementation shown in Figure 2.0, simulation is carried out using Matlab 2019b software to run multi-sensor data fusion algorithms. The fusion algorithm can be developed as a .m file or the simulink model for simulation in Matlab. As compared to the programming languages such as C, C++ or FORTRAN, MATLAB provides an easier and faster way of developing the codes for algorithm implementation

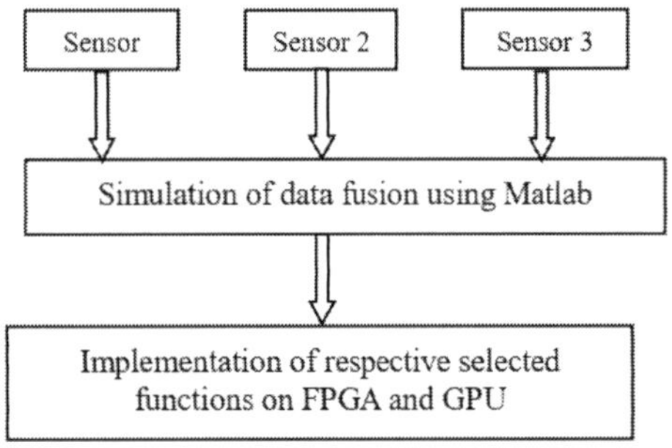

Figure 2.0: Sensor Fusion Process Implementation 1 on Fusion on FPGA

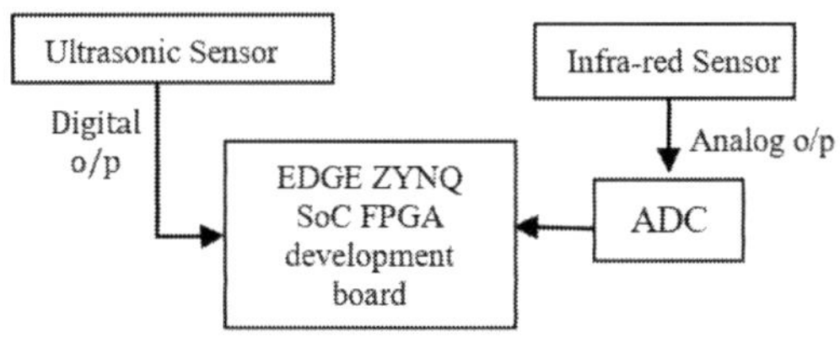

Figure 3.0: Sensor Fusion Process Implementation 2 on FPGA

There are various design options available in Matlab that can be explored to solve the problem in the proposed work and can be checked for feasibility through the simulation of the solution. For the proposed work, the workflow for sensor fusion in mobile robots can be realized by the following steps shown in Figure 4.0. The first step is selecting appropriate sensors as per the requirements of the mobile robot application, followed by the step of choosing the fusion algorithm and deciding on the selection of parameters through several iterations in the simulation process. Further, the fine-tuned algorithm can be fed to an inbuilt HDL(Hardware Description Language) coder in Matlab software, which provides the powerful functions in its HDL toolbox and generates the HDL code VHDL or Verilog as per the user's choice. Thus generated HDL code can now be simulated, synthesized and implemented on the FPGA hardware using the tools, namely, Xilinx ISE and Xilinx Vivado.

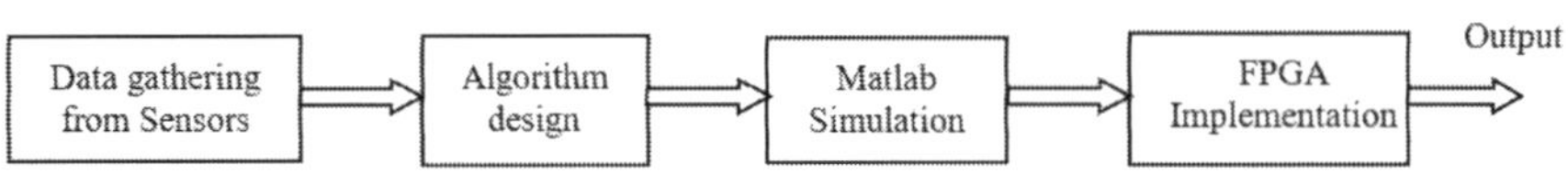

Figure 4.0: Block diagram of the Process flow for Sensor Fusion implementation on FPGA

1)Data Gathering from sensors
The data from various sensors are collected and shaped into the required form. However, some of the sensors produce data/information in analogue form, and hence it has to be converted back to digital form to make it suitable for further process. DAC can be used for this process.
2)Choice of algorithm and its design
The algorithm needs to be chosen based on the proposed mobile robot application. The need is to choose an appropriate fusion algorithm to fuse multiple sensor data for the precise decision making on navigational path planning of the mobile robot in an autonomous fashion. Kalman Filter algorithm is the optimized algorithm used for the sensor fusion process in the proposed work.
3)MATLAB simulation
When the algorithm is simulated on MATLAB, it processes the data and converts it to VHDL or Verilog code. In addition, MATLAB provides various in-built tools such as HDL coder and other additional tools such as state machine charts, image processing tools and others to support the implementation.
4)FPGA implementation
When the VHDL and Verilog codes are obtained, the codes can be implemented on FPGA using XILINX or VIVADO. As per the other implementation shown in Figure3.0, the entire fusion system is implemented on the EDGE ZYNQ SoC FPGA development board and provides the features of a single-board computer built around Xilinx 7-Series FPGA device integrated dual-core ARM Cortex-A9. Here the advantages of utilization of features from the processor logic and reconfigurable logic are exploited. The output from the Infrared sensor is analogue and must be converted to digitized form before sending it as the input to FPGA. As the ultrasonic sensor provides digital data, it can be directly interfaced with FPGA. Once both the sensors are interfaced to FPGA, the fusion of sensor data can be done either using the Central Limit Theorem or the extended Kalman Filter using the processor available on the same board along with FPGA. Thus the sensor fusion algorithm will thus be used for obstacle detection, and further deciding on the navigation path opted by the autonomous mobile robot. This implementation is very widely suitable for boosting robotic applications performance while providing accurate sensing, fast processing, and precise actuation in an absolute sense. The future scope of the proposed study is based on the use of FPGA based reconfigurable computing techniques for robotic solutions. The faculty and the students involved in the proposed research activity got excellent exposure to every research methodology step. The activity provided them with good hands-on experience and also problem-solving skills [21, 22].

5. Conclusion
Here, the focus was on I4.0 standards in developing autonomous systems, which work with better performance and better power efficiency. FPGA based services are of help in providing intelligent system solutions like Auto-guided vehicles or autonomous vehicles in a modern factory environment. The sensor fusion process is an essential process to improve the intelligence of the robot systems through accurate sensing, leading to robust decision-making and precise motion control and actuation. Thus the discussion further emphasizes the need for usage of FPGA based solutions in the sensor fusion process and also on the possible methodologies that can be employed.

Acknowledgement
The authors extend their gratitude to the management of KLE Technological University, Hubballi, Karnataka, India, for the financial support and the provision of a laboratory facility to conduct the research work under the program, Research for undergraduate(REU) at the Department of Automation and Robotics.

References
[1] www.acatech.de/publikationen.
[2] J. Qin, Y. Liu, R. Grosvenor, 2016, A Categorical Framework of Manufacturing for Industry 4.0 and Beyond, Changeable, Agile, Reconfigurable & Virtual Production, Procedia CIRP, 52,173 – 178.

[3] R. Neugebauer, S. Hippmann, M. Leis, M. Landherr, 2016, Industrie 4.0- Form the perspective of applied research, 49th CIRP conference on Manufacturing systems (CIRP-CMS 2016), 2-7.

[4] K.D. Thoben, S. Wiesner, T. Wuest, 2017, Industrie 4.0 and Smart Manufacturing- A Review of Research Issues and Application Examples, International Journal of Automation and Technology, Vol.11, No.1, 4-16

[5] B. Bagheri, S. Yang, H.A. Kao, J. Lee, 2015, Cyber-physical Systems Architecture for Self-Aware Machines in Industry 4.0 Environment, IFAC Conference 38-3, 1622–1627.

[6] Jasprabhjit Mehami, Mauludin Nawi, Ray Y Zhong, 2018, Smart automated guided vehicles for manufacturing in the context of Industry 4.0, Procedia Manufacturing, Volume 26, Pages 1077-1086, ISSN 2351-9789, https://doi.org/10.1016/j.promfg.2018.07.144.

[7] Jyoti S Bali, Anilkumar Nandi, 2017, Simplified Process of Obstructive Sleep Apnea Detection using ECG Signal based Analysis with Data Flow Programming, ICTIS 2017, Ahmadabad 25th and 26th March.2017, Information and Communication Technology for Intelligent Systems (ICTIS 2017), Smart Innovation, Systems and Technologies, vol 84. Springer, Cham, https://doi.org/10.1007/978-3-319-63645-0_18.

[8] Jyoti S Bali, Anilkumar V Nandi, P.S. Hiremath and Poornima G. Patil,2018, Detection of Sleep Apnea in ECG Signal Using Pan-Tompkins Algorithm and ANN Classifiers, 2018, COMPUSOFT, Vol. 7, Issue 11, pp.2852-2861.

[9] Jyoti S Bali, Anilkumar V Nandi, P.S. Hiremath and Poornima G. Patil, 2019, Detection of Sleep Apnea Disorder using an Adaptive Neuro-Fuzzy Classification Method, 2nd International Conference on Recent Innovative Trends in Computer Science and Applications (ICRITCSA-2019), Journal of Computational and Theoretical Nanoscience ISSN: 1546-1955 (Print): EISSN: 1546-1963 (Online)

[10] Bali J., Nandi A., Hiremath P.S, 2020, Efficient ANN Algorithms for Sleep Apnea Detection Using Transform Methods. In: Verma O., Roy S., Pandey S., Mittal M. (eds) Advancement of Machine Intelligence in Interactive Medical Image Analysis. Algorithms for Intelligent Systems. Springer, Singapore, Print ISBN 978-981-15-1099-1 Online ISBN978-981-15-1100-4, DOI: https://doi.org/10.1007/978-981-15-1100-4_5

[11] P.C. Nissimagoudar, A.V. Nandi, Jyoti S. Bali, and H.M. Gireesh, 2020, Convolution neural network-based feature learning model for EEG-based driver alert/drowsy state detection,1, 217-226, Computational Intelligence And Its Applications In Healthcare, Elsevier. ISBN: 978-0-12-820604-1

[12] Jyoti S Bali, Anilkumar Nandi, 2016, An Experience, using Software based Tools for teaching and learning Mathematically intensive Signal Processing Theory Concepts, 2016 IEEE 4th International Conference on MOOCs, Innovation and Technology in Education (MITE), Madurai, pp. 100-104. doi: 10.1109/MITE.2016.029, IEEE Xplore

[13] Jyoti S Bali, Anilkumar Nandi, 2017, ECG Signal based power-aware system for Obstructive Sleep Apnea Detection, *2017 International Conference on Recent Trends in Electrical, Electronics and Computing Technologies*, Warangal, IEEE Xplore, pp. 59-63, doi: 10.1109/ICRTEECT.2017.43.

[14] Suvarna G Kanakaraddi, Karuna C Gull, Jyoti Bali, Ashok K Chikaraddi, Shantala Giraddi, 2021, Disease Prediction Using Data Mining and Machine Learning Techniques, Advanced Prognostic Predictive Modelling in Healthcare Data Analytics,71, Springer Nature.

[15] H. Zhang, W. Sun, H. Ji, X. Wu and G. Lu, 2017, LIU: Localization via Fusion of IMU and Ultrasonic Positioning, 3rd International Conference on Big Data Computing and Communications (BIGCOM), Chengdu, 2017, pp. 272-276, doi: 10.1109/BIGCOM.2017.24.

[16] S. Pérez-Bachiller, D. Gualda, M. C. Pérez, J. M. Villadangos, J. Ureña and R. Cervigón, 2018, Android Application for Indoor Location Using Sensor Fusion: Ultrasounds and Inertial Devices, International Conference on Indoor Positioning and Indoor Navigation (IPIN), Nantes, pp. 1-8, doi: 10.1109/IPIN.2018.8533785

[17] Xiyuan Chen, Yuan Xu, Qinghua Li, Jian Tang, Chong Shen, 2016, Improving ultrasonic-based seamless navigation for indoor mobile robots utilizing EKF and LS-SVM, Measurement, Volume 92, Pages 243-251, ISSN 0263-2241,

https://doi.org/10.1016/j.measurement.2016.06.025.

[18] Li, Jian & Han, Guangjie & Zhu, Chunsheng & Sun, Guiqing, 2016, An Indoor Ultrasonic Positioning System based on TOA for Internet of Things. Mobile Information Systems, 1-10. 10.1155/2016/4502867.

[19] Z. Wang, Y. Wu and Q. Niu, 2020, Multi-Sensor Fusion in Automated Driving: A Survey, in IEEE Access, vol.8, pp. 2847-2868, doi: 10.1109/ACCESS.2019.2962554.

[20] I. Toroslu and M. Doğan, 2018, Effective sensor fusion of a mobile robot for SLAM implementation, 4th International Conference on Control, Automation and Robotics (ICCAR), Auckland, pp. 76-81, doi: 10.1109/ICCAR.2018.8384648.

[21] Jyoti Bali, Ashwini, G. K., Tanvashi, S., & Giriyapur, A. C, 2019, Problem-Solving in Integrated Laboratory using Hackathon Approach, Journal of Engineering Education Transformations, 33(1), 148-152.

[22] Jyoti Bali, Arunkumar Giriyapur, Anilkumar Nandi, 2020, Experiments in Fostering Research skills for Undergraduates in an Inter-disciplinary Engineering Program, WEEF2019, Procedia Computer Science, Volume 172, Pages 609-615, ISSN 1877-0509, https://doi.org/10.1016/j.procs.2020.05.078.

Low Power Memory System Design Using Power Gated SRAM Cell

Srijani Pal[1], Divya S Salimath[1], Banusha Chandran[1], A Anita Angeline[1] and V S Kanchana Bhaaskaran[1]

[1]School of Electronics Engineering, VIT Chennai, Chennai -600127, Tamil Nadu, India

vskanchana@iccc.org

Abstract. Static Random-Access Memory (SRAM) is widely used in cache memory, microprocessors, general computing applications and electronic circuits involving ASIC, FPGA and CPLD. The most commonly used SRAM is the 6T SRAM. However, it incurs higher power consumption and degraded signal to noise margin (SNM) during write and read operations. To overcome these shortcomings, a single ended power gated 11T SRAM for low power operation is proposed. The power consumption reduction is achieved using power gating through virtual VSS (VVSS) signal and transmission gates. Due to the introduction of transmission gates, memory cells realize enhanced write margin characteristics as compared to existing technologies. The proposed cell realizes 33.33% lower power consumption and 50% improvement in read SNM as compared to existing SRAM technologies. To study the impact of technology scaling on our proposed design, the work is carried out in Cadence Virtuoso® tool using both 180nm CMOS technology and BPTM 32nm FinFET technology.

Keywords. Low Power SRAM Design, Power-Gated Memory Design, Single Ended SRAM Cell, SRAM Array, SNM of Memory

1. Introduction

Low power memory design has created a huge impact in SOC design. Since memory is being embedded in the chips, the power consumption needs to be reduced. Low power, high density and high performance have become the most important aspects of state-of-the-art design solutions for memory. The primary focus of this design is to improve area efficiency, reduce power consumption and enhance performance of SRAM cells which are embedded in the devices as an integral part due to various factors.

In lower technology nodes, leakage current of MOSFET devices increases due to second order effects. Since FinFET technology provides improved electrostatic control over the channel, the latter is preferred to planar MOSFET technology. The choice of lower technology nodes enables increased memory density as required in the present times and in order to minimize the power dissipation in the high-density memory chips. However, methods of effectively designing electrically high-performance SRAM cell using FinFET technology has been presented [1][2]. Furthermore, previous works on FinFET technology with 7T SRAM cell validated for performance despite temperature variation and for different FinFET technology nodes, with comparative study analysis was presented in [3]. The various circuits in the power gated SRAM memory are implemented and analysed in both 180nm CMOS technology node and BPTM 32nm FinFET technology models.

The conventional 6T SRAM bit cell has two cross coupled inverters and two access transistors connected to data storage nodes [4]. The inverter pair forms a cross connected latch and provides the necessary information for read and write operation. True and complimentary values of the data are stored on the bit line and bitbar line, respectively. In general, read and write operations require charging and discharging of bitline and bitbar lines. The bit cells are accessed through the access transistors during read and write operations, by asserting the word-line connected to the gates of the access devices high. Care is to be taken while sizing the access transistors with respect to the storage transistor nodes for correct read and write operations of a 6T SRAM cell, and since the same access transistors are used for both read and write operations. For noise free read operation, the width of the NMOS pull down transistors should be greater than the NMOS access transistors and for write operation, on the other hand, the width of the NMOS access transistors should be greater than the PMOS pull-up transistors. Sizing of a conventional 6T SRAM cell shows improvement in the static noise margin (SNM) values, thus achieving better performance of the cell.

A dynamic column-based power supply 8T SRAM structure is discussed by comparing the stability of the 8T SRAM cell with the conventional 6T using N curve analysis [5]. The static voltage noise margin (SVNM) resulted in twice as good result as the conventional 6T SRAM cell. Moreover, the write trip voltage (WTV) was found to be improved as compared to conventional 6T. The differential 10T SRAM implemented in [6] has large parasitic capacitances due to presence of bitline and bitbar lines. Due to the parasitic capacitances, the power consumption is large. Power can be reduced by reducing switching activity. In single ended SRAMs, the switching activity reduces by half, which results in less power consumption when compared with 6T, 8T and 10T SRAM cells [6].

Sense amplifier is required for detection of bitline voltage during read operation. Due to the presence of single bitline, single ended sense amplifier design is required [7]. The sensing power consumption power and delay of PG9T and PG11T. It also includes the simulation results of PG11T for write and read operation alongside the results of 4*4 SRAM array during write and read operation in cells 0 to 15. Section VI includes conclusion.

2. Operation of PG9T SRAM cell

A power gated 6-T SRAM structure has been developed along with various schemes of leakage power reduction, such as fine and coarse power gating techniques [9]. The power gating technique has also been implemented in a 9-T SRAM structure [10].

In PG9T SRAM cell, power gating transistors and Memory Data Read (MDR) transistor are used in addition to the cross coupled inverter pair and access transistor [10]. During write operation, the power gating transistors cut-off the power supply to the bitcell. Hence, the noise during write operation due to the power supply is reduced. During the read operation, the BL is isolated from the bitcell by activating the MDR transistor and deactivating the access buffer (AC2) transistors. Hence, the read current does not flow through the storage nodes, thus eliminating the noise during read operation. The main drawback in PG9T SRAM cells is that it uses NMOS access transistor, which has a weak ability to pass a logic 1. Hence, writing logic 1 in the cell becomes difficult [10].

This paper shows that transmission gates of PG11T SRAM cell are being used instead of access pass transistors to design a more energy-efficient logic 1. It further shows that power consumption is reduced as compared to PG9T.

3. Proposed PG11T SRAM cell

Figure 1 shows PG11T SRAM cell. The cell consists of a positive feedback latch made up of two cross-coupled inverters PU1, PD1, PU2 and PD2 with two power gated transistors PGN and PGP, transmission gates TG1 and TG2, and an MDR transistor.

The transmission gates act like access buffer. During write operation, transmission gates TG1 and TG2 connect the single-ended BL to the internal storage node Q. Read operation is carried out by holding the value of BL or discharging through TG1 and MDR. Row-based signals involve WL, WLB, WLPU and WLPD, while BL, VVSS, WWL and WWLB form the column-based signals. During read operation, the lines WL and WLB are made high. On the other hand, WL, WLB, WWL and WWLB are made high for write operation. The power gated transistors PGN and PGP are made low during write operation by WLPD and WLPU signals, respectively. MDR is high during read operation while it is kept low during hold and write operations.

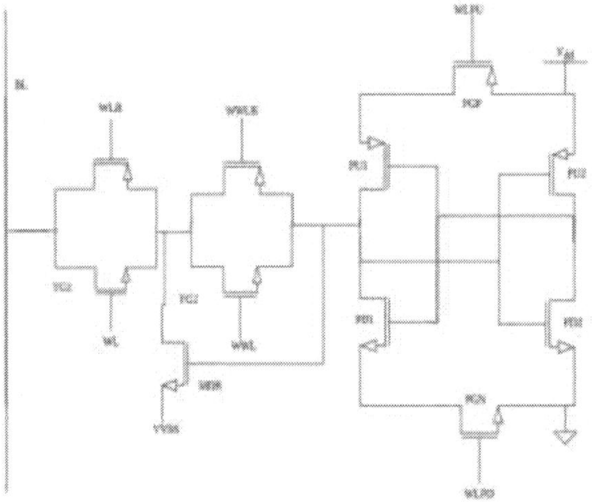

Figure 1. PG11T SRAM cell.

3.1. Write operation.

During write operation, WL and WWL are made 1, and WLB and WWLB are made 0. WLPU is kept high and WLPD is kept low to turn off PGP and PGN, respectively. For writing a logic 1 in the cell, virtual ground VVSS is made high and for writing 0, VVSS is made low. The BL value gets into the SRAM cell and write operation is performed in the SRAM cell.

3.2. Hold operation.

Hold operation is performed by making WL, WWL as 0, and WLB and WWLB as 1, the power gated transistors PGP and PGN are turned on. VVSS is connected to 1. Then, the content of cell is retained.

3.3. Read operation.

For read operation, the WL and WLB are activated by 1 and 0, respectively, and WWL and WWLB are deactivated by 0 and 1, respectively. The virtual ground VVSS is kept at logic 0. Prior to the read operation, BL is pre-charged to Vdd. Whenever bit 1 is stored in the bitcell, MDR transistor turns ON. The pre-charged bitline discharges through the transmission gate TG1 and MDR. for storing bit 1 in the SRAM cell before read operation. Whenever bit 0 is stored in the bit cell, MDR remains off and the bitline does not attempt discharging. Hence, the complementary outputs are obtained during read operation.

4. Design of SRAM array and its peripheral circuits

4.1. Design of pre-charge circuit

The pre-charge circuit is used to speed up the read operation. Figure 2 illustrates the MOSFET-based pre-charge circuit. It uses two PMOS transistors (P1, P2) connected in parallel to drive the bitline high only when the pre-charge activation signal PC is low.

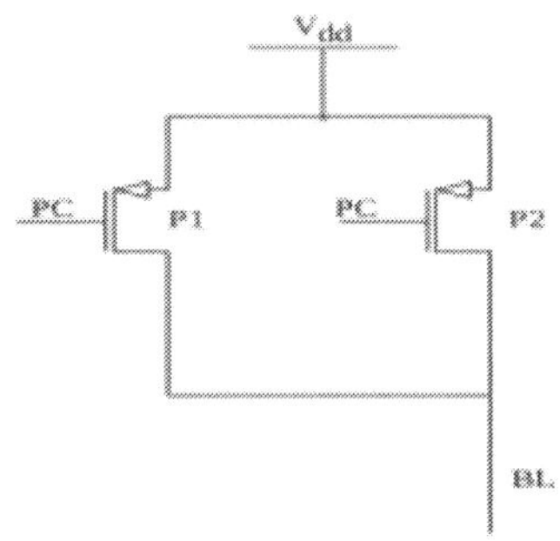

Figure 2. Pre-charge circuit.

4.2. Design of write circuit.

Figure 3 shows the write circuit used for writing onto the cells. It consists of a buffer followed by NMOS N1 transistor to drive the data to bitline when Write Enable (WE) signal is activated. The buffer is used to prevent degradation of Data signal.

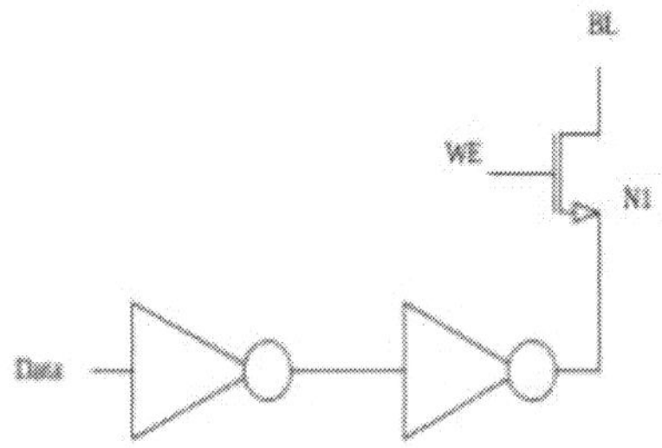

Figure 3. Write circuit.

4.3. Design of sense amplifier

A sense amplifier from [8] has been used in this design with a few modifications. The pull up PMOS transistor has been removed since a separate circuit is used for pre-charging. Furthermore, a footer NMOS N2 is added to prevent the connection to ground unless both the bitline and pre-charge are high. Pre-charging occurs with the PC low, and evaluation occurs with the PC high. The circuit shown in Figure 4 shows the sense amplifier. When PC is high and P2 transistor is turned on, P1 and N1 are operational as inverters with bitline as input. As a result, the complementary bitline is obtained as the output.

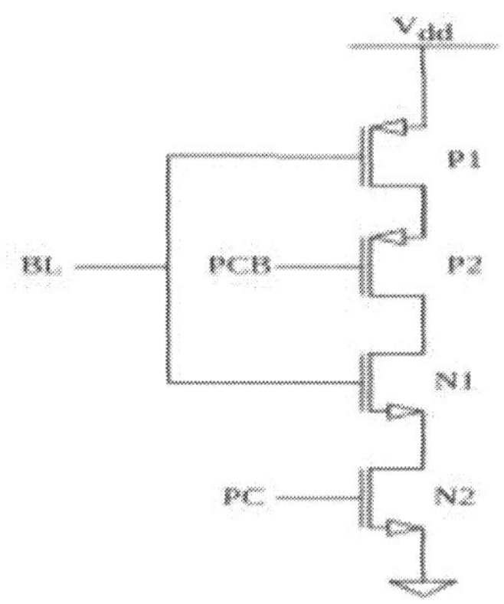

Figure 4. Sense amplifier.

4.4. Design of 4x4 single ended array

The 4x4 array depicted consists of the structure of a PG11T SRAM cell array along with the simple peripheral circuits required. The peripheral circuits include decoder, multiplexer, pre-charge, sense amplifier and write circuits. In the array, the signals WL, WLB, WLPU and WLPD are row-based signals having BL, WWL, WWLB and VVSS as column-based signals. Two 2:4 decoders are instantiated for cell selection. Four-bit address A3, A2, A1, A0 is used for address referencing. The two MSB bits A3, A2 are used for row signal selection through the 2:4 row decoder and for column signal selection, two LSB bits A1, A0 are involved using 2:4 column decoder. A control signal, read (RD) is used which indicates read operation when high and write operation when low. The signals for read and write are switched by a 2:1 MUX. Figure 5 depicts a 4x4 array with the periphery circuits.

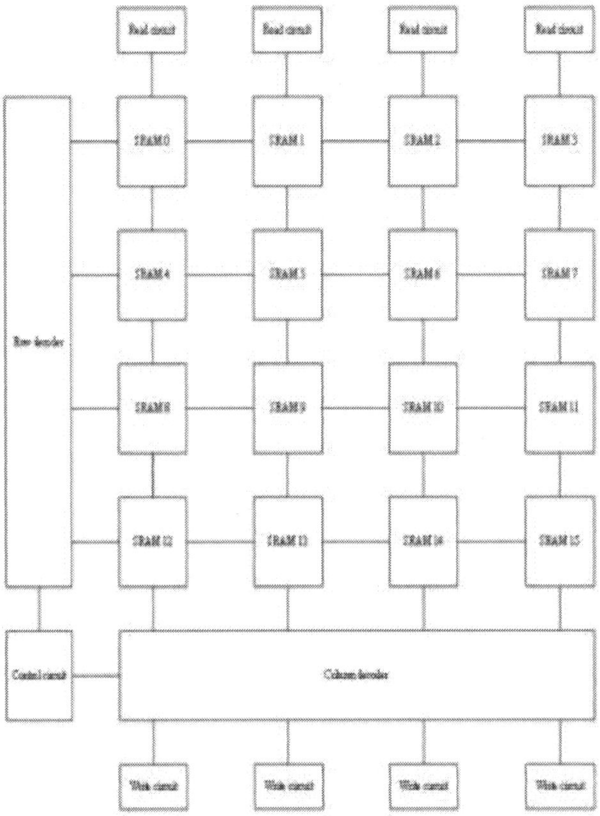

Figure 5. 4x4 array.

4.5. Design of 5:32 decoder

The 5:32 static decoder is implemented as in figure 6. It comprises of a pre-decoder and inverter circuits, which are being replicated in addition to alternate NAND and NOR gate structures. The pre-decoder circuit is used for realizing reduction in the gate count along with smaller number of input wires. It further reduces the number of stages from input to output with reduced delay and power consumption [11].

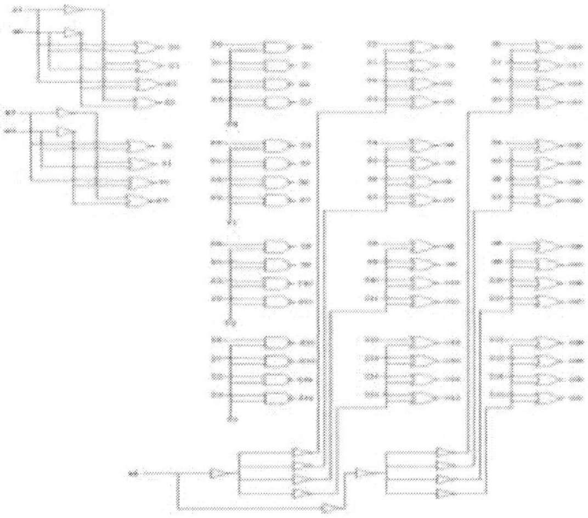

Figure 6. 5:32 decoder.

4.6. Design of 32x27 single ended SRAM array

The single ended SRAM cell is instantiated in the schematic to form an array similar to the 4x4 array. It can store the information in about 864-bit cells. The column-based and row-based signals are used as mentioned in 4x4 Single Ended array. Figure 7 shows a 32x27 SRAM array. The 5:32 decoder in figure 6 is used for address decoding in the 32x27 array.

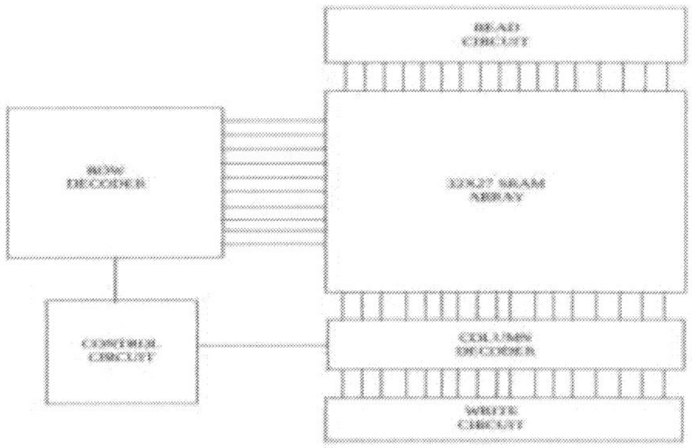

Figure 7. 32x27 array.

5. Results and Discussions

Table 1. Power comparison of PG9T and PG11T in 180nm CMOS technology.

	PG11T (uW)	PG9T (uW)
Average power	5.158	8.41
Write power (logic 1)	18.28	47.75
Read power (logic 1)	5.772	7.99
Write power (logic 0)	10.15	9.526
Read power (logic 0)	0.34	1.261

IOP Conf. Series: Materials Science and Engineering 1187 (2021) 012008 doi:10.1088/1757-899X/1187/1/012008

Table 2. Delay comparison.

	PG11T (180 nm)	PG9T (180 nm)	PG11T (32 nm)	PG9T (32 nm)
Read delay (in ns)	0.188	0.212	0.112	0.0629
Write delay (in ns)	0.231	0.37	0.037	0.071

Table 1 shows the power consumption analysis for read and write operation using logic 1 and logic 0. It can be observed that power dissipation in PG11T is less as compared to PG9T. Table 2. shows the delay comparison of PG9T and PG11T between the values obtained while using 180 and 32 nm technology files.

Figure 8 shows the power consumption values of PG9T SRAM cell while working in 180nm CMOS technology. As can be seen, during 10ns to 20ns, write operations take place, hold operation occurs between the time instances of 20ns to 30ns, and the period between 30ns and 45ns includes read and pre-charge operations. Logic 1 operations are included in this cycle. The same cycle repeats for logic 0 during 55ns to 90ns time period. Figure 9 indicates the power consumption of PG11T SRAM cell while using 180nm CMOS technology during read and write operations. The read and write cycle is similar to PG9T SRAM cell. In 32nm FinFET technology, during read and write operations, figure 10 and figure 11 displays the power consumption of PG9T and PG11T SRAM cells, respectively. PG11T is simulated in transient mode during read and write operation, as shown in figure 12. Initially a logic 1 is written into the cell and it is read. Then a logic 0 is written and read subsequently. Figure 13 indicates the transient simulation of 4x4 array during write operation. The cells are selected using the row, column decoders and multiplexers. Once the cells are selected, data is written into the cells. Logic 1 is written into cells 0 to 15. Figure 14 shows the transient simulation of 4x4 array during the read operation with sense amplifier outputs. Before the cells are selected, all the bit lines are pre-charged and the cells which need to be read are selected using row, column decoders, multiplexers, and sense amplifiers. The logic 1 stored during the write operation in cells 0 to 15 are read. This is confirmed by the bit line discharge. The output of the sense amplifier is same as the data stored in the cells.

From figure 8 to figure 11, it is evident that power consumption during read 1 is higher than during read 0. This is because before the read operation, the bitline BL is pre-charged. During read 1, logic 1 is stored in the cell, which turns on the MDR transistor, creating a discharge path of the pre-charged BL to ground. During read 0 operation, when logic 1 is stored in the cell, the MDR turns off, thus making sure that no discharge path exists for the line BL, so that BL only has to retain the charge it had stored previously on its nodal capacitance. Therefore, read 0 consumes less power than read 1. Spikes are found to be present during the transient analysis, when switching of signals happen, which indicates the fact the dynamic power dissipation happens due to the switching.

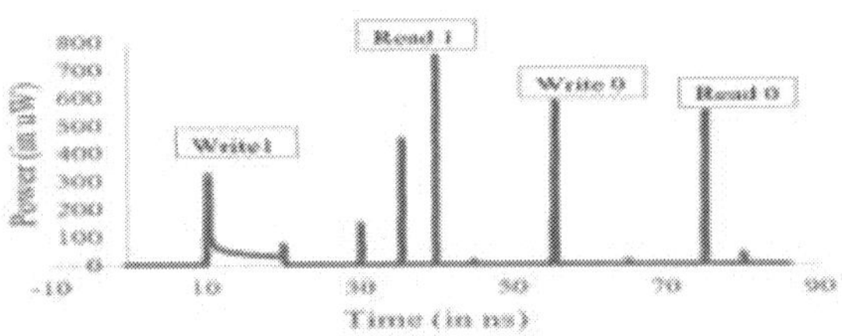

Figure 8. Power analysis in 180nm PG9T SRAM cell.

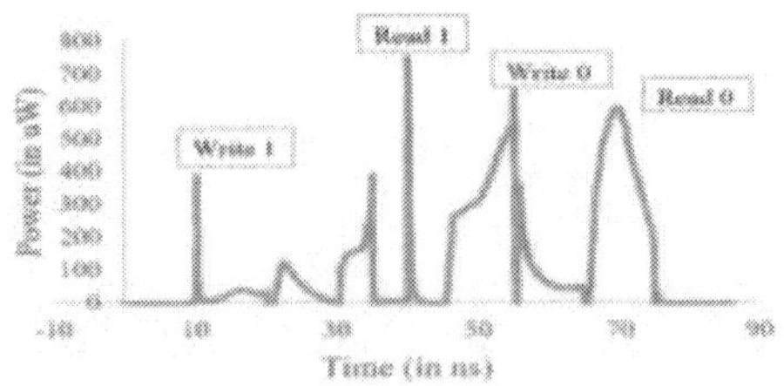

Figure 9. Power analysis in 180nm PG11T SRAM cell.

IOP Conf. Series: Materials Science and Engineering 1187 (2021) 012008 doi:10.1088/1757-899X/1187/1/012008

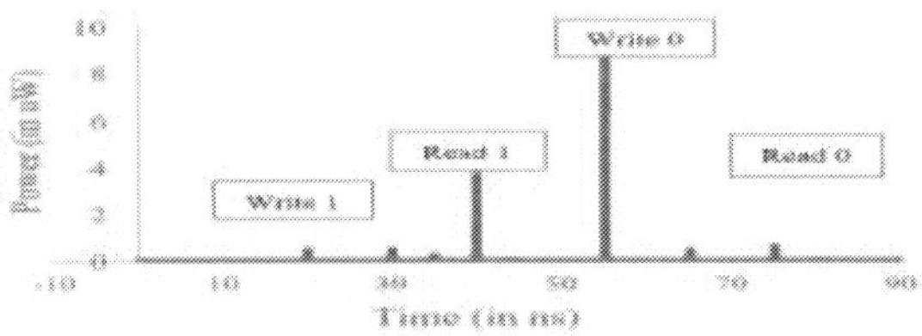

Figure 10. Power analysis in 32nm PG9T SRAM cell.

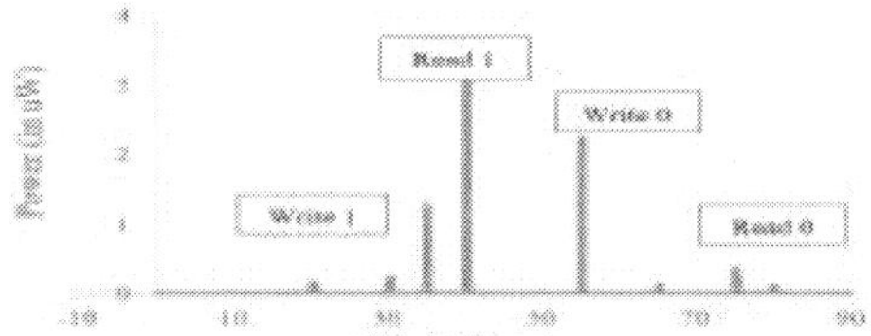

Figure 11. Power analysis in 32nm PG11T SRAM cell.

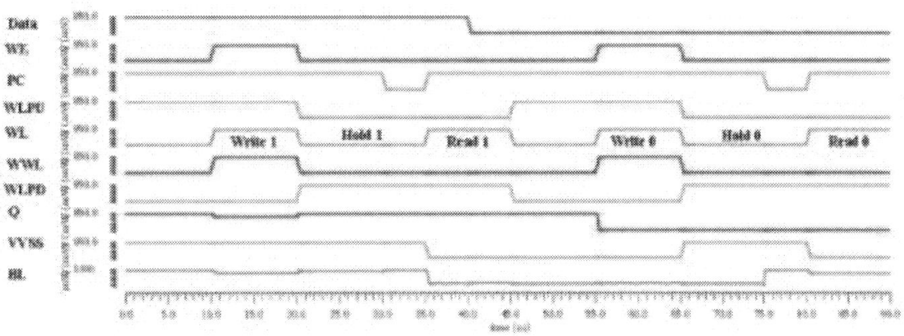

Figure 12. PG11T waveform during read and write operations.

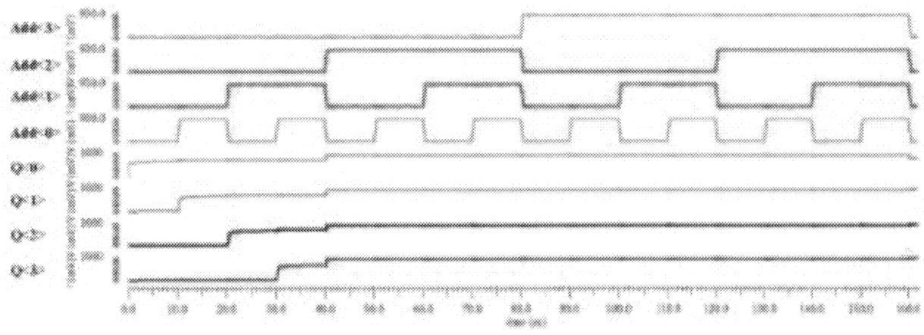

Figure 13. 4x4 array during write operation.

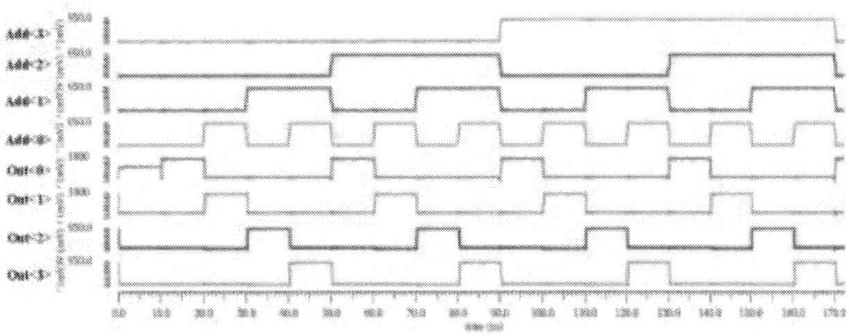

Figure 14. 4x4 array during read operation.

IOP Conf. Series: Materials Science and Engineering 1187 (2021) 012008 doi:10.1088/1757-899X/1187/1/012008

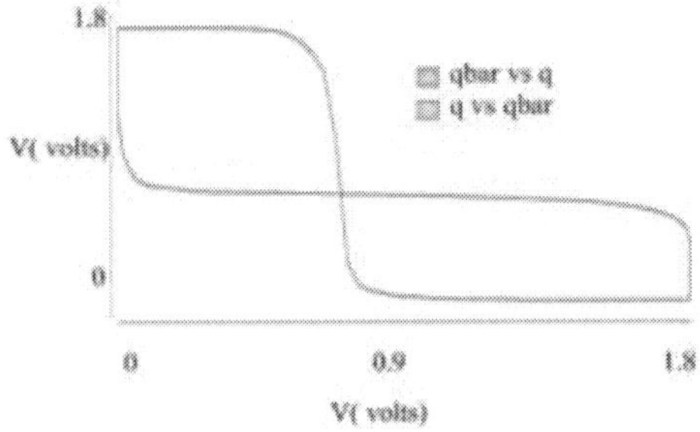

Figure 15. Hold SNM in 6T.

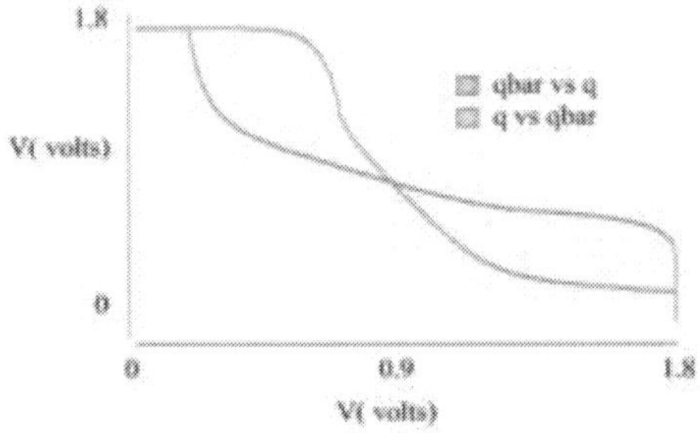

Figure 16. Read SNM in 6T.

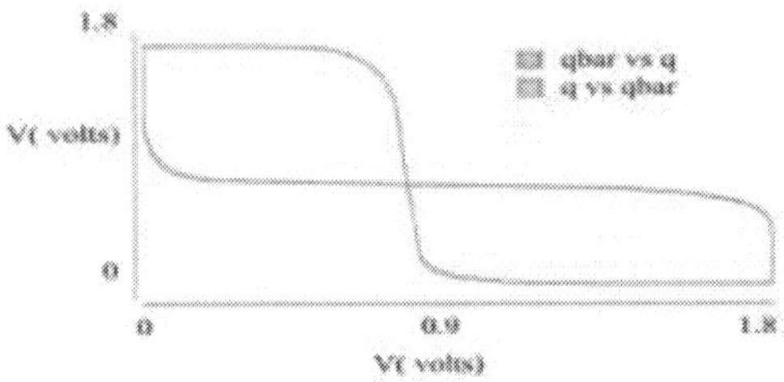

Figure 17. Hold SNM in PG11T.

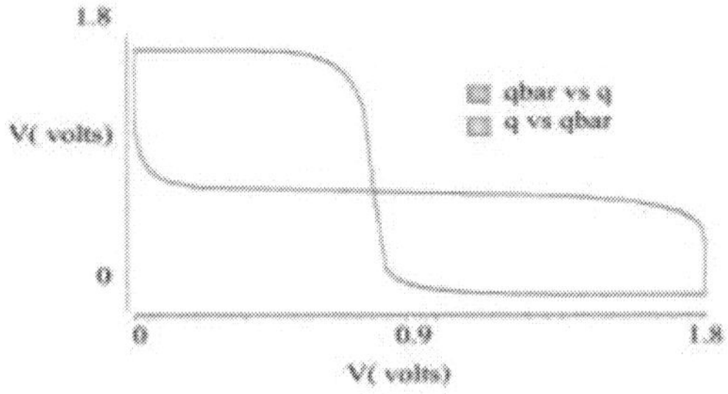

Figure 18. Read SNM in PG11T.

Table 3. SNM comparison.

	PG11T (uW)	PG9T (uW)
Hold SNM (in mV)	848	848
Read SNM (in mV)	424	848

Figures 15, 16, 17 and 18 show the hold and read butterfly curves, respectively, for 6T and PG11T memory cells. The SNM values are tabulated in table 3. The result is obtained by taking the largest square drawn in the two lobes of the butterfly and choosing the minimum of the two values. Then, the diagonal of the corresponding square is taken as the SNM value. The SNM value is then determined by the diagonal of the corresponding square.

It can be observed that the average power results of the proposed SRAM structure realize enhanced characteristics while using 32 nm FinFET technology rather than the 180nm planar MOSFET technology. As lower technology devices require lower voltage supply, they result in reduced power dissipation. Furthermore, the structural configuration of the FinFET with the gate surrounding the channel on the three sides realizes greater control over the channel. Thus, FinFET can drive greater amount of current. The delay comparison results of table 2 shows that the delay is lesser in 32 nm FinFET technology, which is because the FinFET devices have faster switching speed and greater drive strengths. Thus, in 32nm technology the power and delay results are even better when compared to design using 180nm technology.

6. Conclusion

A single ended SRAM cell has been implemented which incurs 33.33% lesser power consumption even while displaying reduced read and write delay values as compared to PG9T. The transient simulations have been carried out for SRAM cell in 180nm CMOS technology and BPTM 32nm FinFET in Cadence

Virtuoso® tool. The power and delay are compared for PG9Tand PG11T SRAM cell in 180 nm and 32nm technology. In 180 nm technology, PG11T has smaller read and write delay as compared to PG9T. The read delay is reduced by 11.32% as compared to PG9T. The write delay is reduced by 37.56% as compared to PG9T. Butterfly curve is plotted for 6T and PG11T. Read SNM improves by 50% as compared to conventional 6T SRAM. In parallel, the hold and read SNM have also been compared with that of PG11T. Simulations are also carried out for 4x4 array. The array stores the given data and the same is verified by the subsequent read operations. Half selection cells are avoided and the same is tested in 4x4 array. It can be seen that necessary bit line discharge has been eliminated with the use of VVSS. A single ended SRAM cell has been proposed which uses power gated transistors and a separate read access buffer for reliable read and write operations.

7. References

[1] S C Song et al, *Systematic approach of FinFET based SRAM bitcell design for 32nm node and below*, in Proc. IEEE Int. Conf. IC Design Technol., May 2009, pp 165–8.

[2] S H Kim and J G Fossum, *Design Optimization and Performance Projection of Double-Gate FinFETs with Gate-Source/Drain Underlap for SRAM Application*, IEEE Trans. Electron Devices, vol 54, no. 8, pp 1934–42, August 2007.

[3] Mohammad Ansaria, Hassan Afzali-Kusha, Behzad Ebrahimi, Zainalabedin Navabi, Ali Afzali-Kush, Massoud Pedram: *A near-threshold 7T SRAM cell with high write and read margins and low write time for sub-20 nm FinFET technologies*, INTEGRATION, the VLSI journal.

[4] Jan M Rabaey, Anantha Chandrakasan and Borivoje Nikolic, *Digital Integrated Circuits*, ISBN 81-7808-991-2, Pearson Education, 2003.

[5] K Dhanumjaya, M Sudha, MN Giri Prasad, K Padmaraju, *Cell Stability Analysis of Conventional 6T dynamic8T SRAM cell in 45nm technology*, International Journal of VLSI design & Communication Systems (VLSICS) vol 3, no.2, April 2012.

[6] C Ik Joon et al, *A 32 kb 10T subthreshold SRAM array with bitinterleaving and differential read scheme in 90 nm CMOS*, IEEE J. Solid-State Circuits, vol 44, no. 2, pp 650–8, Feb. 2009.

[7] L Chang et al, *Stable SRAM cell design for the 32 nm node and beyond*, in Symp.VLSI Technol. Dig. Tech. Papers, June 2005, pp 128–9.

[8] Ameya Chandras, V S K Bhaaskaran, *Sensing Schemes of Sense Amplifiers for Single-ended SRAM*, IEEE, 2017 International Conference on Nextgen Electronic Technologies

[9] D Anandani, A Kumar, V S K Bhaaskaran, *Gating techniques for 6T SRAM cell using different modes of FinFET*, Advances in Computing, Communications and Informatics (ICACCI), 2015 International Conference, pp 483 – 7, 10–3 Aug 2015.

[10] TaeWoo Oh, Hanwool Jeong, Kyoman Kang, Juhyun Park, Younghwi Yang, and Seong-Ook Jung, *Power-Gated 9T SRAM Cell for Low Energy Operation*, IEEE Transactions on Very Large-Scale Integration (VLSI) Systems, vol 25, no. 3, March 2017, 1183

[11] A K Mishra, D P Acharya and P Patra, *Novel Design Technique of Address Decoder for SRAM*, IEEE Int. Conf. on Advanced Communication Control and Computing Technologies 2014, May 2014.

IOP Conf. Series: Materials Science and Engineering 1187 (2021) 012009 doi:10.1088/1757-899X/1187/1/012009

Replacing two As atoms in silicene nanoribbons in the presence of an external electric field

Hoang Van Ngoc

Institute of Applied Technology, Thu Dau Mot University, No 6, Tran Van On Street, Thu Dau Mot city, Binh Duong province, Vietnam

Email:ngochv@tdmu.edu.vn

Abstract. With the continuous development of nanotechnology and science, its life application is extremely large. Silicene nanoribbons (SiNR) are nanostructures with many advantages, replacing another element in SiNR helps to create new materials with many outstanding properties. This work studies the doping of two arsenic (As) atoms at different positions in the unit lattice cell of SiNR, the system is placed in an electric field. The presence of an electric field in the system causes changes in the energy band structure and the density of the state. There are three sites to be doped in here, namely the meta, ortho, and para position. One of these three positions will be the most optimal, the position at which the formation energy is the lowest. The structure diagrams of the energy region and state density will be drawn, compared, and examined.

1. Introduction

The creation of low-dimensional materials in recent years has created momentum for the continued development of science and technology [1-3]. The application of research on low-dimensional materials has resulted in new products with many outstanding features, lighter, smaller, low cost, and high-performance products [4-8]. There is a lot of research on low-dimensional materials such as germanene [5-10], graphene [11,12], and silicene [13-18], making the structure of these structures an important advance in materials science [3]. The study of doping substances in graphene, germanene, and silicene is also of interest to many scientists. Doping these materials with impurity creates structures with more prominent properties and a wider application [3,10,16,18]. The doping of substances to silicene nanoribbons (SiNR) has been studied extensively in recent years [3,16,18]. In this work, the doped substance is arsenic, arsenic is a nonmetal, toxic, and its compound is widely used in life. There are three sites to be studied for arsenic atom doping in SiNR, namely meta, ortho, and para sites. We calculated forming energies for all three positions and came up with conclusions for the most optimal location. Band structure and state density calculations give us an overview of the doping system. The system is placed in a $0.3V/\overset{0}{A}$ electric field, and placing the system in an external electric field induces changes in the band structure and density of states (DOS), which offers applications in controlling the electrical properties of the system. materials using an external electric field. From the research results, we compare the structure of the energy region and the state density of the system when there is an external field and when there is no external field to clearly see the effect of the external field on the system.

2. Configuration

Silicene nanoribbons that we study here have its configuration shown in figure 1.

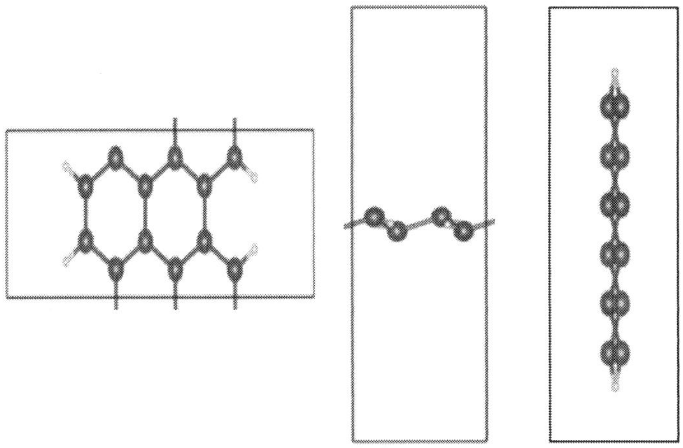

Figure 1. Pristine-configuration of SiNR

The maximum and minimum distances between Si-Si atoms are 2.23934 Å and 2.16769 Å, respectively. Buckling of SiNR is about 0.8 Å. The total energy of pristine system is about -69.408218eV.
Configuration of doping systems are shown in Figures 2, 3, 4.

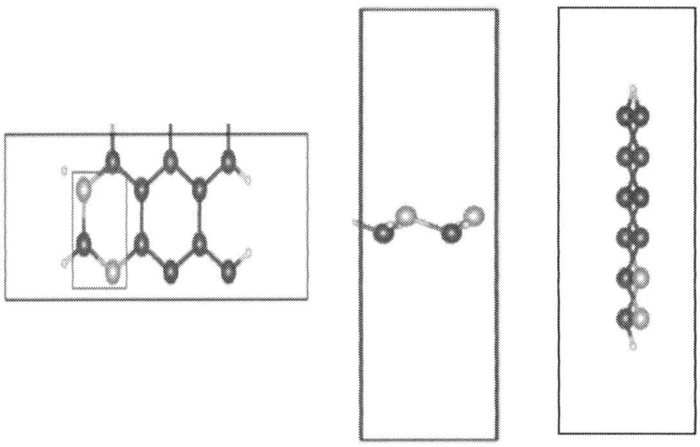

Figure 2. Meta-As doping configuration

The meta doped structure is shown in figure 2. In this configuration, the silicon atom is located between two arsenic atoms. The total energy of meta-configuration is about -69.795159eV. In the figure 3, with an ortho configuration, with two As atoms in adjacent positions, the total energy of the system is -70.218542eV. Figure 4 shows the para doping configuration, in this configuration, two As atoms are placed in two symmetrical positions in the hexagonal cell. One As atom is in top position, the other in valley position. The total energy of system is about -69.991258eV.

IOP Conf. Series: Materials Science and Engineering 1187 (2021) 012009 doi:10.1088/1757-899X/1187/1/012009

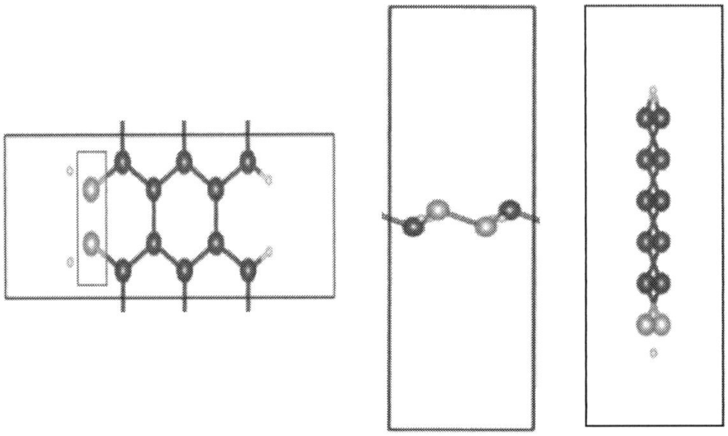

Figure 3. Ortho-As doping configuration

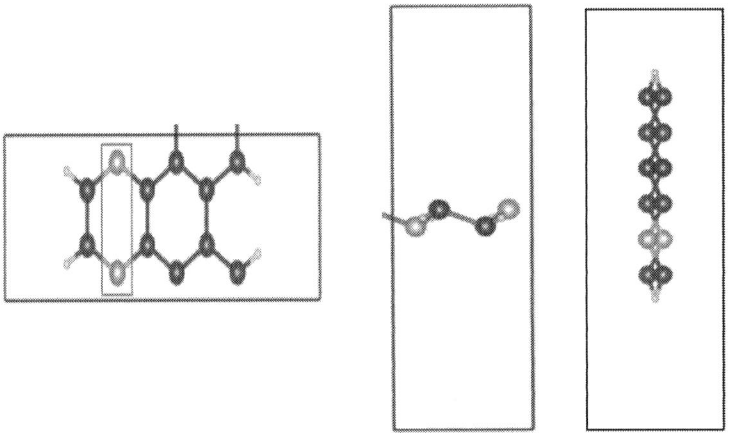

Figure 4. Para-As doping configuration

All three configurations are optimized for the structure and give us the total formation energy to form as follows [4]: $\Delta E_f = E_{0t} - E_{op} + nE_{0Si} - nE_{0As}$

Table 1. The formation energy of the systems in the presence of the external electric field

Doping systems	E_{0t}(eV)	E_{0p}(eV)	E_{0Si} (eV)	E_{0As}(eV)	ΔE_f (eV)
Ortho	-70.218542	-69.408218	-0.1353442	-0.09662854	-0.88775532
Meta	-69.795159	-69.408218	-0.1353442	-0.09662854	-0.46437232
Para	-69.991258	-69.408218	-0.1353442	-0.09662854	-0.66047132

Looking at table 1 we see that the formed energy with the ortho configuration is the smallest (-0.88775532eV), so the ortho configuration is the most optimized configuration corresponding to the As atomic doped SiNR system placed in a the external electric field is about 0.3V/Å.

IOP Conf. Series: Materials Science and Engineering 1187 (2021) 012009 doi:10.1088/1757-899X/1187/1/012009

3. Bandstructure and density of states

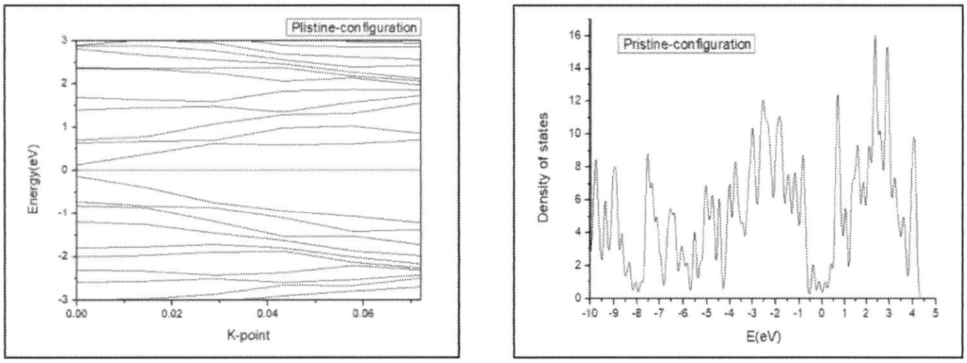

Figure 5. Band structure and DOS of pristine-configuration

The band structure and the state density of the undoped system are shown in Figure 5. The un-doped system has a band gap of 0.285eV, which is a semiconductor.

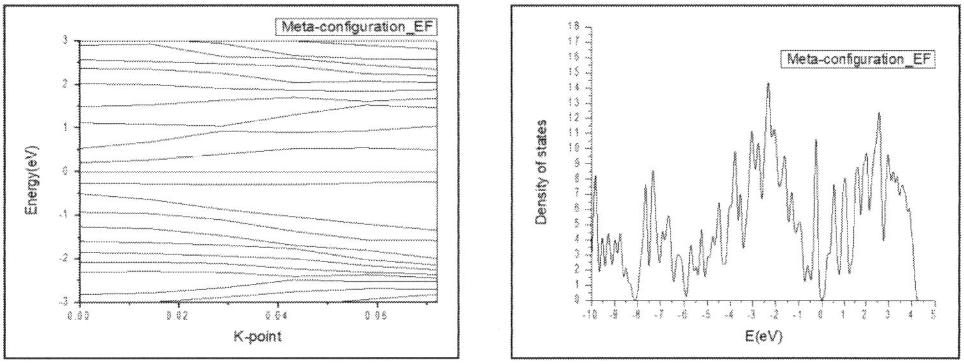

Figure 6. Band structure and DOS of meta-configuration in the presence of the external electric field

Figure 2 shows band structure and DOS of the meta-doped structure in the presence of an external electric field, the band gap still exists, the band gap energy is about 0.472eV. Compared to the pristine configuration we see that the band gap is wider, this is a semiconductor, compared with the pristine configuration, these states moved closer to the Fermi level.

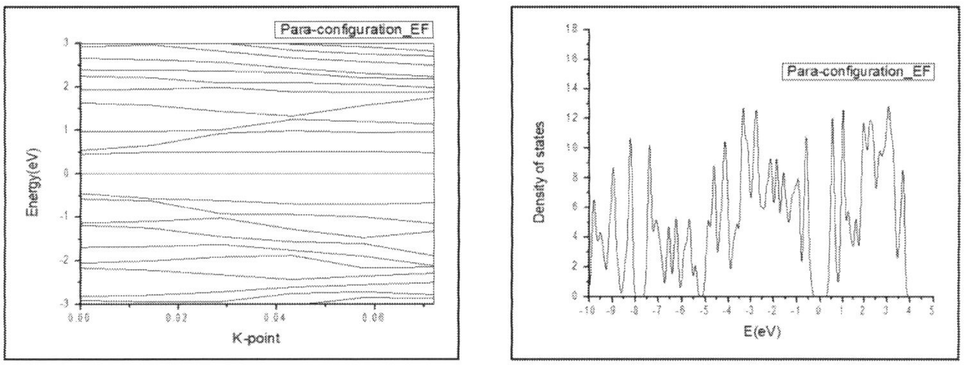

Figure 7. Band structure and DOS of para-configuration in the presence of the external electric field

IOP Conf. Series: Materials Science and Engineering 1187 (2021) 012009 doi:10.1088/1757-899X/1187/1/012009

Figure 7 shows the energy band structure and the state density of the para doped system in the presence of an external electric field. Comparing with pristine system, we can see that the energy gap is now wider, the size of the energy gap is about 0.921eV, the doping system is a semiconductor. The states are more concentrated around the Fermi level.

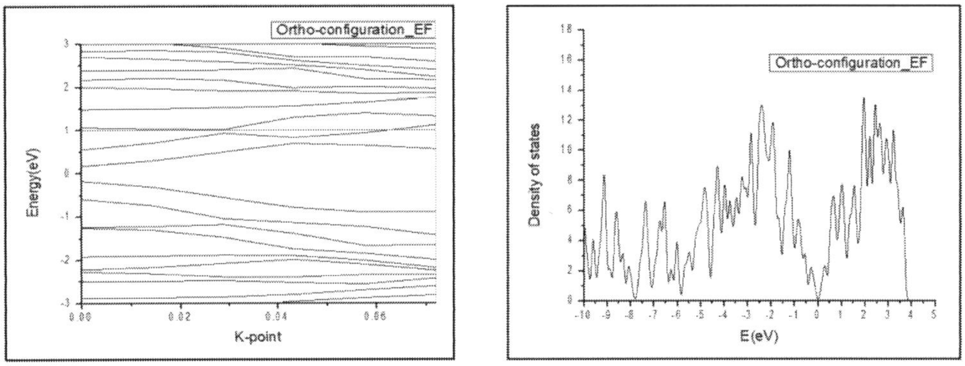

Figure 8. Band structure and DOS of ortho-configuration in the presence of the external electric field

The energy gap disappears in figure 8, the semiconductor becomes a conductor of electricity, the doped structure is metallic. The collective states are more concentrated at the bottom of the conduction band.

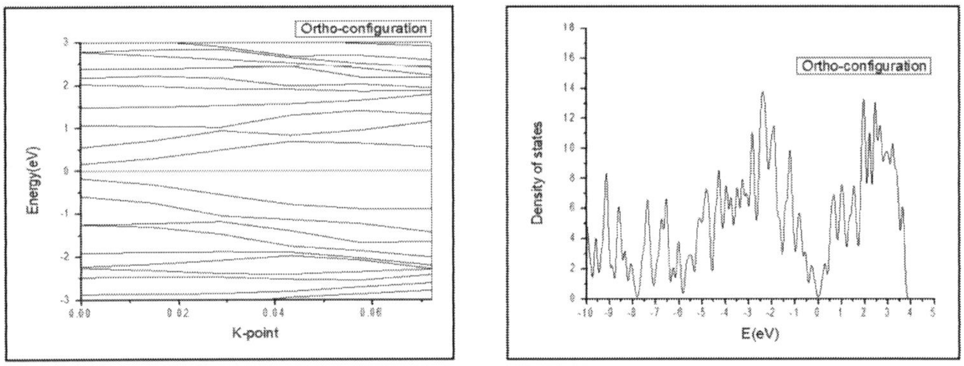

Figure 9. Band structure and DOS of ortho-configuration without the external field

In the absence of an external electric field (figure 9), the ortho doped configuration is still a semiconductor with an energy gap of 0.388eV, compared to the pristine system, the energy gap has been wider.

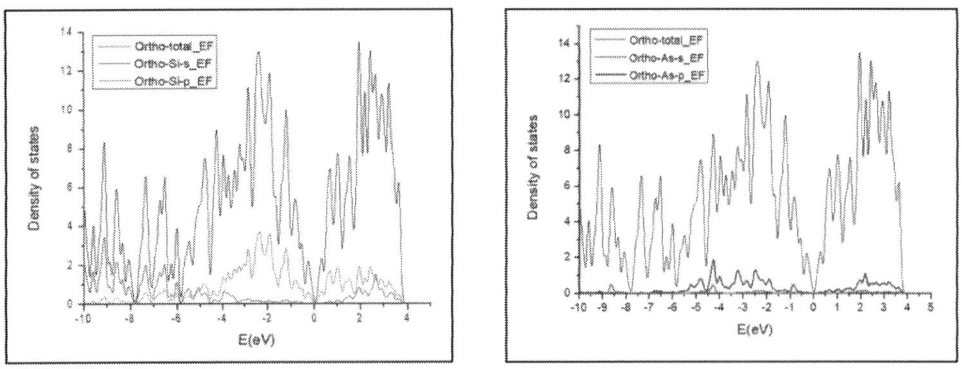

Figure 10. Total and partial DOS of ortho-configuration in the presence of the external electric field

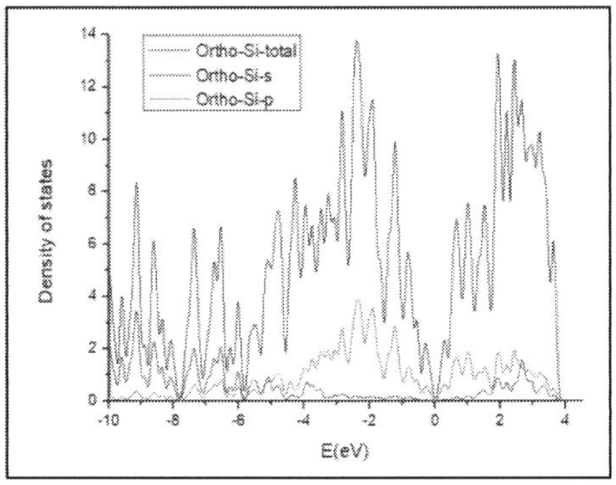

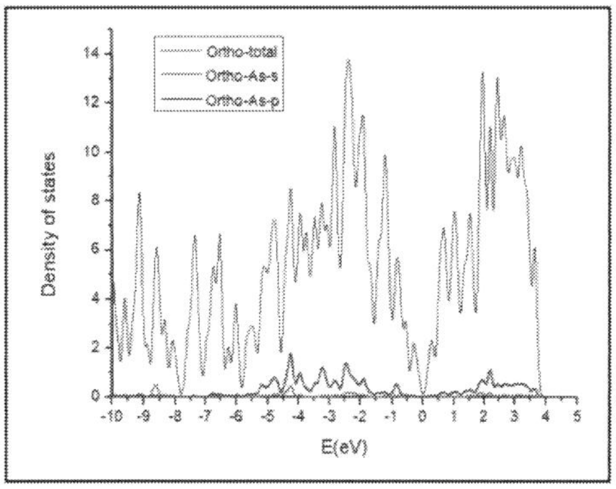

Figure 11. Total and partial DOS of ortho-configuration without the external field

The total state density and the partial state density in the ortho doped configuration are shown in figure 10 (when the external electric field is present) and figure 11 (when there is no external electric field). The contributions of the states Si-s, Si-p, As-s, As-p are almost unchanged in the presence of an external electric field and when there is no external electric field. The contributions of the Si-p state are essential to the formation of the conduction and valence bands.

4. Conclusions

Nanotechnology science research has many great applications. Doping another atom in a material helps to create materials of scientific value. Doping two arsenic atoms in silicene nanoribbons in the presence of an external electric field has been studied and analyzed. There are three doping configurations studied here, namely meta, ortho, and para configuration. By calculating the forming energy, we have found that the most optimal structure when there is an external electric field of 0.3V/ $\mathring{A}$ is the ortho configuration with the forming energy of -0.88775532eV. The pristine configuration is a semiconductor with a narrow energy gap, when doped with arsenic we find that both meta and para configurations are semiconductors with wider energy gap, especially with ortho, the band gap is lost, and for a conductor of electricity. The contributions of the partial states Si-s, Si-p, As-s, As-p have been studied here, the Si-p state is the state that contributes mainly to the formation of the band structure energy of the ortho optimal system.

Acknowledgement

This research is funded by Thu Dau Mot University, Binh Duong Province, Vietnam.

References

[1] P. Vogt, P. De Padova, C. Quaresima, J. Avila, E. Frantzeskakis, M. C. Asensio, A. Resta, B. Ealet, and G. Le Lay 2012 Silicene: compelling experimental evidence for graphenelike two-dimensional silicon *Physical review letters*, vol. 108, no. 15, pp. 155501.

[2] S. Cahangirov, M. Topsakal, E. Aktürk, H. Şahin, and S. Ciraci 2009 Two-and onedimensional honeycomb structures of silicon and germanium *Physical review letters*, vol. 102, no. 23, pp. 236804.

[3] D. K. Nguyen, N. T. T. Tran, Y.H. Chiu, G. Gumbs and M.F. Lin 2020 Rich essential properties of Si-doped graphene *Scientific Reports,* vol. 10, no. 12051.

[4] Qing Pang, Yan Zhang, Jian-Min Zhang, Vincent Jib and Ke-Wei Xuc 2011 Electronic and magnetic properties of pristine and chemically functionalized germanene nanoribbons *Nanoscale,* vol. 3, pp.4330-4338.

[5] L. Li, S. z. Lu, J. Pan, Z. Qin, Y. q. Wang, Y. Wang, G. y. Cao, S. Du, and H. J. Gao 2014 Buckled germanene formation on Pt (111) *Advanced Materials*, vol. 26, no. 28, pp. 4820-4824.

[6] M. Derivaz, D. Dentel, R. Stephan, M. C. Hanf, A. Mehdaoui, P. Sonnet, and C. Pirri 2015 Continuous germanene layer on Al (111) *Nano Letters*, vol. 15, no. 4, pp. 2510-2516.

[7] P. Liang, Y. Liu, S. Xing, H. Shu, and B. Tai 2016 Electronic and magnetic properties of germanene: Surface functionalization and strain effects *Solid State Communications*, vol. 226, pp. 19-24.

[8] E. Bianco, S. Butler, S. Jiang, O. Restrepo, W. Windl, and J. Goldberger 2013 Stability and exfoliation of germanane: a germanium graphane analogue *ACS Nano,* vol. 7, no. 5, pp. 4414.

[9] C. J. Rupp, S. Chakraborty, R. Ahuja, and R. J. Baierle 2015 The effect of impurities in ultrathin hydrogenated silicene and germanene: a first-principles study *Physical Chemistry Chemical Physics,* vol. 17, no. 34, pp. 22210-22216.

[10] B. van den Broek, M. Houssa, E. Scalise, G. Pourtois, V. Afanasev, and A. Stesmans 2014 First-principles electronic functionalization of silicene and germanene by adatom chemisorption *Applied Surface Science*, vol. 291, pp. 104-108.

[11] V. O. Özçelik, H. H. Gurel, and S. Ciraci 2013 Self-healing of vacancy defects in single-layer graphene and silicene *Physical Review B*, vol. 88, no. 4, pp. 045440.

[12] N. J. Roome and J. D. Carey 2014 Beyond graphene: stable elemental monolayers of silicene and germanene *ACS Appl Mater Interfaces,* vol. 6, no. 10, pp. 7743-7750.

[13] M. M. Monshi, S. M. Aghaei, I. Calizo 2017 Doping and defect-induced germanene: A superior media for sensing H_2S, SO_2, and CO_2 gas molecules *Surface Science,* vol. pp. 96-102.

[14] M. Dávila, L. Xian, S. Cahangirov, A. Rubio, and G. Le Lay 2014 Germanene: a novel two dimensional germanium allotrope akin to graphene and silicene *New Journal of Physics*, vol. 16, no. 9, pp. 095002.

[15] Z. Ni, Q. Liu, K. Tang, J. Zheng, J. Zhou, R. Qin, Z. Gao, D. Yu, and J. Lu 2011 Tunable bandgap in silicene and germanene *Nano Letters*, vol. 12, no. 1, pp. 113-118.

[16] S. M. Aghaei, M. M. Monshi, and I. Calizo 2016 A theoretical study of gas adsorption on silicene nanoribbons and its application in a highly sensitive molecule sensor *RSC Advances,* vol. 6, no. 97, pp. 94417-94428.

[17] S. M. Aghaei, I. Torres, and I. Calizo 2016 Structural Stability of Functionalized Silicene Nanoribbons with Normal, Reconstructed, and Hybrid Edges *Journal of Nanomaterials,* vol. 2016, p. 8, Art. no. 5959162.

[18] S. M. Aghaei, M. M. Monshi, I. Torres, and I. Calizo 2016 Edge functionalization and doping effects on the stability, electronic and magnetic properties of silicene nanoribbons *RSC Advances,* vol. 6, no. 21, pp. 17046-17058.

IOP Conf. Series: Materials Science and Engineering 1187 (2021) 012010 doi:10.1088/1757-899X/1187/1/012010

Efficient Waste Management Models for Smart Cities: A Case Study

K. Reji Kumar [1]. and E. N. Satheesh[2].

[1]Department of Mathematics, N. S. S. College, Cherthala, India.
[2]Department of Mathematics, N. S. S. Hindu College, Changanacherry, India.

rkkresearch@yahoo.co.in

Abstract. Waste management is a matter of the greatest concern in the modern world. We do not have an efficient waste management system in India. Developing a comprehensive network of waste management is the biggest challenge that we are facing today. Research on smart cities brings together the advances in the field of science and technology to ensure sustainable development of cities ensuring a healthy, wealthy, and peaceful life for all. In this paper, we discuss a mathematical modelling method that can effectively be implemented to build a network of waste management in which human being, organizations, institutions, and government agencies play their role. We discuss the waste management system of Thiruvananthapuram, the capital city of Kerala, as an example. We show that the waste management system of the city is not and sufficient and efficient. We also give suggestions to improve the system.

1. Introduction

The smart city is a modern concept, which is a contribution of the technologically advanced world. New research in smart cities aims at improving personal and social life using advanced digital technologies and artificial intelligence. Using machine learning techniques we capture data, process, inspect, and analyze the collected data, which are related to human life, and then applied to manage the system efficiently [1].

The objects and materials that we possess in our life have some value for us. When its value becomes null the object becomes waste in the judgment of an individual. But for a different person, the same thing may be valuable. So the meaning of the word "waste" is highly subjective. We do not go much deep into this philosophic aspects of the meaning of the word because it requires a detailed discussion and hence it is set aside for another occasion. Many objects are commonly called waste by all. In this study, we mainly focus on such things.

If an object has some reuse value in the evaluation of some people, that object could be given to them. This is called waste reuse. Many objects labeled waste can be recycled to produce similar kinds of objects or different kinds of new objects. Some waste materials are very rich sources of energy. If an object cannot be used in any one of the forms mentioned so far, it should be disposed of permanently. The general approach to the concept of waste management has three components.

Waste Reuse Management (WReuM), Waste Recycle Management (WRecM), and Waste Disposal Management (WDM). An efficient waste management system must effectively handle all the three components. The first component consists of the marketing or free transfer of second-hand materials from the owner to the needy person. The percentage of the third component in the waste management system must be reduced to zero. Unfortunately in India, the percentage of the third component is much higher than the first two. The percentage of the second component is extremely low. In the following section we discuss the problems that prevail in the waste management system in India.

2. Waste Management in India

Mathematical modelling is an important method that can effectively be used to represent real-life problems. It can help us to find some solution to the problem. In this section, we briefly review the mathematical modelling techniques used by researchers for modelling waste management problems. The majority of the municipal waste management models identified in the literature are decision support models. Morrissey et al. [2] divided the models into three categories. First type is based on cost-benefit analysis. The second type is based on life cycle assessment and the last is based on multi-criteria decision making. In his view, the current models have many shortcomings. Some models emphasize the refinement of evaluation steps, instead of addressing the decision-making process itself. A waste management model must consider environmental, economic, and social aspects together to be sustainable. Many of these models do not consider all the three important factors together in their model.

Mathematical modelling of waste management systems is not a new concept. The period 1970 – 90 mathematical modelling techniques were applied in the waste management problems. An outline of these models is given by Morrissey et al. [2]. The approach to the problem of waste management in India is extremely different from that of other countries in the world. We do not have a centralized and collaborative system for waste management. Ours waste management system is highly localized and not networked. We are backward in applying modern technology and scientific advancements to improve the efficiency of the system of waste management. Planning and implementation of waste management techniques completely rest on the local authorities. State Government and Central Government are devoid of their responsibilities in building a network of waste management system. We have underestimated the scope and opportunities of a well networked WMS. An efficient WMS can make the country self sufficient in many ways. Generation of energy, recycling and reuse of used materials, creation of new job opportunities, control over the scarcity of materials for the development etc are the main issues that can be addressed by tan efficient waste management system. The most pathetic condition of Indian waste management systems is discussed by Sunil Kumar et al. [3] and Reji Kumar et al. [5] as well as the opportunities associated with it. In this paper, the authors initiate a study of WMS based on network analysis. Our waste management system is a complete failure due to inadequate waste collection, transport, treatment, and disposal. Waste generation in India is growing fast due to fast growth in population, rapid urbanization, and indifferent behaviour of citizens. The diverse culture of India is making the situation extremely dangerous and impossible to get corrected naturally.

The quantity of waste generated varies from place to place. It is very high in megacities and very low in rural areas. It is almost proportionate to the total population that inhibits in an area. The Ministry of Environment and Forest (MoEF) has issued MSW (Management and Handling) Rule 2000 which intends to ensure proper waste management in India [4]. According to this Rule municipal authorities are responsible for implementing the law and developing infrastructure for solid waste management. This is the main drawback of this Rule. Waste management is not a simple task that can be effectively planned and implemented by local bodies. It needs national-level planning the strategy and centralized control over its implementation. All activities must be monitored and reviewed for their effectiveness at different levels. At the grass-root level, every citizen of the country must take part effectively in the process. Duties and responsibilities of an ordinary individual must be defined in clear terms and the same must be done at all levels in the system.

The waste management in India is mainly the informal sector. The informal sector consists of small-scale, labor-intensive, largely unregulated and unregistered agents who use outdated methods for collection and processing of waste. The informal sector is run by the contribution of waste pickers, who collect household, commercial, and industrial waste. Hundreds of thousands of waste pickers and waste collecting agents in India depend on waste for an income. These waste pickers and collection agents do not have any training in waste management. The existing network of waste management systems in India consists of waste pickers, waste collectors, and waste recycling centres. This is a natural network in which the nodes are the waste pickers, waste collectors, and collection centres. Waste pickers collect only those waste materials which have some monetary value. They leave the other kind of waste in the environment. The waste materials that have no monetary value are usually sent to landfills. Landfills are not an acceptable solution to the problem. Complete recycling of all materials in the form of waste is the ultimate aim of a sustainable waste management system.

This situation leads to an inefficient and incomplete waste management system. Thus we need a new efficient network of waste management, in which each individual discharges his predetermined duties to the maximum possible extent for the effective management of waste materials. In the following section, we describe how to change the current system by applying the techniques of social network analysis.

Social network is a network of individuals together with the relations existing among them. World Wide Web, Internet, friendship network, collaboration network etc are examples of social networks. Social network analysis is the scientific study of the role and importance of the members of the network along with the characteristics of the relation among the members using the methods of graph theory. Importance of some actors in spreading information in complex networks is studied in [7, 8].

Centrality measure of a network is a well studied parameter which helps us to understand the importance of nodes in the network. According to the relative importance of actors in a network the members are ranked. Degree centrality, Closeness centrality, Betweenness centrality, K-core decomposition, Mixed degree decomposition method etc are some important centrality measures that can be used to find rank of the members according to their importance in the network. Improved k-shell decomposition method is a modified version of the k-shell decomposition method proposed by Kumar et al. [13]. A new measure of centrality named the m-ranking method was proposed by Kumar et al. [6]. This method is modified to handle the situation of directed network [10] and the influence in the network due to personal variations [9]. A method to find the m-ranking approximately is discussed in [14]. The methods are applied to find the ranking of nodes in two real world networks [11, 12].

3. Mathematical Modelling of Waste Management Systems

The system of waste management in a country can be modeled using graph theoretic networks and its efficiency and effectiveness can be studied using the techniques in social network analysis. Mathematical modeling techniques are not effectively being used in the field of waste management. Researchers in the field have turned their attention to apply mathematical modeling techniques to solve the problems of waste management only in the recent past [5]. So application of social network in the field is completely new and offers a wide spectrum of openings.

A network representing the WMS can consist of all the members of the society at the root level. These members can be connected to various collection centers located in the neighborhood. The local collection centers play an important role in the management of waste. The centers must be strengthened by providing sufficient workers and technological knowhow to separate and collect the waste materials according to the future use the procedure through which it passes. Separation and collection must start at the point of the generation of the waste. So a very pleasant relation between the members of the society and the first level collection centers are essential for the proper functioning of the entire system of waste management. We propose a two way relation between the first level collection centers and the members of the society. First level centers must be connected to the second level collections centers or waste processing and recycling centers. The responsibility of building and maintaining of the second level centers rests mainly on the state and central government. Currently in India the task of establishing the second level centers is done by local self government. This is one of the main reasons of inefficient system of waste management in India. The entire system and all levels of waste management must be monitored and policy for decision making and evaluation for continuous improvement of the system must be done by department of the state and central government.

From the above discussion it is clear that the state and central government must act as center of all activities that control the entire system. In other words they are the hubs in the network. So the centrality measure of the nodes representing the state government and the central government in the network must be high when compared to the values of the nodes representing the first or second level. Majority of networks studied in social network analysis are natural and dynamic. It originates by itself and undergoes changes over time. But in the current situation, we propose to build an artificial (man-made) network named waste management network. This network is very comprehensive and must contain all individuals in a geographical area under study. Next we proceed to discuss more about a network that can be developed artificially. As we have mentioned earlier waste collection centers are the second level nodes in the network. Waste recycling centers and engineered landfills are nodes at the third level and the state and central government have representation in the network at the apex level. The proposed network is a multi layered system. At the bottom level we include the Households, offices and small business centers. The collection and processing centers belong to the middle level and the final processing centers and recycling points become the top level in the hierarchy.

Next, we discuss a network model, which represents the waste management system in the capital city of Kerala, Thiruvananthapuram. In 2011 there was only one waste dumping area (called Vilappinsala) in the city limit. It was forced to shut down due to the protest from the public over the mismanagement of waste at the site. As a solution to the problem of waste, the Thiruvananthapuram Municipal Corporation (TMC) adopted a decentralized system for the management of waste.

Total area of Thiruvananthapuram city, which is divided into 100 wards has land area 214.86 square kilometers. The population of the city is approximately 0.9 million. According to government statistics the city produced 250 tones of municipal solid waste. Following the protest in 2011 the TMC adopted a policy called segregated collection of waste. Segregation of waste at the origin and resource recovery at the root level is the main feature of the policy. Additionally participation of the individuals and institutions is ensured uniformly all over the city. It is based on the principle of proximity, which ensures that the waste collection agents or collection centers are in the easy reach of every individual of the society. The responsibility of waste management of bulk generators and commercial establishments rests basically on them. But the facilities of processing will be provided by the local self government. The TMC has established a system of collecting the segregated waste from the houses by imposing a monthly charge of Rs 800. Residential flats and gated communities are encouraged to set up organic waste management facility offering 50 per cent subsidy. TMC also established collection centers at about 25 places in the city, where people can drop the separated and waste materials [15, 16].

According to the records of TMC about 50% percent of the total households in the city and 350 bulk generators compost their wet waste. Approximately 100 bulk generators and 4000 households have setup biogas plants for the treatment of waste. Pipes, kitchen-bin, pot or bio composting, aerobic bins etc are various technologies used for composting. Using these methods approximately 67% of the waste has been diverted from the landfills [15]. In Table I, we sum up various infrastructure facilities that are effectively being used in the waste management and the percentage of waste that each facility can handle [15]. The quantity of waste is given in tone in the third column and the percentage utilized in the fourth column.

In the Figure 1, a graph theoretic representation of the waste management system in TMC is given. There are thousands of households, business enterprises, hotels, etc which generate solid waste. They are together represented by the red ellipse in the figure. It is a huge network of thousands of nodes which are connected to 15 yellow coloured nodes that represent the collection points of waste materials. Unfortunately, there are no major recycling centres or disposal centres in the city. So, level 3 is empty in the first figure.

IOP Conf. Series: Materials Science and Engineering 1187 (2021) 012010 doi:10.1088/1757-899X/1187/1/012010

Table 1. Infrastructure and percentage of waste it can handle at Thiruvananthapuram

Infrastructure	Units	Capacity: tons per day	Utilization: % tons per day
Kitchen Bins	19000	19.00	80%
Bio Bins	109	2.72	60%
Biogas Plants	3982	3.98	60%
Pipe compost Units	87000	87.00	87%
Organic Waste Converters	2	0.5	100%
Aerobic bins	383	11.49	100%
Mobile composting units	154	4.62	100%
Community biogas plants	23	23.00	80%
Dry waste collection bins	2	0.2	100%
Dry leave collection bins	3	0.3	100%
Material recovery facilities	44	44.00	100%
Resource recovery facilities	2	10.00	100%
Facilities in the private sector	19	150.00	60%
Total		356.8	67.89%

This shows very clear laxity from the side of the corporation in setting up modern waste processing and recycling centres. The corporation must have started waste processing centres to manage effectively the waste materials collected at the collection points. Instead, the segregated materials are either buried or sent to other states for further processing. It is a clear evidence of the lack of interest from the side of the municipality.

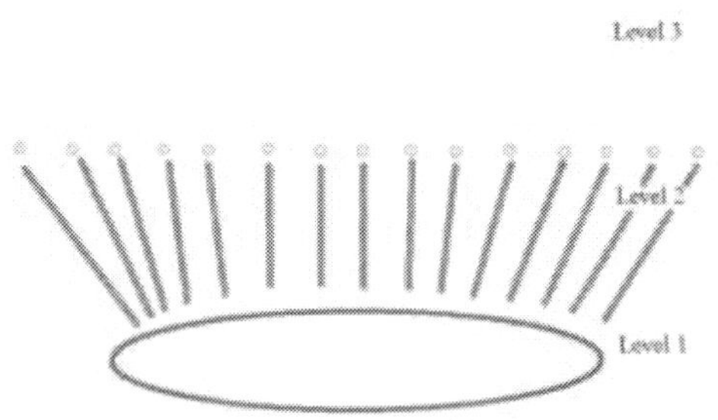

Fig. 1. Current SWM of Thiruvananthapuram

The Figure 2 is a modified representation of the network by including some waste processing centres. Waste to energy generation centre, material recycling centre and permanent disposal centre are the three centres proposed. In the new imaginary network, the three new processing centres are given green colour. The first collection centre is connected to all three processing centres. Similarly, all the remaining collection centres are connected to the processing centres. It is represented by the blue-coloured dots in the figure. It is clear that the current system which exists in Thiruvananthapuram city is not efficient.

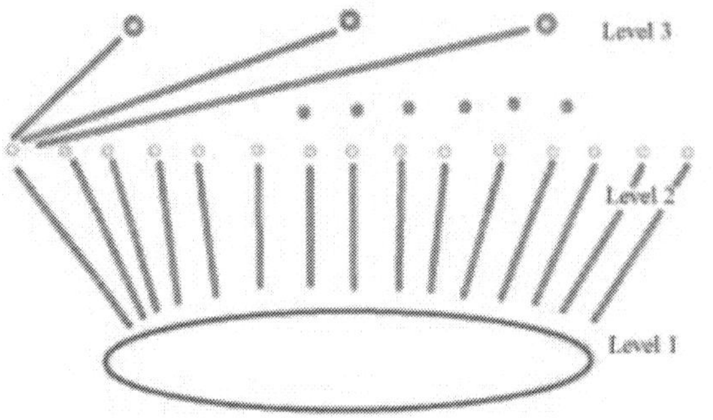

Fig. 2. An Improved SWM for Thiruvananthapuram

We can call a waste management system sufficient if it can handle the entire quantity of all type of waste materials without making problems to the society and environment during a particular time period. So a system which can process only a portion of waste is not sufficient. A system is consistent if it can manage the waste generated constantly for a long time. A WMS, which is efficient, must be sufficient also. But a sufficient system may not be efficient always. Consistency checks whether a system shows steady performance for a long time. This measure evaluates the performance of the system based on time. From the data presented in the Table 1, it is clear that the WMS of TMC is not sufficient to meet the requirement because only 67.89 % of the waste generated is processed currently. In addition to this no data is available regarding the income generated from waste processing and recycling, job opportunities created by the current system etc. So the system is highly inefficient.

In order to improve the waste management system in India, we need to develop an efficient waste management network artificially. Waste collection and post collection procedures must be modified using modern technology. Every individual taking part in the waste management procedure must be properly trained and must be equipped to meet the challenges of waste management in the future. The best and most appropriate methods for waste collection and disposal must be implemented. A professional approach to the problem is a pressing need. There is also a lack of accountability in current SWM systems throughout India. The responsibility to implement the WMS in the country must be shifted from local authorities. It must be reassigned to the State government and the Central government. The central government must take a pivotal position in the whole process.

4. Conclusion

This paper is a step forward to improve the waste management system in India. It can be attained by redefining the role of all members in the society in the big task of waste management. Three important measures of a WMS is defined and discussed. The situation is analyzed for the waste management system of Thiruvananthapuram Municipal City. The network representing the current WMS of the city is given along with the suggestions to improve the performance of the system.

Acknowledgments

We are very grateful to the anonymous referees for their appropriate and constructive suggestions that helped me to improve this paper

References

[1] Larissa R. Suzuki and Anthony Finkelstein Edotore, Machine learning and AI – ensuring fairness in smart cities, Data as Infrastructure for Smart Cities.

[2] A.J. Morrissey and J. Browne, Waste management models and their application to sustainable waste management, Waste Management 24 (2004) 297–308

[3] Sunil Kumar and Stephen R. Smith. Geoff Fowler, Costas Velis, S. Jyoti Kumar, Shashi Arya1, Rena1, Rakesh Kumar1, and Christopher Cheeseman, Challenges and opportunities associated with waste management in India, rsos.royalsocietypublishing.org R. Soc. open sci. 4: 160764

[4] Ministry of Environment and Forests (MoEF). 2015, The Gazette of India. Municipal solid waste (Management and Handling) Rules, New Delhi, India.

[5] K. Reji Kumar and E. N. Satheesh, Developing an Efficient Waste Management System: An Approach by Using Some Methods of Mathematical Modelling, Proceedings of the ICMSMT 2020, Vol. 4, IOP Conference series, Material science and Engineering, 1654 – 1660.

[6] K. Reji Kumar, Shibu Manuel and Deepu Benson, The m-ranking of nodes in complex networks, Proceedings of COMSNETS 2017. 9th International Conference on Communication Systems & Networks, https://doi.org/ 10.1109/COMSNETS.2017.7945419

[7] K. Reji Kumar K and Shibu Manuel, Spreading information in complex networks: A modified method, Proceedings of International Conference on Emerging Technological Trends 2016. ICETT 2016, **IEEE Xplore,** https://doi.org/ 10.1109/ICETT.2016.7873636.

[8] Reji Kumar Karunakaran, Shibu Manuel and Edamana Narayanan Satheesh, Spreading information in complex networks: An overview and some modified methods, Chapter in Graph Theory - Advanced Algorithms and Applications, InTech Publishers, http://dx.doi.org/ 10.5772/intechopen. 69204

[9] K. Reji Kumar, and Shibu Manuel, Personal influence on the spreading of information – a network based study. Int. J. Math. Trends Technol. 52 (8) (2017).

[10] K. Reji Kumar, and Shibu Manuel, A centrality measure for directed networks: m-ranking method. Zyer, T., et al. (eds.) Social Networks and Surveillance for Society. Lecture Notes in Social Networks. Springer, Heidelberg (2019), https://doi.org/10.1007/978-3-319-78256-0_7

[11] K. Reji Kumar, and Shibu Manuel, Collaborations of Indian institutions which conduct mathematical research: a study from the perspective of social network analysis. Scientometrics. (2018) 117: 1041–1051 https:// doi.org/10.1007/s11192-018-2898-0

[12] K. Reji Kumar, and Shibu Manuel, A network analysis of the contributions of Kerala in the field of mathematical research over the last three decades, Proceedings of the International Conference on Computer Networks, Big Data and IoT, 2018, https://doi.org/ 10.1007/978-3-030-24643-3_70

[13] K. Reji Kumar, and Shibu Manuel, An improved k-shell decomposition for complex networks based on potential edge weights. Int. J Appl. Math. Sci. 163–168 (2016)

[14] K. Reji Kumar and Shibu Manuel, (2020) An Approximation to m-Ranking Method in Networks. In: Balaji S., Rocha Á., Chung YN. (eds) Intelligent Communication Technologies and Virtual Mobile Networks. ICICV 2019. Lecture Notes on Data Engineering and Communications Technologies, vol 33. Springer, Cham. https://doi.org/10.1007/978-3-030-28364-3_33.

[15] https://www.downtoearth.org.in/news/waste/ten-zero-waste-cities-how-thiruvananthapuram-cleaned-up-its-act-68539, Accessed on 29th March 2021.
https://zerowasteworld.org/wp-content/uploads/India.pdf, Accessed on 29th March 2021.

Design and simulation of Sierpiński Triangular Fractal Antenna for S-Band

Jahanvi H. Vyas[1], NB Gohil[2], Ved Vyas Dwivedi[3]

[1]*MEEC Department, Shantilal Shah Engineering College, Bhavnagar*
vyasjahanvi@gmail.com
[2]*Assistant Professor, EC Department, Shantilal Shah Engineering College, Bhavnagar*
nbgohil1@gmail.com
[3] *Vice Chancellor, Gokul Global University, Siddhpur, Gujarat*
provost@gokuluniversity.ac.in

Abstract: -In current wireless communication scenarios, you need to reduce the size of the antenna and achieve better performance without affecting other antenna settings. A fractal antenna is an antenna that uses a fractal and similar design to maximize the effective length of a material that can transmit and receive electromagnetic radiation in a specific total area, or to augment its environment (internal components or external structures) or the volume. Fractal antennas offer better parameters while providing better performance at many different frequencies. The Sierpinski triangle is a fixed set of fractal attractions. Its general shape is an equilateral triangle, which is recursively divided into smaller equilateral triangles. Sierpinski's triangular fractal geometry is used to reduce antenna size and improve antenna performance at required frequencies. In this proposed task, a simple technique is used to design a Sierpinski triangular fractal antenna. These antennas are lightweight and small in size. This design covers the S band and is suitable for wireless applications.

Keywords: *fractal Antenna; Sierpinski triangle; dual-frequency; lightweight; Wi-MAX.*

1. INTRODUCTION

Antennas are a necessary and important component for wireless communication. Most of the antennas are imported to my country due to poorly designed antennas. There are not enough models to manufacture in our country. Once the antenna is designed, it will be relatively easy to manufacture [1]. Wireless applications require a small antenna. This demand can be met by using fractal microstrip antennas. Recently, the design of fractal antennas has attracted many researchers due to its low cost, light weight and small size. Sierpinski's triangular fractal geometry is used to reduce antenna size and improve antenna performance at required frequencies [2]. The fractal antenna uses the fractal, Self-similar design to maximize the effective length of a material that can receive or transmit electromagnetic radiation in a given surface or total volume, or stretch its volume, or stretch its surroundings (in internal parts or external structure). Fractal antennas can operate with excellent performance at many different frequencies at the same time [3]. The main objective of this article is to design a lightweight and compact antenna. This design covers the S band and is suitable for wireless applications. Fractal antennas can operate with excellent performance at many different frequencies at the same time.

IOP Conf. Series: Materials Science and Engineering 1187 (2021) 012011 doi:10.1088/1757-899X/1187/1/012011

2. ANTENNA DESIGN

The proposed antenna is designed with microwave studio computer simulation software. This antenna design uses low cost FR4 material as the 1.6mm high card. The dielectric constant of the panel is 4.4. Microstrip line power supplies are used because of their reliability and ease of manufacture [5]. This design uses the shape of the Sierpinski joint until the second iteration. The permittivity of the substrate (εr) and the lowest resonant frequency (fr) are taken into account to calculate the length of the sides of the triangular patch antenna. Equation (1) indicates the side length of the triangular patch [6].

$$a = \frac{2C}{3fr\sqrt{\varepsilon r}} \quad (1)$$

In equation (1), C indicates the speed of light in free space. Substituting all the parameters of equation (1), the side length of the triangular patch is 47.33 mm. Figure 1 shows the proposed antenna design. Equations (2) and (3) show the calculation of the number Nn of triangles and the fractional area Fn of each iteration [7] [8].

$$Nn = 3^n \quad (2)$$

$$Fn = \left(\frac{3}{4}\right)^n \quad (3)$$

Figure 1 (a) is an initiator made up of triangles fed by microstrip lines. Figure 1 (b) is the first iteration with the inverted triangle removed from the center of the patch. Figure 1 (c) is the second iteration of the three triangles removed from the remaining patches. These three triangles are scaled-down versions of the triangles that were removed in the first iteration.

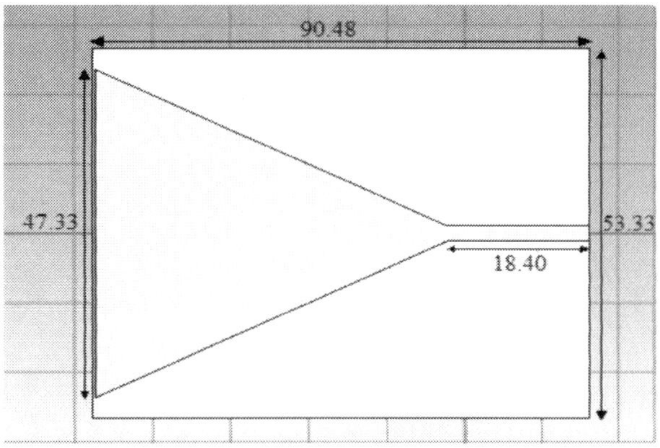

Figure 1 (a) Proposed antenna design, 0th iteration

IOP Conf. Series: Materials Science and Engineering 1187 (2021) 012011 doi:10.1088/1757-899X/1187/1/012011

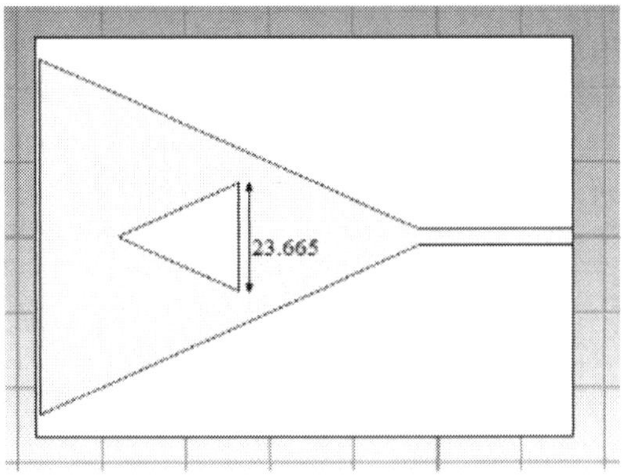

Figure 1 (b) Proposed antenna design, first iteration

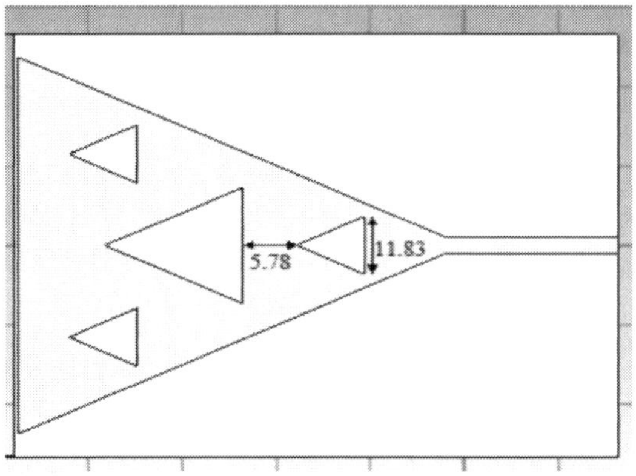

Figure 1 (c) Proposed antenna design, second iteration

3. RESULTS OF THE SIMULATIONS

The proposed antenna design is simulated with CST Studio software. This antenna design uses low cost FR4 material as the substrate with a height of 1.6mm. The dielectric constant of the substrate is 4.4. Microstrip line power supplies are used because of their reliability and ease of manufacture. Copper (annealed) is used as a patch with a ground 0.035mm high. Figure 2 (a) is a graph of the amount of reflection attenuation at the 0th iteration. Figure 2 (b) is a graph of the amount of reflection attenuation at the first iteration. Figure 2 (c) is a graph of the amount of reflection attenuation of the second iteration.

A. Mitigation of reflection until the second iteration: -

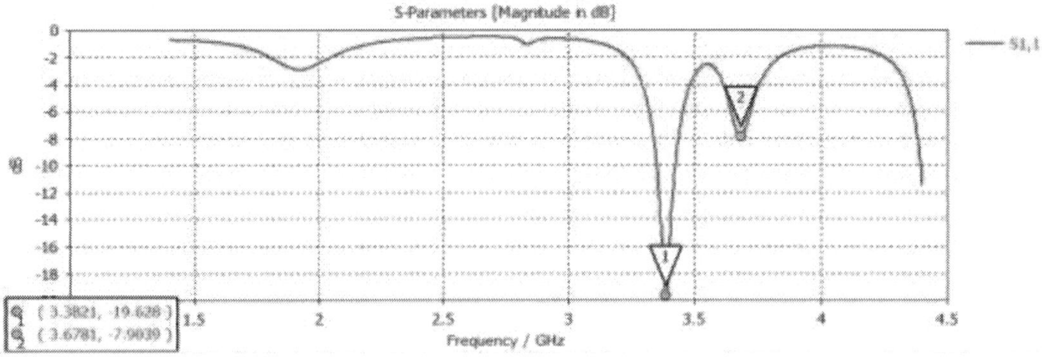

Figure 2 (a) Plot of the antenna's return loss proposed in the 0th iteration

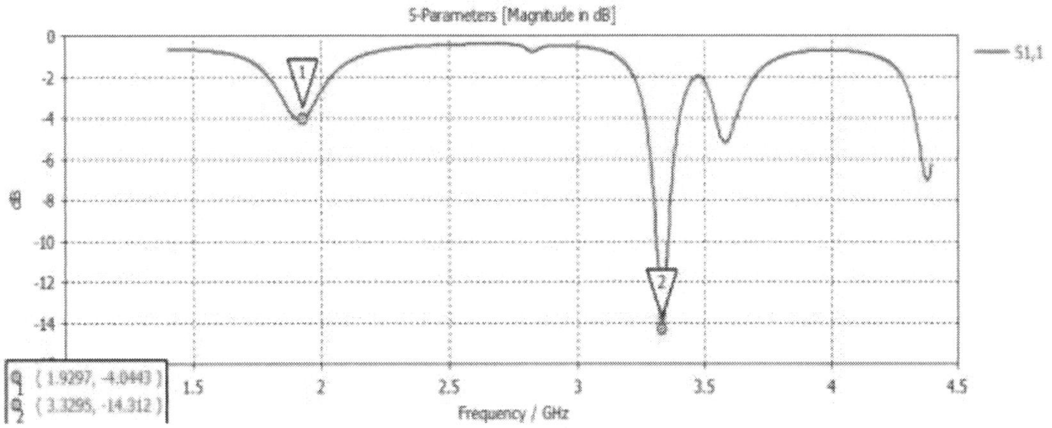

Figure 2 (b) Plot of the antenna reflection loss proposed in the first iteration

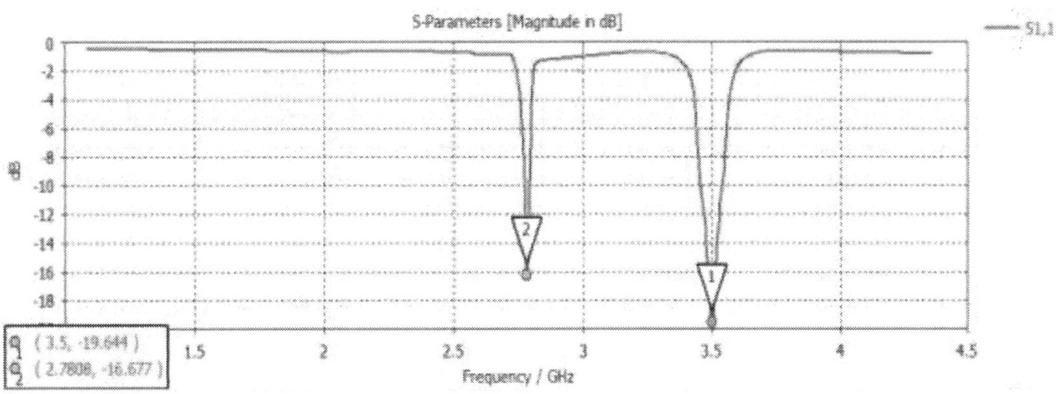

Figure 2 (c) Plot of the antenna reflection loss proposed in the second iteration

In the simulation results, the initiator shows a resonant frequency of 3.3 GHz and the reflection attenuation is -19.628 dB. The first iteration gives a resonant frequency of -14.31 dB of reflection attenuation at 3.32 GHz. The second iteration gives a dual resonant frequency of 2.7 GHz and 3.5 GHz, which is responsible for Wi-MAX applications and has a reflection attenuation of -16.667 dB and -19.644 dB, respectively. With this proposed antenna, the SWR in the 2.7 GHz frequency range is 1.5416 dB and the SWR in the 3.5 GHz frequency range is 1.2356 dB.

IOP Conf. Series: Materials Science and Engineering 1187 (2021) 012011 doi:10.1088/1757-899X/1187/1/012011

B. Proposed antenna radiation pattern: -

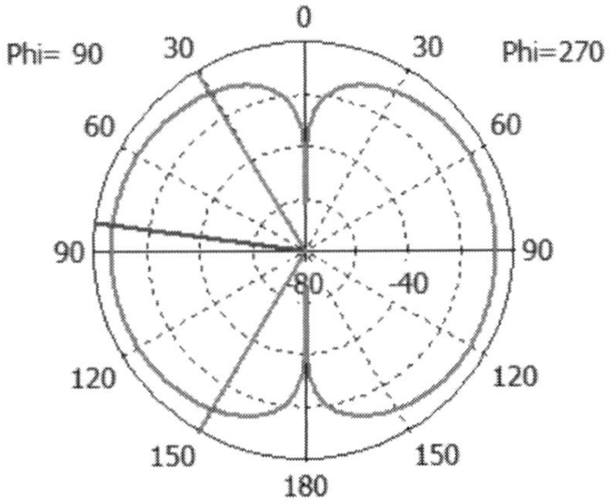

Figure 3 (a) Proposed antenna radiation 2.7 GHz model

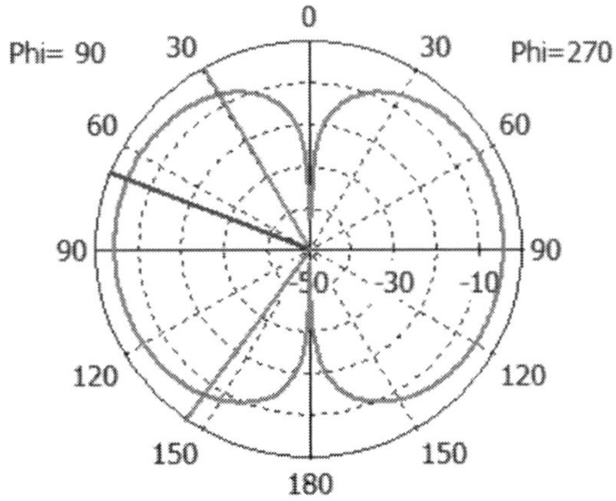

Figure 3 (b) Proposed antenna radiation 3.5 GHz model

Figure 3 (a) shows the proposed antenna radiation pattern at the desired frequency of 2.7 GHz. Figure 3 (b) shows the radiation pattern of the proposed antenna at the desired frequency of 3.5 GHz.

4. CONCLUSION

Triangular Fractal Antennas for 2.7 GHz and 3.5 GHz (S-band) frequencies were originally designed by CST Studio and performed best in these frequency bands. The combination of fractal antennas and microstrip foam produces compact antennas while providing the superior performance required. Originally designed Sierpinski triangular fractal antenna. It operates at 2.7 GHz and 3.5 GHz (S band) and performs best at these frequencies. The original design of the Sierpinski Triangular Fractal Antenna acquires the parameters of -16.677 dBi and -19.644 dBi at 2.7 GHz and 3.5 GHz, respectively, by combining the frequencies of 2.7 GHz and 3.5 GHz (band S) in one card. I go. This antenna is suitable for Wi-MAX applications for S-band wireless communication.

REFERENCE

[1] Sheetal Bukkawar and Vasif Ahmed, "Square Fractal Antennas for Multiband Applications", IEEE Publications.

[2] Yu-wei ZHANG, Shu LIN, Shang YD, Ling LID, Wei-yuan ZHANG, Cai-tian YANG, Alexander Deni sov, "Simulation of a Minkowski double polarized fractal microstrip antenna for the S-band "Design", 2018 IEEE.

[3] N. Srikanta and M. Pachiyannan, "Wideband, dual frequency CPW-powered fractal antennas for wireless applications", 2019 IEEE.

[4] R. Yogamathi, S. Banu, A. Vishwapriya, "Design of Fractal Antennas for Multiband Applications", 4th ICCCNT, 2013.

[5] Fajar Wahyu Ramadhan, Nanang Ismail, Innel Lindra, Eki Ahmad Zaki Hamidi, M. AliRamdhani, Emerson Pascawira Sinulingga, IEEE Publications, "Multiband Microstrip Fractal Antennas for S-Band and C-Band Radar Applications", 3rd international conference 2019 on electrical engineering, telecommunications and IT (ELTICOM).

[6] Shashank S. Kulkarni and Sisir Kumar Nayak, IEEE Publications, "Ring Fractal Patch Antennas for GPS Applications", 2019 5th International Conference on Technological Convergence (I2CT), India. March 29-31, 2019.

[7] Djafri, K., Challenge, M., Aksas, R., Mouhouche, F. Dehmas, M., "Monopoly antenna based on miniaturized concentric fractal rings for WLAN / WiMAX applications" Radio engineering, 28 (1), pp. 39-44, April 2019.

[8] Josue B. Benavides, Rafael A. Lituma, Paul A. Chasi, Luis F. Guerrero, IEEE publication, "New modified hexagonal fractal antenna with multiband notch characteristics for UWB applications", 2018 IEEE National Conference on communication, signal processing and networking (NCCSN).

[9] Narinder Sharma, Ramandeep Kaur, Dr. Posted by Vipul Sharma, IEEE Publication, "Analysis and Design of Rectangular Microstrip Patch Antennas Using Fractal Technology for Multiband Wireless Applications", 2016 International Conference on Microelectronics and telecommunications engineering.

[10] Amit Ranjan, Sunil Kumar Singh, "Multifractal Triangular Microstrip Patch Antennas for UWB Applications", International Journal of Engineering Trends and Technology (IJETT), published October 2015.

Deep Neural Network for Electroencephalogram based Emotion Recognition

Shruti Garg[1*], Soumyajit Behera[2], Rahul K Patro[2] and Ashwani Garg[3]

[1]Department of Computer Science and Engineering, Birla Institute of Technology, Ranchi-JH 835215, India
[2]Department of Mathematics, Birla Institute of Technology, Ranchi-JH 835215, India
[3]Faculty of Health, Biomedical Science and Medical Science, Griffith University, Queensland-4122, Australia
*Corresponding Author: Shruti Garg. Email: gshruti@bitmesra.ac.in

Abstract. Emotion recognition using electroencephalogram (EEG) signals is an aspect of affective computing. The EEG refers to recording brain responses via electrical signals by showing external stimuli to the participants. This paper proposes the prediction of valence, arousal, dominance and liking for EEG signals using a deep neural network (DNN). The EEG data is obtained from the AMIGOS dataset, a publicly available dataset for mood and personality research. Two features, normalized and power and normalized wavelet energy, are extracted using Fourier and wavelet transform, respectively. A DNN with three different activation functions (exponential linear unit, rectified linear unit [ReLU] and leaky ReLU) has been applied for single and combined features. The result of combined features with leaky ReLU is found to be the best, with a classification accuracy of 85.47, 81.87, 84.04 and 86.63 for valence, arousal, dominance and liking, respectively.

Keywords: Deep neural network, Emotion Recognition, EEG.

1. Introduction

Neural Networks (NNs) [1] are used in many research studies. Pattern recognition [2], image processing [3] and computer vision [4] are important applications of NNs. The electroencephalogram (EEG) is a tool used to record electrical activity in the human brain by providing external stimuli [5], as shown in figure 1.

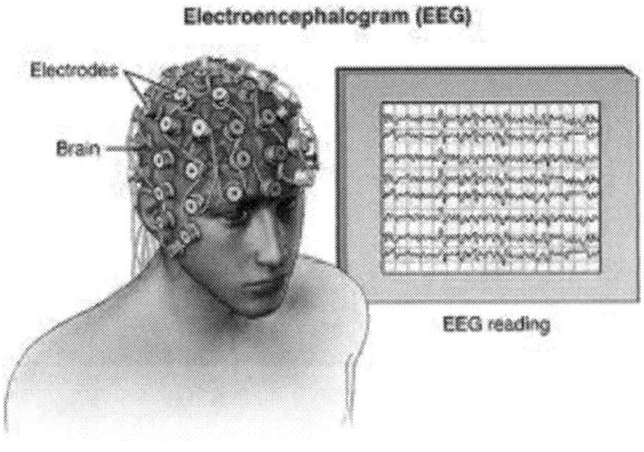

Figure 1. Procedure of EEG recording [6]

These EEG signals are used in the medical field for diagnosing mental diseases and abnormalities, as well as in research for identifying moods and personalities [7]. Moods cannot be identified by the different modalities of data because they do not produce a significant change in data [8]. There are several emotions generated in the human brain at any one stage of a mood. Thus, emotions are the basic unit to identify moods from images, texts, voices and physiological signals [9]. The EEG signal is a kind of physiological signal and is the preferred tool to identify emotions because it is a direct response of the human body, whereas voice and facial signals are acted/elicited responses [10]. Thus, an EEG-based emotion recognition (ER) model using a deep neural network (DNN) is presented in this paper.

Designing an EEG-based ER system involves the following steps:

- Data collection
- Pre-processing
- Feature extraction
- Classification

Data collection has been done in [10–17], in which [10–14] are publicly available EEG datasets, and [11, 14] provides pre-processed dataset for research. The DEAP [11] dataset is the most popular among researchers. In contrast, very few studies have been conducted using the AMIGOS [14] dataset because it is new and has a more complicated structure than DEAP. Moreover, it provides low accuracy compared to DEAP because of its imbalanced instances of data. AMIGOS provides a variable-length row according to the length of the video shown to different users, which makes the application of classification models more complicated [18]. This paper uses the AMIGOS dataset for emotion classification.

There are several literature studies on ER-based machine learning (ML), deep learning (DL) and other artificial intelligence systems [19]. These techniques are applied at the feature extraction and classification stages. The important features extracted from EEG signals for ER are entropy, energy and power from Fourier, wavelet and short-time Fourier transforms [20]. This paper studies ER and only discusses classification. A variety of ML methods, such as decision tree, support vector machine (SVM), k-nearest neighbour, naïve Bayes and random forest (RF), are used to identify emotions. Furthermore, DL methods, such as long-short term memory and convolution neural network, are used for ER [21]. A summary of ML and DL methods applied on four important datasets (DEAP [11], DREAMER [12], SEED [13] and AMIGOS [14]) is presented in Table 1.

Table 1. Summary of studies conducted on ER

Ref., year	Dataset	Feature Extraction	Algorithm Category	Method	Accuracy
[22], 2017	DEAP	Wavelet Entropy, Energy	ML	KNN, SVM	86.75%.
[23], 2020		Power, Entropy, Fractural Dimension, Statistical features, Wavelet Energy	ML	KNN, SVM	78.96%
[24], 2020		Multi-level feature Capsule Network (End to End network)	DL	Multi-level feature Capsule Network	98.32%
[25], 2021		Differential Entropy	DL	Graph Convolution Network+ LSTM	90.60%
[12], 2017	DREAMER	Power Spectral Density	ML	SVM	62.49%
[26], 2018		Power Spectral Density	DL	Dynamic Graph CNN	86.23%
[27], 2020		Region asymmetric convolution neural network (RACNN) (End to End network)	DL	RACNN	95.00%
[28], 2021		Tomographic and Holographic 3D feature map constructed by DL	DL	CNN	90.43%

[29], 2019	SEED	Differential Entropy	ML	DT, KNN, RF	75.00%
[30], 2020		Local and global inter channel relation	DL	Regularized Graph Neural Network	85.30%
[31], 2020		High order statistics	DL	LSTM	90.81%
[32], 2020		Electrode Frequency Distribution Map +STFT	DL	CNN	90.59%
[14], 2018	AMIGOS	All band, PSD, spectral power asymmetry between 7 pairs of electrodes in the five bands.	ML	SVM	57.7%
[33], 2018		Conditional Entropy (CF) feature, CNN based feature using EEG topography	ML	Extreme learning machine (ELM)	72.36%
[34], 2019		PSD, Conditional Entropy, PSD image based Deep learning features	DL	LSTM	83.02%
[35], 2020		Spectrogram Representation	DL	Bidirectional LSTM	83.30%

Table 1 shows that DL methods give better accuracy than ML methods. The highest accuracy was achieved on the DEAP dataset, and the lowest accuracy was achieved on the AMIGOS dataset, revealing that the AMIGOS dataset requires an efficient method.

Thus, a DNN for ER in the AMIGOS dataset using EEG signals has been proposed in this paper.

The rest of this paper comprises four more sections. Section 2 provides details of the emotion model used in the research. Section 3 details the methodology and the experiment. Section 4 comprises the results and discussions. Finally, section 5 provides a conclusion.

2. Emotion Model

There are two ways to recognise emotions in the field of affective computing. The first is in terms of discrete emotions, such as sadness, happiness, fear, anger and disgust [36]. The second is in terms of emotions models. Emotion models consist of emotion in terms of valence, arousal and dominance (VAD) [37]. All VAD are assessed between 1–9, where valence represents positivity of emotions, arousal represents excitement, and dominance represents domination towards content shown. Fig. 2 shows the change in VAD according to how much the subject likes the audio-visual content presented.

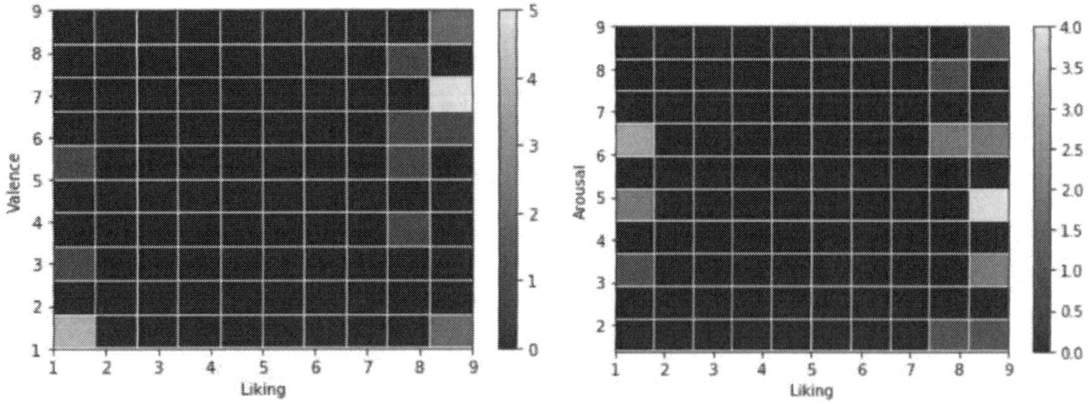

(a) (b)

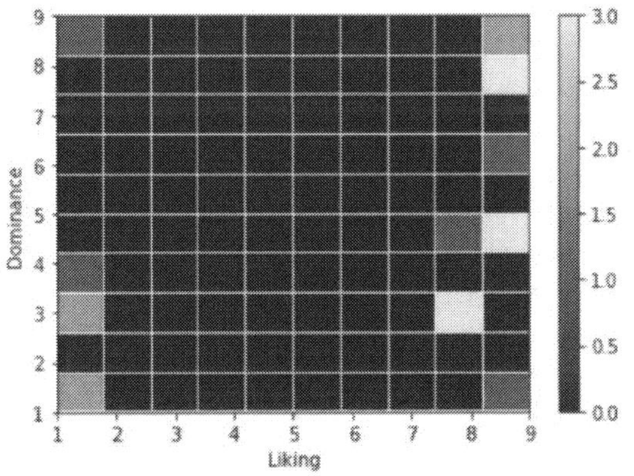

(c)

Figure 2. Changes in Valence, Arousal and Dominance according to liking (a) Valance vs liking (b) Arousal vs liking (c) Dominance vs liking

It is evident from Fig. 2 that emotions are influenced by the liking of content. Thus, liking is also predicted along with VAD in this work.

3. Methodology

Emotion recognition has significant research importance in affective computing. It resembles human recognition if done by NNs, as NNs are known as simulations of the human brain [38]. Neural networks are utilised for recognising emotions in the AMIGOS dataset in the VAD model along with liking. The following are the ER steps:

1. Extraction of EEG data from the AMIGOS dataset and preparation of the CSV file.
2. Balancing the VAD emotions
3. Extraction of bands from the EEG data
4. Feature extraction and fusion
5. Application of DNN
6. Model evaluation

Fig. 3 depicts the block diagram of this paper's work.

3.1 Dataset Description

AMIGOS is a multi-model dataset for personality and mood. This dataset consists of 40 subjects. All 40 people were shown 20 movies. Of the 20 movies, 16 were short videos (51 to 150 seconds long), and four were long videos (14:06 to 23:35 minutes long). Participants' EEG, ECG, GSR and video signals were recorded while watching the movies. Each participant was assessed in terms of VAD, familiarity and liking, using both self-assessment and annotation by an external expert.

There are 17 columns present in the dataset. The first 14 represent the data recorded by 14 EEG electrodes, the next two gives the ECG signal recording, and the last one is the GSR signal data. The signals were pre-processed at a sampling frequency of 128 Hz. The five dimensions of emotions were measured on a scale of 1–9, where 1 is the lowest, and 9 is the highest.

Pre-processing:

For EEG:

1. The data has been downsampled to 128 Hz.
2. The reference has been averaged to a common reference.
3. The 4.0–45.0 Hz bandpass frequency filter has been pertained.

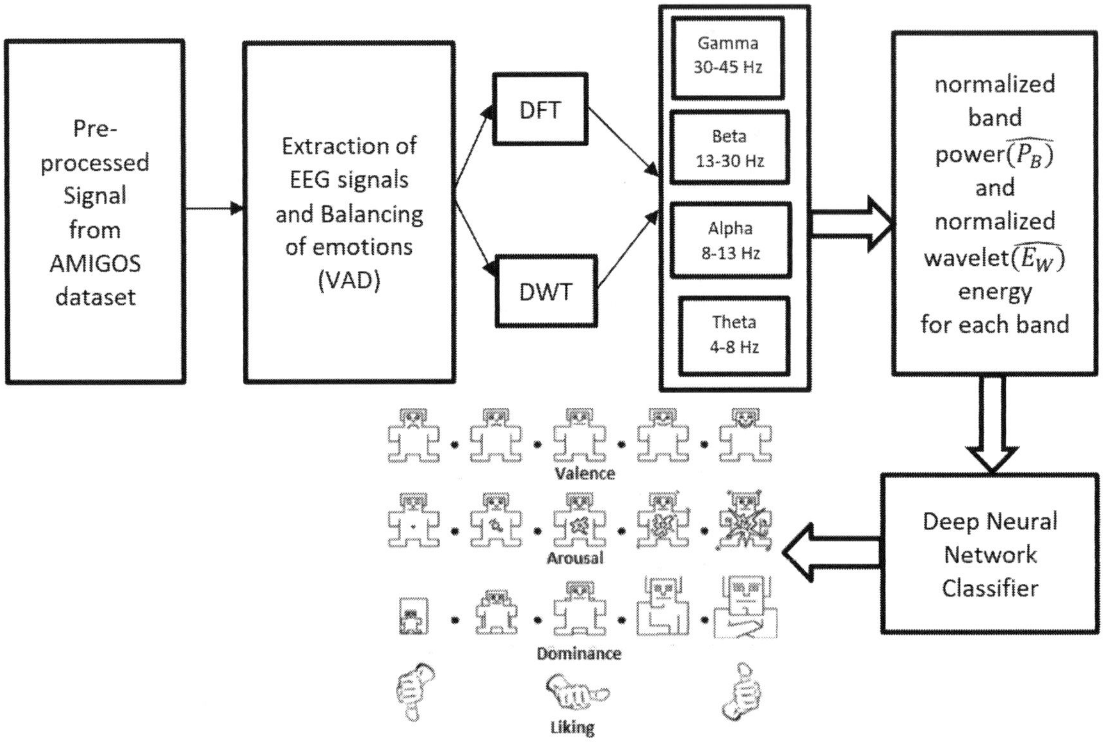

Figure 3. Proposed Emotion Recognition System

Here, VAD and liking are taken for classification as high and low classes, and their encoding is shown in Table 2.

Table 2. Encoding of VAD and liking

Emotion	Low	High
Valence	≤ 4.5	≥ 4.5
Arousal	≤ 4.5	≥ 4.5
Dominance	≤ 4.5	≥ 4.5
Liking	≤ 4.5	≥ 4.5

3.2 Feature Extraction:

Discrete Fourier transform and discrete wavelet transform have been used here for feature extraction. The feature taken from the Fourier transform is the normalized band power (NBP), and the normalized wavelet energy (NWE) was taken from the wavelet transform.

3.2.1 Normalized Band Power

To calculate the NBP, the Fourier transformX_k was first calculated using Equation 1.

$$X_k = \sum_{n=0}^{N-1} x_n e^{-2\pi nk/N} \quad (1)$$

$where, N\ is\ length\ of\ vector\ x\ and\ 0 \leq k \leq N-1$

$where, N\ is\ the\ length\ of\ vector\ x\ and\ 0 \leq k \leq N-1$

Then, the signal in the frequency domain was further decomposed into five frequency bands (4–8 Hz, 8–13 Hz, 13–16 Hz, 16–30 Hz and 30–45 Hz) to extract features. Beta bands were decomposed into two (Beta1 and Beta2) to equalise the dimensions with wavelet transform. The band power and NBP were calculated for each band from Equations 2 and 3.

$$P_B = \sum_0^k |X_k|^2 \tag{2}$$

Where P_B represents power of band B, k is length of each band.

$$\hat{P}_B = \frac{P_B}{\sum_B P_B} \tag{3}$$

Where $\hat{P}_B$ is called normalized band power.

3.2.2 Normalized Wavelet Energy (NWE)

Wavelet energy is calculated here using multilevel Daubechies-4 wavelet transform. The multilevel wavelet transform decomposes the signal into different frequency bands and is obtained by successive low-pass filter h[n] and high-pass filter g[n] of the time domain signal x[n] as shown in Fig. 4.

Since the EEG signal provided in the dataset ranges from 4 Hz to 45 Hz, a five-level decomposition is sufficient for the band information, as shown in Figure 4.

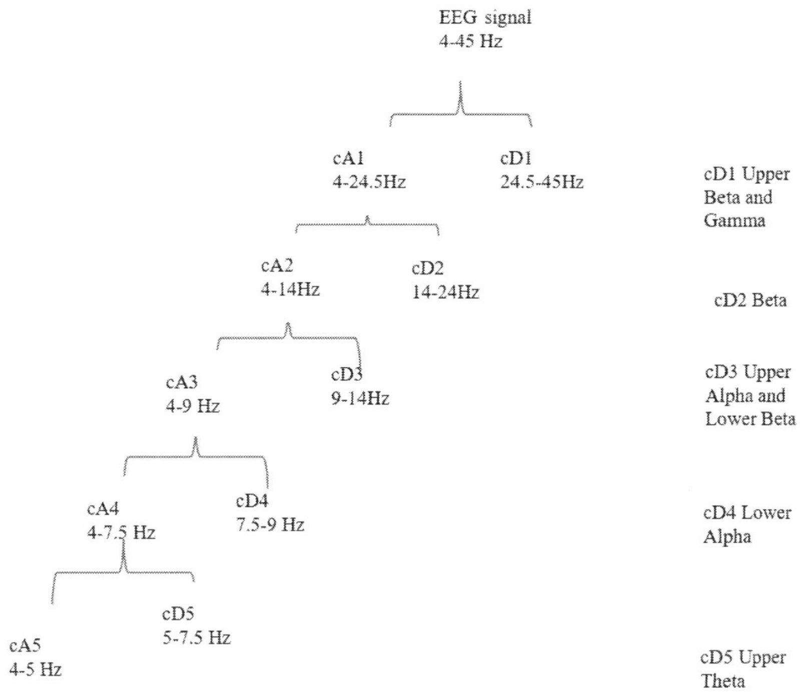

Figure 4. Wavelet Decomposition of different bands

Finally, the features provided to the classifiers are NBP obtained from Fourier transform, NWE obtained from wavelet transform and {NBP, NWE}.

3.3 Neural Network

The NNs [38] are particularly designed to simulate the functioning of the human brain and consist of three basic units:

- Neuron
- Activation function
- Number of layers

The neuron is connected at different layers, as shown in Fig. 5.

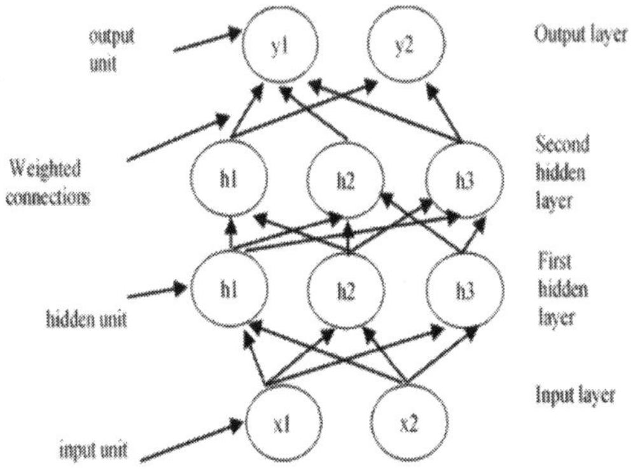

Figure 5. Architecture of Neural Network

The NN classifier uses layered architecture, a collection of multiple layers called input layer, hidden layer and output layer. Each layer consists of multiple neurons. The calculation done at each neuron is shown in Fig. 6.

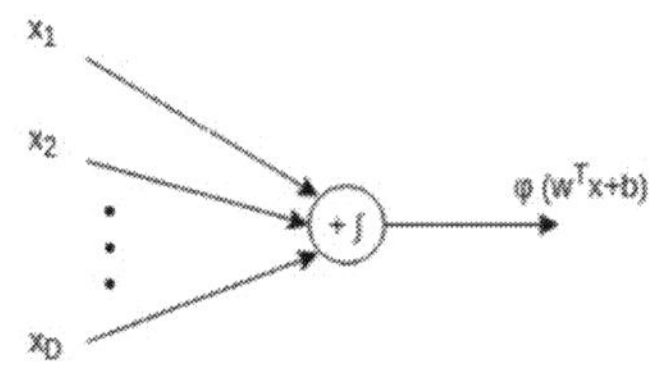

Figure 6. Output calculation at single neuron

Where input vector X consists of m features $x_1, x_2, \ldots, x_m$, and y represents an output of a single neuron. φ represents the activation function, which has been taken as the exponential linear unit (ELU) [39], rectified linear unit (ReLU) [40] and leaky ReLU [41] here. The activation function serves to convert the output of neurons into non-linear forms. The selection of an appropriate activation function is critical to obtaining better classification outcomes. The ELU activation function uses a log curve to transform the NN output into a non-linear form. The ReLU activation function diminishes all negative values of outcome to zero, which creates a problem of vanishing gradients [41]. This problem has been resolved by the leaky ReLU. The leaky ReLU does not make negative gradients as zero; rather, it assigns a small value to negative gradients, as shown in Fig. 7.

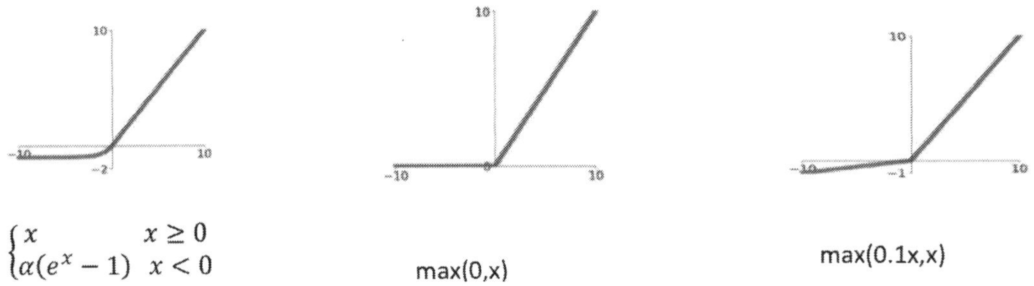

Figure 7. Activation Functions (a) ELU (b) ReLU (c) Leaky-ReLU

The output of every neuron work is the input for the next neuron. The same process is repeated for every neuron at each layer. Once the output of all output neurons is calculated, it will be compared with the actual output, and the weight change will be calculated by Equation (4).

Updated weight = old weight + learning rate * (expected output – predicted output) * input (4)

4. Results and Discussions

The experiments conducted in this paper were done in Python 3.7 on a computer with an Intel i5 processor, 8 GB of RAM and an Nvidia graphics card. A DNN with three different activation functions (ELU, ReLU and leaky ReLU) was applied to determine VAD and liking in EEG signals. The architecture of the DNN used here is shown in Table 3.

Table 3. Architecture of DNN

Layer	**Size**
Input	70x128
Hidden	128x256
Hidden	256x512
Output	512x4

The batch size taken here is 32, and the binary cross-entropy loss function is used to calculate the loss.

The feature extraction is done by the NBP and NEW, along with the combined feature vector {NBP, NWE}. The accuracies obtained by different activation functions are shown in Table 4, with the highest accuracy in bold.

Table 4. Accuracies obtained by experiments

S. No.	**Feature Extraction**	**Activation function**	**Valence(%)**	**Arousal(%)**	**Dominance(%)**	**Liking(%)**
1.	NBP	ELU	65.00	55.60	60.32	79.95
		ReLU	74.40	62.01	66.48	81.86
		Leaky ReLU	**75.50**	**68.37**	**66.76**	**81.88**
2.	NWE	ELU	81.80	75.65	78.97	83.95
		ReLU	83.66	78.93	80.94	84.66
		Leaky ReLU	**83.89**	**78.97**	**80.96**	**84.73**
3.	NBP, NWE	ELU	84.41	78.76	81.09	85.45
		ReLU	85.28	81.53	83.44	86.56
		Leaky ReLU	**85.47**	**81.87**	**84.04**	**86.63**

Table 4 shows that the leaky ReLU with combined features gives the highest accuracy for all emotions. The validation vs training loss is shown in Figure 8 for the leaky ReLU activation function.

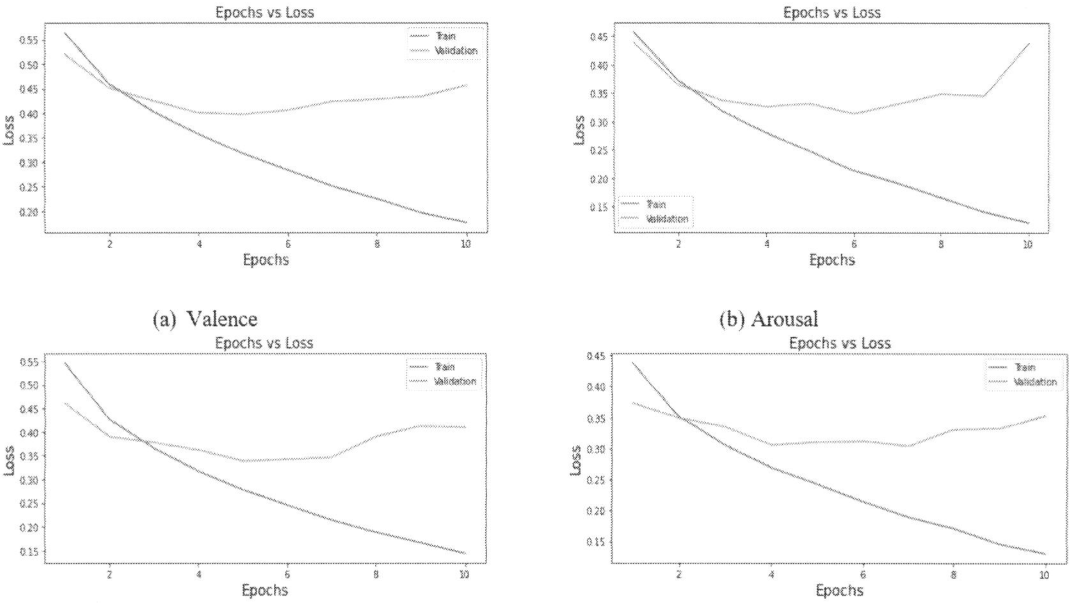

(a) Valence (b) Arousal

(c) dominance (d) liking

Figure 8. Training vs validation loss curve for (a) valence (b) arousal (c) dominance and (d) liking

The outcome of the DNN is also compared with state-of-the-art methods by the applied stacking classifier. The ML methods ensembled in the stacking classifier are RF, Gaussian naïve Bayes, SVM and logistic regression, as shown in Table 5.

Table 5. Comparison of present work with existing method

Ref, Year	Emotions				Modality	Features	Classifier
	A	V	D	L			
[14], 2018(Original paper)	57.7	56.4	-	-	EEG	All band, PSD, spectral power asymmetry between 7 pairs of electrodes in the five bands.	SVM
[33], 2018	71.54	66.67	72.36	-	EEG	Conditional Entropy (CF) feature, CNN based feature using EEG topography	Extreme learning machine (ELM)
[42], 2018	68.00	84.00	-	-	EEG+ECG +GSR	Time, frequency and Entropy domain features	GaussianNB, XGBoost
[43], 2019	83.02	79.13	-	-	EEG	PSD, Conditional Entropy, PSD image based Deep learning features	LSTM
[44], 2020	83.3	79.4	-	-	EEG+ECG +GSR	Spectrogram Representation	Bidirectional LSTM
[45], 2020	75.00	87.50			EEG	Spectrogram Representation	Deep Convolution Neural Network
[46], 2021	87.39	90.54	-	-	EEG	Features extracted from topographic and holographic feature map	CNN+SVM
Our Method	80.77	69.89	71.75	78.40	EEG	NBD + NWE	Stacking Classifier
	85.47	**81.87**	**84.04**	**86.63**	EEG	NBD + NWE	DNN

Table 5 shows that the research problem of identifying emotions in the AMIGOS dataset was initially posed in 2018 by Correa et al. The accuracy achieved in their initial analysis was 57.7%. Then, different solutions using ML and DL were proposed in [33, 42–46] and have achieved accuracies up to 90.54%. However, only valence and arousal were predicted in [42–46], whereas VAD was predicted in [33]. Liking and VAD were predicted in the current work. Since VAD is highly dependent on liking, as shown in Fig. 2, the feature extraction is done in the time and frequency domain in [14]. On the other hand, 3D features were extracted in [44–46]. The combined features {NBP, NWE} were used to predict

VAD and liking, which helped to achieve higher accuracy in the present work. Another observation made in this paper is that the leaky ReLU activation function of a DNN with combined features gives better accuracy than state-of-the-art methods.

The research on ER obtained from EEG signals using DL is not popular due to the limited availability of large datasets [47]. Although the dataset used in this present work is large, it consists of variable length instances, which restrict the ease of applicability of models [43]. The collection of EEG signal is a clinical process in which many individuals are not comfortable; thus, DL is applied in various research for speech and facial ER [48]. Thus, research on ER using DNNs from EEG signals is highly recommended. Two datasets [17, 49] for EEG-based ER have been proposed recently and could be explored in the future.

5. Conclusions

An EEG-based ER system was proposed in this paper. AMIGOS, a mood and personality dataset, was used for ER. The features extracted for classification are NBP and NWE. The combined features were used in the DL classifier to identify VAD and liking. It was observed that DNNs provide better accuracy than ML methods. The proposed system could be used to recommend songs/movies to viewers according to their liking.

The present work can be extended to identify emotions in other physiological signals, such as GSR, ECG and electrodermal activity. Furthermore, a multimodal system could be proposed for ER.

Reference:

[1] Oh JW, Lee IW, Kim JT and Lee GW 1999 Application of Neural Networks for Proportioning of ACI Materials Journal Jan 96(1).

[2] Bishop CM 1995 Neural networks for pattern recognition Oxford university press Nov 23.

[3] Egmont-Petersen M, de Ridder D and Handels H.2002 Image processing with neural networks—a review Pattern recognition Oct 1;35(10):2279-301.

[4] Zhou YT and Chellappa R 2012 Artificial neural networks for computer vision Springer Science & Business Media Dec 6.

[5] Barlow JS 1993 The electroencephalogram: its patterns and origins MIT press.

[6] Siuly S, Li Y and Zhang Y 2016 Electroencephalogram (EEG) and its background In EEG Signal Analysis and Classification pp. 3-21 Springer, Cham.

[7] Alarcao SM and Fonseca MJ 2017 Emotions recognition using EEG signals: A survey IEEE Transactions on Affective Computing 10(3):374-93.

[8] Schnall S 2010 Affect, mood and emotions Social and emotional aspect of learning pp. 59-64.

[9] Castellano G, Kessous L and Caridakis G 2008 Emotion recognition through multiple modalities: face, body gesture, speech In Affect and emotion in human-computer interaction pp. 92-103 Springer, Berlin, Heidelberg.

[10] Soroush MZ, Maghooli K, Setarehdan SK and Nasrabadi AM 2017 A review on EEG signals-based emotion recognition International Clinical Neuroscience Journal 4(4):118.

[11] Koelstra S, Muhl C, Soleymani M, Lee JS, Yazdani A, Ebrahimi T, Pun T, Nijholt A and Patras I 2011 Deap: A database for emotion analysis; using physiological signals IEEE transactions on affective computing 3(1):18-31.

[12] Katsigiannis S and Ramzan N 2017 DREAMER: A database for emotion recognition through EEG and ECG signals from wireless low-cost off-the-shelf devices IEEE journal of biomedical and health informatics 22(1):98-107.

[13] Zheng WL and Lu BL 2015 Investigating critical frequency bands and channels for EEG-based emotion recognition with deep neural networks IEEE Transactions on Autonomous Mental Development 7(3):162-75.

[14] Correa JA, Abadi MK, Sebe N and Patras I 2018 Amigos: A dataset for affect, personality and mood research on individuals and groups IEEE Transactions on Affective Computing.

[15] Khare SK and Bajaj V 2020 Time-frequency representation and convolutional neural network-based emotion recognition IEEE transactions on neural networks and learning systems.

[16] Gao Q, Wang CH, Wang Z, Song XL, Dong EZ and Song Y 2020 EEG based emotion recognition using fusion feature extraction method Multimedia Tools and Applications 79(37):27057-74.

[17] Joshi VM and Ghongade RB 2020 IDEA: Intellect database for emotion analysis using EEG signal Journal of King Saud University-Computer and Information Sciences.

[18] Siddharth S, Jung TP and Sejnowski TJ 2019 Utilizing deep learning towards multi-modal bio-sensing and vision-based affective computing IEEE Transactions on Affective Computing.
[19] Gu X, Cao Z, Jolfaei A, Xu P, Wu D, Jung TP and Lin CT 2021 EEG-based brain-computer interfaces (bcis): A survey of recent studies on signal sensing technologies and computational intelligence approaches and their applications IEEE/ACM transactions on computational biology and bioinformatics.
[20] Supratak A, Wu C, Dong H, Sun K and Guo Y 2016 Survey on feature extraction and applications of biosignals In Machine learning for health informatics pp. 161-182 Springer, Cham.
[21] Zhang J, Yin Z, Chen P and Nichele S 2020 Emotion recognition using multi-modal data and machine learning techniques: A tutorial and review Information Fusion. 59:103-26.
[22] Mohammadi Z, Frounchi J and Amiri M 2018 Wavelet-based emotion recognition system using EEG signal Neural Computing and Applications.28(8):1985-90.
[23] Nawaz R, Cheah KH, Nisar H and Yap VV 2020 Comparison of different feature extraction methods for EEG-based emotion recognition Biocybernetics and Biomedical Engineering 40(3):910-26.
[24] Liu Y, Ding Y, Li C, Cheng J, Song R, Wan F and Chen X 2020 Multi-channel EEG-based emotion recognition via a multi-level features guided capsule network Computers in Biology and Medicine 123:103927.
[25] Yin Y, Zheng X, Hu B, Zhang Y and Cui X 2021 EEG emotion recognition using fusion model of graph convolutional neural networks and LSTM Applied Soft Computing 100:106954.
[26] Song T, Zheng W, Song P and Cui Z 2018 EEG emotion recognition using dynamical graph convolutional neural networks IEEE Transactions on Affective Computing 11(3):532-41.
[27] Cui H, Liu A, Zhang X, Chen X, Wang K and Chen X 2020 EEG-based emotion recognition using an end-to-end regional-asymmetric convolutional neural network Knowledge-Based Systems 205:106243.
[28] Topic A and Russo M 2021 Emotion recognition based on EEG feature maps through deep learning network International Journal of Engineering Science and Technology.
[29] Qing C, Qiao R, Xu X and Cheng Y 2019 Interpretable emotion recognition using EEG signals IEEE Access. 7:94160-70.
[30] Zhong P, Wang D and Miao C 2020 EEG-based emotion recognition using regularized graph neural networks IEEE Transactions on Affective Computing.
[31] Sharma R, Pachori RB and Sircar P 2020 Automated emotion recognition based on higher order statistics and deep learning algorithm. Biomedical Signal Processing and Control 58:101867.
[32] Wang F, Wu S, Zhang W, Xu Z, Zhang Y, Wu C and Coleman S 2020 Emotion recognition with convolutional neural network and EEG-based EFDMs Neuropsychologia 146:107506.
[33] Siddharth, Jung TP and Sejnowski TJ 2018 Multi-modal approach for affective computing In 2018 40th Annual International Conference of the IEEE Engineering in Medicine and Biology Society (EMBC) pp. 291-294 IEEE.
[34] Siddharth S, Jung TP and Sejnowski TJ 2019 Utilizing deep learning towards multi-modal bio-sensing and vision-based affective computing IEEE Transactions on Affective Computing. .
[35] Li C, Bao Z, Li L and Zhao Z 2020 Exploring temporal representations by leveraging attention-based bidirectional LSTM-RNNs for multi-modal emotion recognition Information Processing & Management 57(3):102185.
[36] Lench HC, Flores SA and Bench SW 2011 Discrete emotions predict changes in cognition, judgment, experience, behavior, and physiology: a meta-analysis of experimental emotion elicitations Psychological bulletin 137(5):834.
[37] Barrett LF 1998 Discrete emotions or dimensions? The role of valence focus and arousal focus Cognition & Emotion 12(4):579-99.
[38] Abdi H 1994 A neural network primer. Journal of Biological Systems 2(03):247-81.
[39] Clevert DA, Unterthiner T and Hochreiter S 2015 Fast and accurate deep network learning by exponential linear units (elus) arXiv preprint arXiv:1511.07289.
[40] Arora R, Basu A, Mianjy P and Mukherjee A 2016 Understanding deep neural networks with rectified linear units arXiv preprint arXiv:1611.01491.
[41] Kolbusz J, Rozycki P and Wilamowski BM 2017 The study of architecture MLP with linear neurons in order to eliminate the "vanishing gradient" problem In International Conference on Artificial Intelligence and Soft Computing pp. 97-106 Springer, Cham.
[42] Tung K, Liu PK, Chuang YC, Wang SH and Wu AY 2018 Entropy-assisted multi-modal emotion recognition framework based on physiological signals In 2018 IEEE-EMBS Conference on Biomedical Engineering and Sciences (IECBES) pp. 22-26 IEEE.
[43] Siddharth S, Jung TP and Sejnowski TJ 2019 Utilizing deep learning towards multi-modal bio-sensing and vision-based affective computing IEEE Transactions on Affective Computing.

[44] Li C, Bao Z, Li L and Zhao Z 2020 Exploring temporal representations by leveraging attention-based bidirectional LSTM-RNNs for multi-modal emotion recognition Information Processing & Management. 57(3):102185.
[45] Singh G, Verma K, Sharma N, Kumar A and Mantri A 2020 Emotion Recognition using Deep Convolutional Neural Network on Temporal Representations of Physiological Signals In 2020 IEEE International Conference on Machine Learning and Applied Network Technologies (ICMLANT) pp. 1-6 IEEE.
[46] Topic A and Russo M 2021 Emotion recognition based on EEG feature maps through deep learning network International Journal of Engineering Science and Technology.
[47] Santamaria-Granados L, Munoz-Organero M, Ramirez-Gonzalez G, Abdulhay E and Arunkumar NJ 2018 Using deep convolutional neural network for emotion detection on a physiological signals dataset (AMIGOS) IEEE Access 7:57-67.
[48] Hossain MS and Muhammad G 2019 Emotion recognition using deep learning approach from audio–visual emotional big data Information Fusion 49:69-78.
[49] Arnau-González P, Katsigiannis S, Arevalillo-Herráez M and Ramzan N 2021 BED: A new dataset for EEG-based biometrics IEEE Internet of Things Journal.

Real-time Flow Visualization using Projection Mapping Technique

T Nishanka[1], Prajwal D[1], Nishmitha M[1], and Vinayambika S Bhat[1]
[1]Department of Electronics and Communication Engineering, Mangalore Institute of Technology and Engineering, Badaga Mijar, Mangaluru - 574225, Karnataka, India.

Abstract. Flow visualization is a fundamental topic in scientific visualization and has been the subject of active research for a long time. Usually, information begins from numerical simulations, for example, those of computational fluid dynamics, and should be investigated by methods for perception to acquire a comprehension of the flow. With the rapid rise of computational power for simulations, the interest in further developed visualization strategies has thrived. The idea is to attain advanced visualization through the means of projection mapping. When incorporated with the properties of the fluid, whose flow is to be studied, we can create a visual simulation by extracting the real-time variations data through sensors and feed it to the projector, which results in a visualization method in the 3D plane. The complete process is made visible as the image is projected through light and also in space. Real-time Flow Visualization using Projection Mapping methods will help attain the same results as the conventional methods at less cost. As this process is crucial for many industries in the testing and design sectors, the desired output must be obtained within the given time constraint. This method will help to optimize the complete process. The result can be used to derive the necessary conditions for the best design and performance of the object or vehicle.

1. Introduction

Flow visualization is a technique that makes the particle flow visible to our bare eye as most of them are invisible to get their numerical information [1]. Mathematical computations and experiment can examine the flow of the particles. Flow visualization methods are used to deliver analytic data near the movement around the model [2]. Flow visualization provides information on the entire subject instantly clear without the need for data processing.

In tentative methods, flow is envisioned using three methods. Surface flow visualization, particle tracer method, optical method. These flow visualization methods have been quite helpful in making the data clear, but the experimental setup is usually time-consuming and very expensive.

Projection mapping is related to movie mapping, is a mapping method used to convert things, often asymmetrical shapes, into a plane where a movie projection can occur [3]. These things may be composite industrial surroundings, such as structures, small indoor items, or acting stages. With the help of specific software and tracking sensors, a two- or three-dimensional body is spatially mapped on the simulated program, which imitates the actual environment it is projected on. Projection mapping intends to deliver a new involving experience for the audience through a visual imitation. The overview uses technology to control lighting onto the changeable exterior type and turn usual objects into collaborating 3D displays.

Flow visualization using the projection mapping technique can help us understand the flow of the medium along the surface in real-time through visual interpretation. The projection mapping adapts to the real-time input given to the system, it is easier to create an interactive environment susceptible to the immediate changes in the flow of the medium and can further be used to produce a real-time flow visualization, through which necessary data can be derived with regards to the medium for further studies and calculation on the same field.

2. Methodology

The methodology involves four major components; Program/Content, Mapping, Visualization/Display, Calculation.

The programs for the particular particle or medium we want to study the flow visualization will be executed and projected onto space using projectors (from more than one place). The code visualizes the water body as a mesh. Each element of the mesh has a particular vector associated with it. The waves are created using the sine wave piling equation based on the basic sine equation. The Bump Mapping Technique achieves the depth of the wave and the detailing. The sine wave piling algorithm is used to design the water simulation [4]. It is a Fourier composing method to pile up several sine waves with different frequency and amplitude.
The basic sine wave equation is given as:

$$Y\ (x,\ y,\ t) = A*cos\ (w * (x,\ y) + wt*t + FI; \quad (1)$$

Where A = Amplitude of the waves
w = Spatial angular frequency
wt = S*2*PI/L; Temporal angular frequency
FI = Initiatory phase

The sine wave piling equation, which is derived from the equation mentioned before, is:

$$H(x,y,t) = \sum_{i=1}^{n} Ai \times \sin((Dix \times x + Diy \times y) \times \frac{Li}{2\pi} + t \times Si \times \frac{2\pi}{Li}) \quad (2)$$

In equation (2), Ai is the amplitude of number i sine wave;
Li is the wavelength of that wave;
Dix, Diy are the direction of it;
Si is the speed of it;
t is time.

Bump mapping is a computer graphics technique where at each pixel; a heightmap is considered as a basis to look up a perturbation that is normal to the surface of the object being rendered, and the same is applied before calculating the contrast and the illumination at different parts and positions of the object [4].

First, for each pixel, the bumps are converted into vectors present on the bump map. In left figure 1, a zoomed-in view of the bump-map is shown; the lighter pixels stand out compared to the dark ones. A vector is computed for each pixel. These vectors represent inclines of the surface at a pixel. In figure 1, on the right side, the red vectors point in the downhill direction.

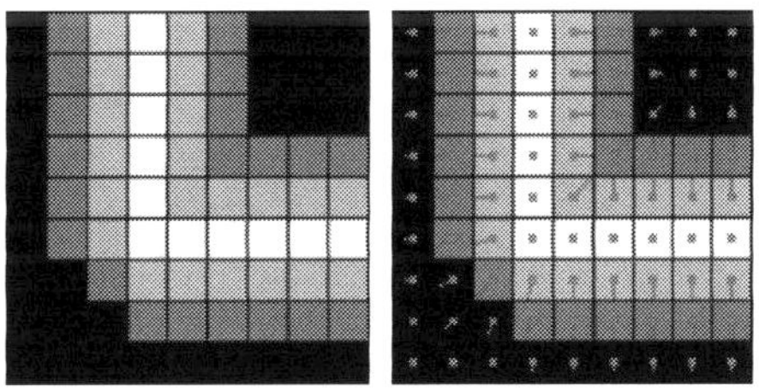

Figure 1. Details of Bump Mapping

One widely used method to calculate these vectors is to calculate the X and Y gradient of a pixel:

$$x\_gradient = pixel(x-1, y) - pixel(x+1, y); \tag{3}$$

$$y\_gradient = pixel(x, y-1) - pixel(x, y+1); \tag{4}$$

The x_gradient and the y_gradient is used to adjust the normal vector of the polygon at particular points. In the figure 2 a bump map and polygon are shown. They all show U and V vectors. After the simple adjustment, we can see the new Normal vector is:

$$New\_Normal = Normal + (U * x\_gradient) + (V * y\_gradient); \tag{5}$$

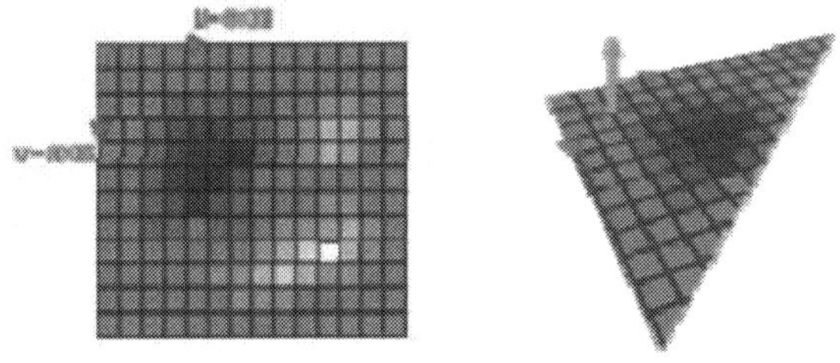

Figure 2. Bump Map and the Polygon

The brightness of the polygon at a given time is calculated using the New_Normal [4].

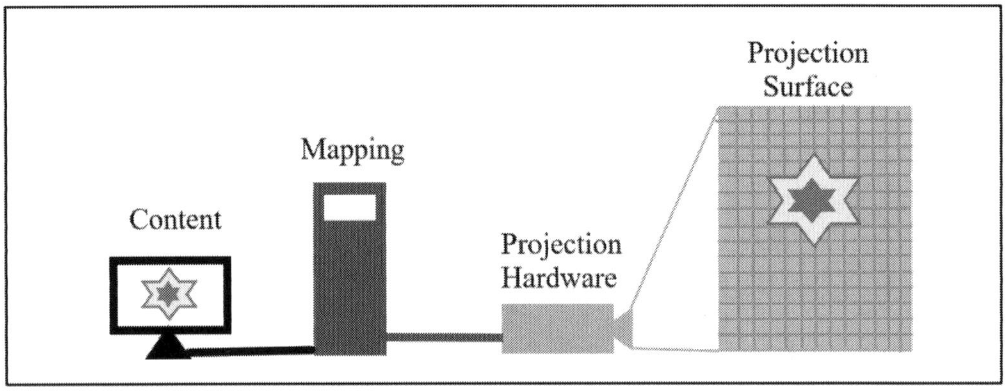

Figure 3. Projection-Mapping Flow

The object, be it a plane or a vehicle, will be brought into the frame, mapped by the Kinect sensor. The program is updated with real-time data to display the variation in the flow due to the object. The projection of simulation data is achieved using projection mapping; the complete process flow is shown in figure 3. Projection-Mapping is also called Spatial Augmented Reality and is used to create illusions in the real world [4]. After that, the process of visualization takes place. TouchDesigner Software is used to map the display onto the screen. The program's executable file is exported to the software, which is then projected onto the designed setup with the help of a projector. The sensor will capture the movement in the frame, and changes are made in the program accordingly. The result of the program is being projected virtually in the space. Thus, we can see the flow variation which will take place with our bare eyes.

When it comes to calculation, depending on the parameters required, it can be point-based or surface-based. The streamlines of the flow can be observed. The tangential stress along with the transition of the boundary layer can be found out. Along with these, the diffusion of mass and change in density can be calculated. Flow visualization is a classic sub-field of scientific visualization, and we can use it in studying the vortices or the turbulent nature of fluid [5]. As we are considering light as the source to depict the motion of a fluid, delay in terms of data acquisition and processing is less or non-

IOP Conf. Series: Materials Science and Engineering 1187 (2021) 012013 doi:10.1088/1757-899X/1187/1/012013

existent. The idea is to attain information on unsteady motion by considering various fluids and cases. This will also help in understanding the flow characteristics of fluids.

3. Real-time Implementation (System Description)

The complete system can be divided into 5 steps, and the flow of the process is shown in Figure 4.

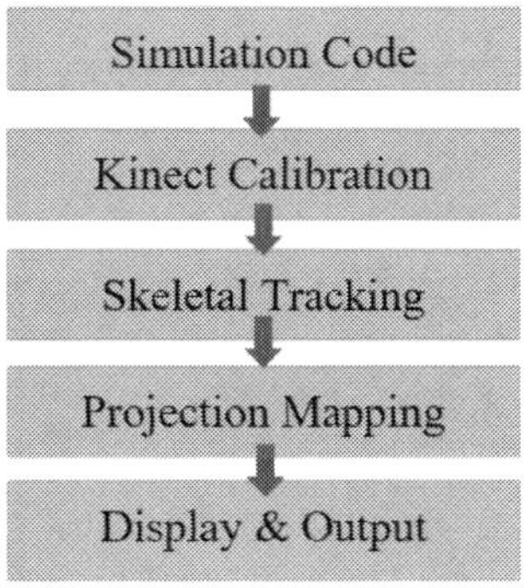

Figure 4. The flow of Implementation Process

3.1. Simulation Code

The simulation code is a processing code. Processing is software that helps visualize the code. Using the techniques mentioned in the methodology, water simulation code is executed. The input to this code is the data from the Kinect sensor. The software uses Kinect library to utilize the various functions of the Kinect v2 and handpick the features relevant to the program. The user interface of the software is shown in Figure 5.

Figure 5. User Interface of Processing Software

IOP Conf. Series: Materials Science and Engineering 1187 (2021) 012013 doi:10.1088/1757-899X/1187/1/012013

3.2. Kinect Calibration

The Kinect SDK user interface is figure 6. Kinect calibration is done with the help of the Kinect SDK, which Microsoft provides. In this process, check for the configuration and verification of the sensor. Also, test the sensor with regards to depth readings, color readings, previews and resolutions. This enables us to understand the input capacity of the sensor, which is to be noted for later use in the base code. The Kinect and Projector are calibrated to have the same focal center. Gray codes are projected onto the screen. Given the coordinate mapping functions of the Kinect SDK, this Gray code pattern is used to establish the precise mapping from a 3D point in the Kinect camera's coordinate frame to the corresponding point in the projector's image [6].

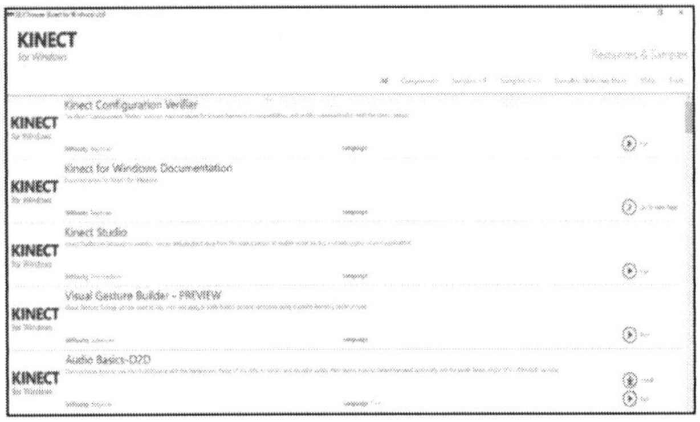

Figure 6. Kinect SDK User Interface (UI)

3.3. Skeletal Tracking and Depth Mask

Skeleton Tracking Depth Mask code from Kinect v2 library is used to retrieve the data of the joints and the positions of the parts of the body. Kinect SDK can also be used to make use of this feature of the Kinect sensor. Further concentrate on a particular part of the body (in this case, hands) and proceed with the code and calculations. Figure 7 presents the output of the skeleton tracking program and joint chart for Kinect v2

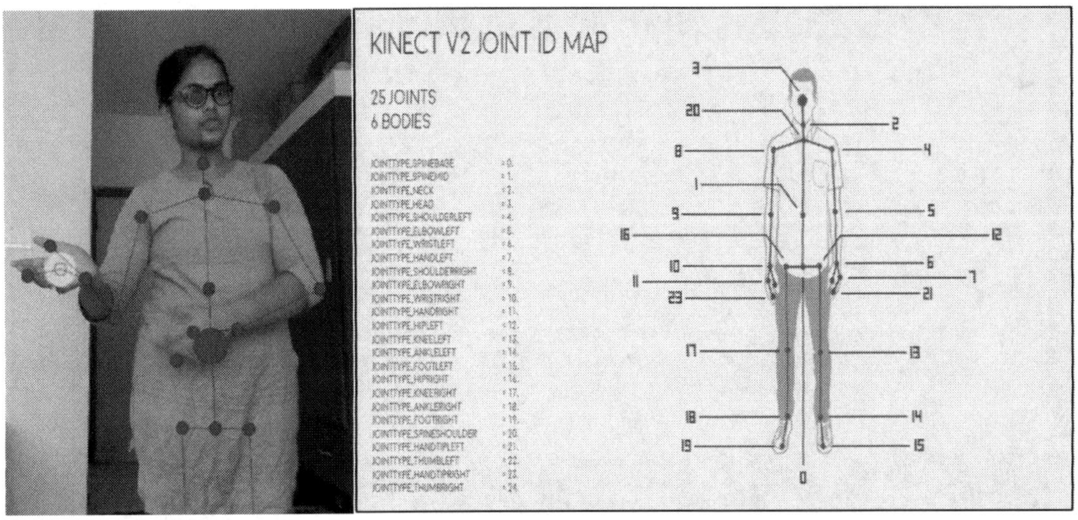

(a) (b)

Figure 7. (a) The output of skeleton tracking program, **(b)** Joint chart for Kinect v2

3.4. Projection-Mapping

Projection-Mapping is achieved using the TouchDesigner Software and the projector. The projection mapping concept was established to accurately map the world, utilizing more than one dimension [7]. The software maps the object to be used as a screen, and then the display is projected onto the screen. Without any screens or lenses, the projection mapping presents the virtual layer directly onto the physical object [8]. The projected images are bound within the limits of the screen. Thus spaces that lie beyond the screen are not projected by the projector. This technique helps in concentrating only on a particular part of the screen. It also enables us to use various regular and irregular shapes as a screen, which, when used properly, helps attain a 3D effect of the projected images. Figure 8 and 9 present an overview of TouchDesigner Software UI and Mapping and projecting onto an irregular object, respectively.

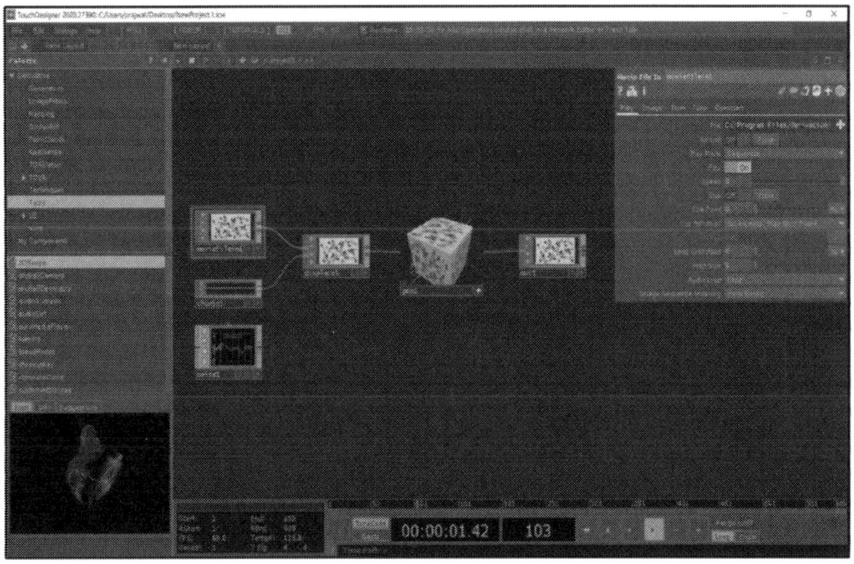

Figure 8. TouchDesigner Software UI

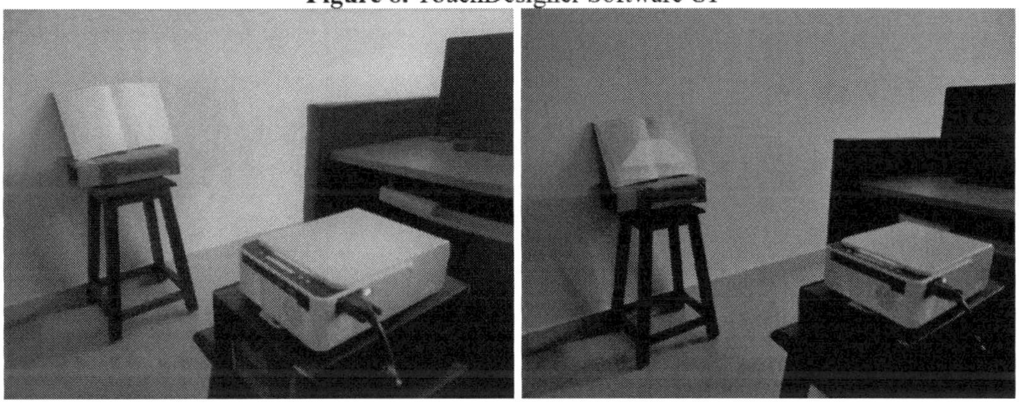

Figure 9. Mapping and projecting onto an irregular object

3.5. Display and Output

The setup includes three components. Kinect sensor, projector and the screen, as shown in figure 10. Among these, the Kinect and the projector are connected to the system. When the code is executed, the water simulation is displayed by the projector. The Kinect captures the changes, and the same can be seen on the screen. Figure 11 shows the water flow output and projected display when the flow occurs due to interaction between the projection and sensor.

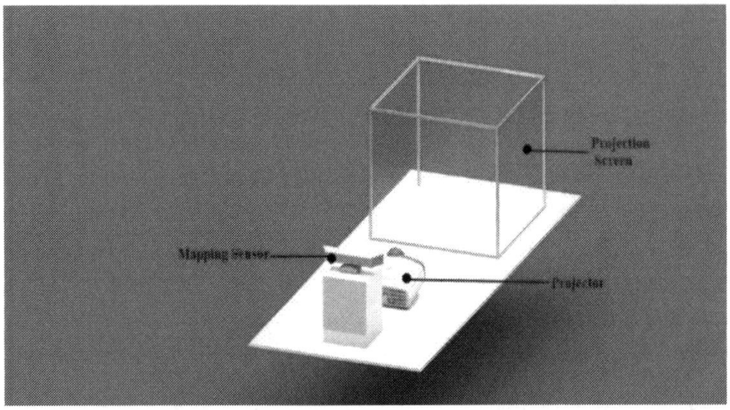

Figure 10. Setup of the project

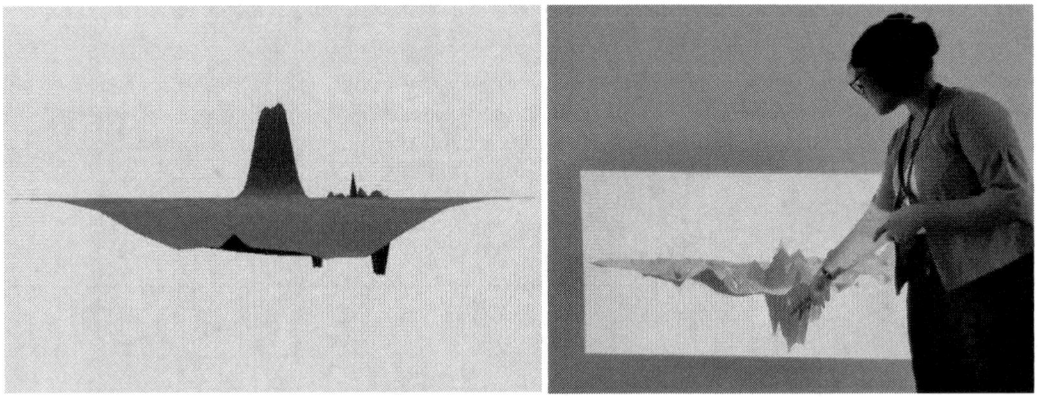

Figure 11. Simulation code output and projected display of interaction between the fluid and the user through the sensor resulting in flow visualization.

4. Hardware Description

A brief description of the projector and the Kinect v2 sensor, along with its adapter, is given below. The basic requirements for the project were to keep the whole setup cost-effective. The products mentioned below were cheaper than the other alternatives present in the market.

4.1. Projector

Figure 12 IBS T5 smart projector HD 3D Wi-Fi mira cast 3200 lumens home cinema portable projector. It has an LCD chipset and has a lamp life of 60000 hrs. The maximum projection distance is 40ft. It has a resolution of 800x480 pixel and a maximum brightness of 4700 lumens.

Figure 13. IBS T5 Smart Projector

4.2. Kinect v2 Sensor

The Microsoft Xbox One Kinect Sensor features an exceptional 1080p Full HD camera with advanced infrared technology for capturing accurate motion capture data. The Kinect Sensor features a microphone array with built-in noise isolation. This feature can be used to command and control the Xbox and a TV with our voice and gestures. This is one of the methods to achieve a projector depth camera system to execute the process of projection mapping [9]. The outlook of the Microsoft Xbox One Kinect Sensor is in figure 14.

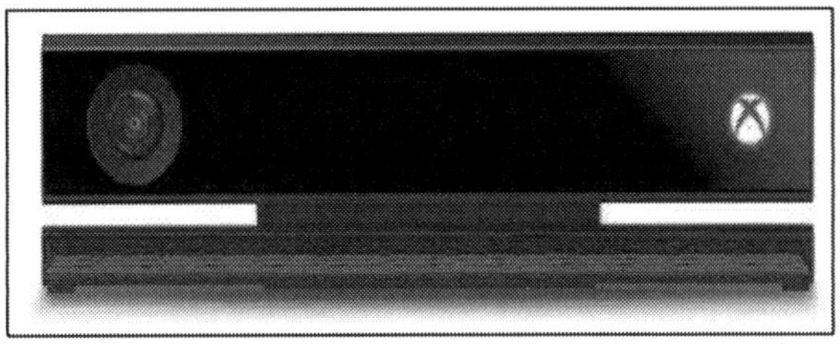

Figure 14. Microsoft Xbox One Kinect Sensor

4.3. Adapter

The adapter is compatible with the Xbox One S and Windows 10 PC. The adapter comes with a USB 3.0 cable which enables the user to use it with both the systems which are mentioned before. As Microsoft has discontinued manufacturing the adapter as a part of the sensor, the adapter is an individual component. The Xbox One Kinect Adapter is shown in figure 15.

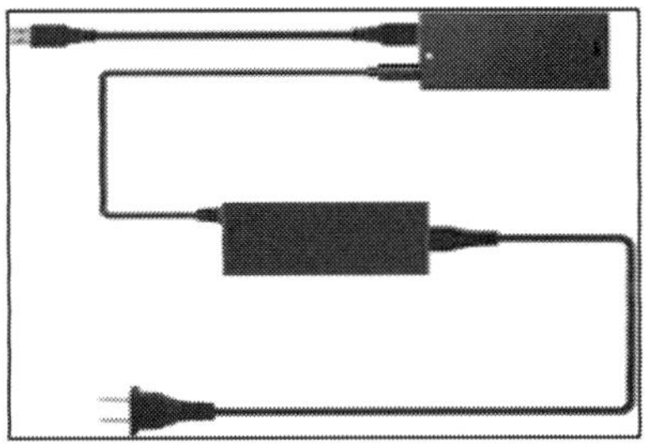

Figure 15. Xbox One Kinect Adapter

5. Other Approaches

The sine summing method has been considered to achieve the fluid motion for the simulation code. During our research, we came across various other methods to achieve fluid simulation. These methods are currently used in computer graphics. The theory and methodology can be applied to the present code to transform and obtain fluid simulation. Two of these approaches briefly discussed.

5.1. Linear Wave Theory for water simulation

A topic that is mainly used in the field and study of ocean surface waves and coastal engineering. The initial assumption in this theory includes that the height of the wave is small compared to the depth and the wavelength of the water [10]. The water height of a single wave is given as:

$$\eta(x,t) = Acos(k \cdot x - \omega t + \theta_0) \tag{6}$$

Where A is the wave amplitude,
x is a point in the plane,
k is the wave vector defining the direction of the wave,
k=$\|k\|$ Where k is the wavenumber and $\lambda = \frac{2\pi}{k}$ is the wavelength
ω is the frequency andθ_0 is the phase shift

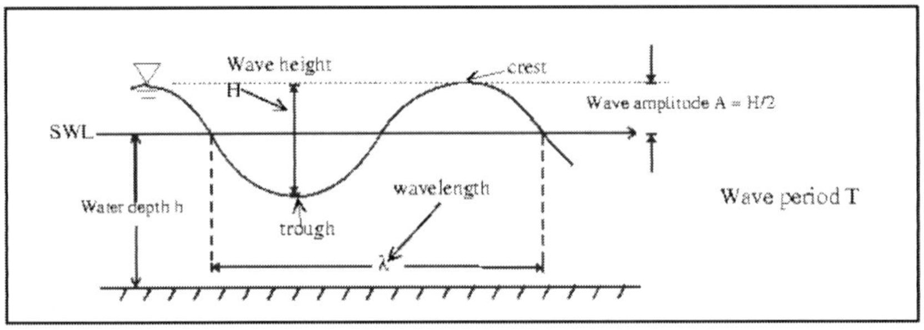

Figure 16. Linearized (Airy) Wave Theory

Relation between the wavenumber k with the frequency ω as per the dispersion relation is

$$\omega = \Omega(k) = \sqrt{\left(g + \frac{\gamma}{\rho}k\right)k} \tag{7}$$

where g is the gravitational acceleration
γ is the surface tension,
ρ is the water density.
The energy density of a wave is

$$E = \frac{1}{2}(\rho g + \gamma k^2)A^2 = \frac{\rho\Omega^2}{2k}A^2 \tag{8}$$

which is transported with the group speed

$$c_g = \frac{\partial\Omega}{\partial k} = \frac{\partial\Omega}{\partial k}\hat{k} \tag{9}$$

The transportation energy describes how the propagation of the amplitude takes place over the surface.

5.2. Stabilized ISPH-Based Method

ISPH is an acronym for incompressible smoothed particle hydrodynamics. Primarily used in Computer Graphics and Realistic Animation, this is one among the many approaches that provide users with the accurate result at the same time considering the variable involved in random flows [11].

IOP Conf. Series: Materials Science and Engineering 1187 (2021) 012013 doi:10.1088/1757-899X/1187/1/012013

Based on the Lagrangian description, the continuity and Navier-Stokes equation for incompressible fluids can be written as,

$$\nabla \cdot u = 0, \tag{10}$$

$$\frac{Du}{Dt} = -\frac{1}{\rho}\nabla P + \upsilon\nabla^2 u + g + \frac{1}{\rho}\nabla \cdot \tau \tag{11}$$

where ρ is the density,
υ is the kinematic viscosity of the fluid,
u is the velocity vector,
P is the pressure of the fluid,
t indicates time,
g is the gravity acceleration,
τ is the turbulence stress

Extending the above equation further, we derive an equation for smoothed particle hydrodynamics, which further extends to incompressibility using the final equation. To obtain this equation, various steps are to be followed depending on the parameters taken into consideration and the fluid.

The method has been used to simulate complex scenes with millions of sampling points. ISPH method is proven to be robust when it comes to non-linear flow. This method would only be suggested when a high processing system and large data storage capacity are available. The millions of sampling points are considered to render the final product, and large storage is a must.

6. Discussions

We have considered the boundary of the fluid body to be small. Also, the wavelength and amplitude of the wave are initially taken to be small. We can expect inaccuracies when the environment for the fluid changes. Large amplitudes of the wave will lead to the propagation of particles at the wrong speed, giving a very unrealistic, choppy water surface. The executable code is indirectly connected to the projection mapping software, and this might lead to delay in times when we don't have enough storage or processing power. Thus, looking for an alternative or converting the complete code to suit the TouchDesigner software would help in such cases.

Flow visualization has many techniques direct, texture-based, geometric and feature-based [12]. These methods can be used as stepping stones to understand and evaluate flow visualization under different parameters and categories [13]. NASA uses other methods such as an Infrared flow visualization system to visualize transonic and supersonic flow regimes. They observe the shock impingement, boundary layer transition and surface temperature gradients through this method [14].

Computational Fluid Dynamics is used to compute the fluid simulations in the present time. The problem that users and researchers encounter while using this process is huge unstructured datasets [15]. Used in automotive industries, oceanography, meteorology and medicine, flow visualization is observed through CFD [15].

The scope for advancement in the project completely depends on the parameters to which it is exposed. Depending on the kind of flow visualization required, the model and program can be modified. This project can be converted into an automated system by choosing the suitable microcontroller compatible with Processing and TouchDesigner software. The final code can be dumped into the microcontroller. In such a case, high processing capacity and high storage space are supposed to be considered.

7. Conclusions

Real-time flow visualization using projection mapping technology will be cost-effective and less space-consuming. As light is being used as a medium of projection and the sensor keeps updating the programs every time it captures a change or variation in the environment, this method is expected to have no latency and no error in the final output. When the output is compared with the theoretical values, there should be a minimum to no error. The objective is to attain the replica of the fluid flow and visualization of the same for an analytical and observational study.

The results of the initial simulations of the complete setup have recorded no latency. The interaction between the human and the sensor to acquire real-time data is accurate. To further precision, an auto-calibration model for the projector and the Kinect sensor can be devised. Projection-Mapping is executed manually. The automap feature of the software is also utilized to map the object. For future developments in the project, simulation code can be developed to plot charts and help calculate velocity and density coefficients. These results can be compared with the traditional methods to gauge the accuracy and identify the project's shortcomings.

Digitization of flow visualization will help all the domains that study the flow to determine various features and characteristics of the fluids and gases. These studies directly or indirectly affect the designs and different mathematical models devised to apply and use them to reference future observations.

8. References

[1] Kaur Gurjot 2016 Wake Visualization Behind Different Body Shapes *Int. J. of Engineering Research & Technology* IJERT05(2016)08 p 16
[2] Zhou Yi Shuangjiu Xiao Ning Tang Zhiyong Wei and Xu Chen 2016 Pmomo: Projection mapping on movable 3D object In *Proc. of the 2016 CHI Conf. on Human Factors in Computing Systems (CA, USA)* pp. 781-790
[3] Costa Nuno and Artur Arsenio 2015 Augmented reality behind the wheel-human interactive assistance by mobile robots *6th Int. Conf on Automation, Robotics and Applications (ICARA)* pp 63-69
[4] Kai Li and Liming Wu 2007 *Water simulating in Computer Graphics* (Sweden: MSI Vaxjö University)
[5] Weiskopf Daniel and Gordon Erlebacher 2015 Overview of flow visualization *The Visualization Handbook* (Utah: Academic Press) pp 261-278
[6] Benko Hrvoje Andrew D Wilson and Federico Zannier 2014 Dyadic projected spatial augmented reality *Int. Proc. of the 27th annual ACM symposium on User interface software and technology* (New York: Association of Computing Machinery) pp. 645-655.
[7] Motta T Loaiza M Raposo A and Soares L 2014 Kinect projection mapping *SBC Journal on Interactive Systems* 5(2014)03 pp 4-14
[8] Soyoung Jung Frank Biocca and Daeun Lee 2015 Effect of 3D Projection Mapping Art: Digital Surrealism *Virtual, Augmented and Mixed Reality 2015 (Lecture Notes in Computer Science vol 9179)* eds R Shunmaker and S lackey (Switzerland: Spring International Publishing) pp. 361-367
[9] Anselm Grundhöfer and Daisuke Iwai 2018 Recent advances in projection mapping algorithms, hardware and applications *Computer Graphics Forum* 37(2018)02 pp 653-675.

[10] Skrivan Tomas Andreas Soderstrom John Johansson Christoph Sprenger Ken Museth and Chris Wojtan 2020 Wave curves: Simulating lagrangian water waves on dynamically deforming surfaces *ACM Transactions on Graphics* TOG39(2020)04 pp 65-1

[11] Hassaballah M Abdelraheem M Aly and A Abdelnaim Interactive fluid flow simulation in computer graphics using incompressible smoothed particle hydrodynamics 2020 *Computer Animation and Virtual Worlds* 31(2020)02 p e1916.

[12] Laramee Robert S Helwig Hauser Helmut Doleisch Benjamin Vrolijk Frits H Post and Daniel Weiskopf 2004 The state of the art in flow visualization: Dense and texture-based techniques *In Computer Graphics Forum* 23(2004)02 (Oxford: Blackwell Publishing Ltd.) pp 203-221

[13] Brambilla Andrea Robert Carnecky Robert Peikert Ivan Viola and Helwig Hauser 2012 Illustrative flow visualization: State of the art, trends and challenges *Proc. of Eurographics 2012 State of the Art Reports* (Italy: The Eurographics Association) pp 74-95

[14] Garbeff Theodore J and Jennifer K Baerny 2017 Recent advancements in the infrared flow visualization system for the NASA Ames Unitary Plan Wind Tunnels *55th AIAA Aerospace Sciences Meeting* (Texas: AIAA Inc.) p 1051.

[15] Laramee Robert S Gordon Erlebacher Christoph Garth Tobias Schafhitzel Holger Theisel Xavier Tricoche Tino Weinkauf and Daniel Weiskopf 2008 Applications of texture-based flow visualization *Engineering Applications of Computational Fluid Mechanics* 2(2008)03 pp 264-274.

Acknowledgements

Authors would like to acknowledge the Mangalore Institute of Technology and Engineering (MITE), Moodabidri for providing an ecosystem to carry out the research work.

Restoration of degraded underwater images with attenuation coefficients as a clue

Dikshith C S, Sameeksha M Vernekar, Chaitra Desai, Ujwala Patil, Uma Mudenagudi

Centre of Excellence in Visual Intelligence(CEVI), KLE Technological University, Hubballi, KA, IN

E-mail: dikshithcs123@gmail.com, samikshavernekar99@gmail.com, chaitra.desai@kletech.ac.in, ujwalapatil@kletech.ac.in, uma@kletech.ac.in

Abstract. Underwater image restoration methods require the knowledge of wideband attenuation coefficients per color channel. In this paper, a Generative adversarial network(GAN) framework is proposed for underwater image restoration using the attenuation coefficients as an input. Attenuation coefficients like atmospheric light and transmission map are also estimated in order to restore an underwater image. For atmospheric light estimation, a combination of Res-net and Dense-net architecture is modelled and making it a learning based Residual-Dense Network. Through literature it is evident that, the transmission map describes about the portion of the light that is not scattered and reaches the camera. Transmission map is generated using the underwater image formation model. Further, these attenuation coefficients are given as an input to the proposed GAN architecture for underwater image restoration. The GAN architecture is having a generator block and a discriminator block. The result is demonstrated on synthetic and real underwater data-set and then compared with state-of-the-art method.

1. Introduction

Underwater image restoration is one of the demanded areas of research due to the challenges involved. Underwater images are more prone to various distortions like poor contrast, colour deviation or refractions of light. The main reasons for the degrading of an underwater image are light scattering and absorption of light. Scattering of visible light gradually reduces the sharpness of the image whereas the varying degrees of light attenuation travelling in water results in the change of colour. There are various techniques and methodologies are being used to process and restore an underwater image. As the visible light energy propagates deeper and deeper, the colour drops gradually depending upon the different wavelengths. And as the visible light coming from the image travels towards the camera, a part of this is visible light energy encounters the suspended particles in water. Therefore, results in scattering of light energy by these particles. Simultaneously, some kind of uneven illumination is caused, leading to spots that are brighter at the centre of the underwater image.

According to the Jaffe-McGlamery model of linear superposition, an underwater image can be represented by few parameters namely forward scattering, direct scattering as well as backward scattering parameters. The Jaffe-McGlamery model will cover a multiple range of imaging conditions and also which possesses few of the complex numerical methodologies, hence cannot be utilized for a single image restoration design. The underwater image is split into two segments

IOP Conf. Series: Materials Science and Engineering 1187 (2021) 012014 doi:10.1088/1757-899X/1187/1/012014

namely the light transmitted which is undeviating and the transmission of light by the hazy aqua medium and also the minute particles, which is namely the veiling light. The mathematical equation is expressed as,

$$I(a,b) = J(a,b) * t(a,b) + A * (1 - t(a,b)) \tag{1}$$

The RGB model is defined using the above variables, where I(a,b) represent the image captured through the camera, J(a,b) represent the haze-free image, t(a,b) will represent the transmission of the medium across the beam which will describe the part of light which is never scattered back and will reach the camera and A will represent global atmospheric light. The idiom J(a,b)t(a,b) will be a direct attenuation and the idiom A*(1-t(a,b)) represents water light luminescence. The t(a,b) which is the transmission of medium is a rapidly decaying activity as,

$$t(a,b) = exp(-\beta * d(a,b)) \tag{2}$$

where, β represent the attenuation coefficient and d(x) represents the distance from the scene to the camera, also called the scene depth.

In recent years, the deep learning frameworks for underwater image restoration have developed rapidly, especially the CNN (convolutional neural network) , mainly used in classification of image, object detection, and motion detection, and they perform much better compared to the traditional methods. But in the case of underwater image restoration using CNN is limited due to the lack of real-world underwater scenes. Therefore, using a synthetic underwater dataset would be an effective approach. Some generative adversarial networks (GAN) are used for generating real underwater images.

In this paper, a GAN framework is put forward for Underwater image restoration using the attenuation coefficients like the atmospheric light A of the equation and the transmission map t(x) of the equation as an input. Estimation of atmospheric light is carried out by modeling a combination of Res-net and Dense-net architecture and proposing a Residual-Dense Network model. The transmission map is known as the portion of the light that is not scattered and reaches the camera. A transmission map is generated using the image formation model. The two attenuation coefficients that are, the atmospheric light and transmission map are used as an input for the proposed GAN underwater restoration model. The GAN architecture mainly has two blocks, a generator block; U-Net is used for the generator network, and a discriminator block; the proposed GAN having a discriminator block is executed as a Patch-GAN. The proposed technique successfully dehazes an underwater image when tested on quantitative and qualitative synthetic data and real data also surpass various state-of-the-art techniques of dehazing.

2. Related work

The attenuation coefficients for the ocean show many temporal variations as it depends on many organic concentrations and substances in water. The contrast will have much significance on underwater-computer-vision, here the details of the attenuation coefficients is used for the color restoration algorithm. Derya Akkaynak[1] has set a goal in the paper to grasp the base knowledge of visual hydrography in order to enhance the underwater-computer-vision techniques. Initially, the loci are derived that constitute an appropriate search space in underwater image restoration techniques, along with the reference for realistic computer-generated images. Through the wideband coefficients, it is shown that it relies on both the properties of scene and the water, a possession which is overlooked on the recently utilized for the image formation. Then verify the preferred loci model using in-situ analyses using 2 non-identical optical water bodies.

The wideband coefficient becomes sensitive to the real color and the range of the object. This relatively invalidates the image formation model and hence the error is analyzed from this discrepancy. Phillip Isola[2] investigates that the image-to-image translation problem can be addressed using a conditional adversarial network as a general-purpose solution. Here the network not only learns how to map from input image to output image but also learns a corresponding loss function to train the mapping. In the manner, GAN will attain generative model and a conditional GAN will attain a conditional generative model of data, hence making the conditional GAN fit the image to image interpretation. The author demonstrates that this approach is efficient at synthesizing images from label maps, image colorizing, and reconstructing objects.

The key global atmospheric light is ignored as the prsesent learning techniques which will anticipate only transmission medium using the CNN(convolutional-neural-network). Chongyi Li[3] proposes an adaptive cascaded CNN for single image dehazing, by considering the attenuation coefficients, that is atmospheric light globally and transmission medium together by 2 task based sub-networks. Moreover, a common feature is shared between the two cascaded subnetworks. Ultimately, using the parameters which will be estimated by the model, a restored image is attained. More accurate and effective restoration performance is achieved by the atmospheric scattering model.

3. Methodology

In this section, an architecture is been proposed for underwater image restoration. The framework mainly consists of 2 parts,namely Discriminator and Generator. A Generator network will learn a function to map from an input image to an output image. A generative adversarial network will be an appeal to train generator, which will be used to generate images using generator. The Discriminator network classifies if the image generated is Fake or Real. We have used a synthetic data-set which is been prepared using the NYU dataset. We have used this synthetic data-set for training the architectures.

generative adversarial networks

3.1. Atmospheric light estimation

A learning-based approach for the estimation of atmospheric light is proposed. Initially, the input image is passed through the convolutional layer and further passed through the dense blocks to estimate the atmospheric light like AR, AG, AB for the respective red, green and blue channels. The main block of our dense residual network is dense blocks. Here each and every layer of the network is connected to each and every other layer and we have added residual connections in-between the dense blocks.

Apart from adding more layers of CNN, adding residual networks will increase the performance and also accuracy. The residual block is adding output from the previous layer is added to the output of some forward layer. ResNet uses an additive method (+) that merges the previous layer (identity) with the future layer. Here we have added 2 dense blocks to our network. In CNN layers due to more layers, there may be chances of losing the information before reaching its destination. In DenseNet, each and every previous layer is connected to the next layers to avoid losing the information. DenseNet concatenates the previous layers with the future layers.

3.2. Transmission map estimation

We have estimated transmission map by using 2 methods. Firstly, by using underwater image formation model.

The transmission map $t(x, y)$ is given as,

$$t(x, y) = exp(-\beta * d(x, y)) \tag{3}$$

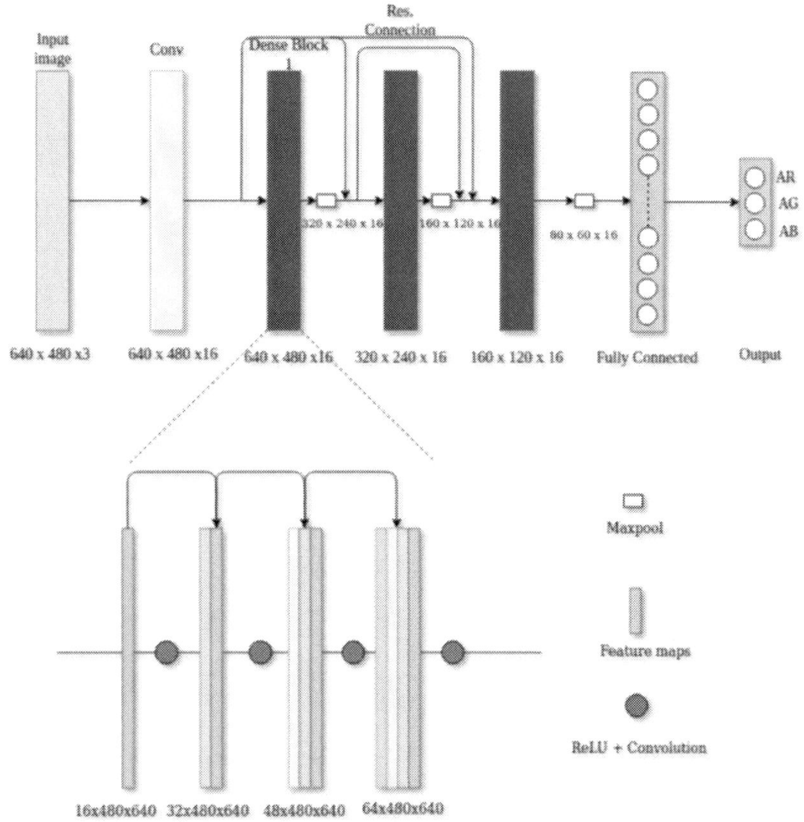

Figure 1. Learning based estimation of atmospheric light using Residual Dense Network

- β is the attenuation coefficient.
- $d(a, b)$ is the distance from camera to the scene.

Here beta value varies from [0.4,1.6].
Secondly by using CNN layer. The architecture of the CNN layer is shown below.

We have proposed our model on estimation of transmission map as shown above. Here the input image is passed through several layer layer of CNN to train on the transmission map data set which is estimated using the equation as stated above and we have tested on the testing data set which is added in the result section.

3.3. Proposed architecture for Underwater image restoration

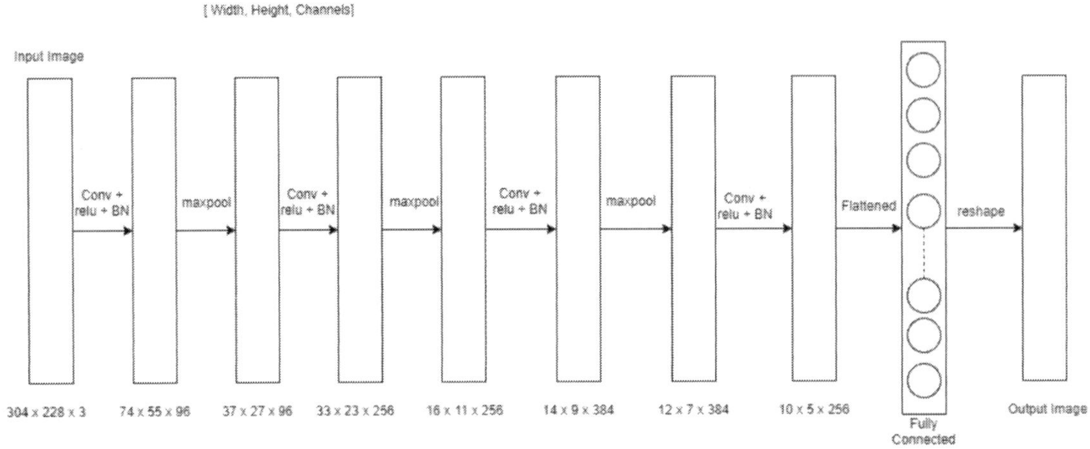

Figure 2. Architecture of Transmission map

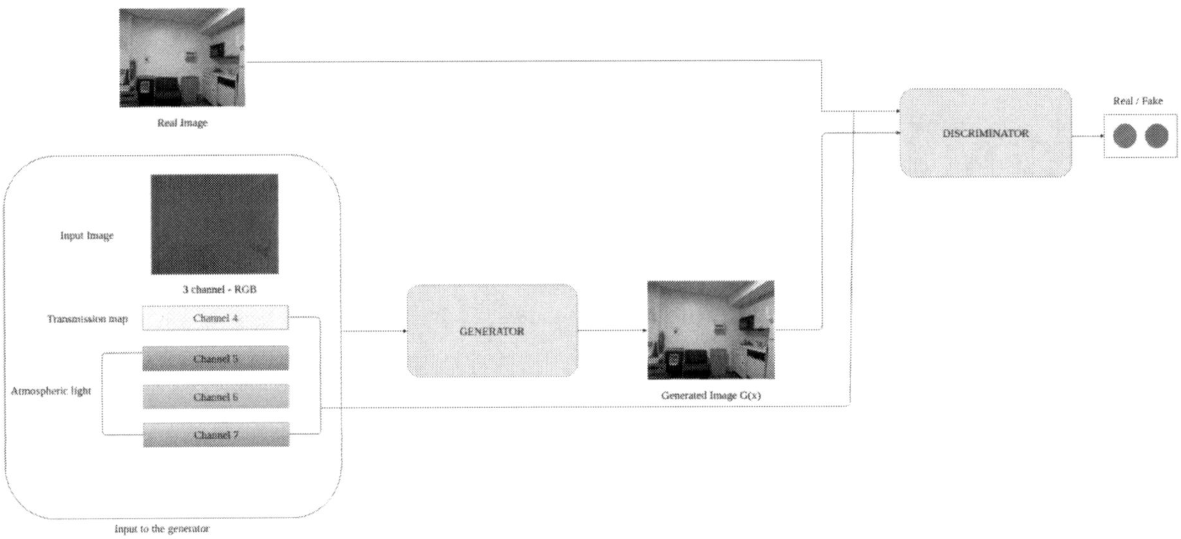

Figure 3. Underwater restoration model

Generator :

For generator architecture, we have used U-Net architecture for the model to learn and generate images. We have given a hazy image, transmission maps, and atmospheric light as input for the generator network. A Generator network uses the U-Net architecture which is an encoder and decoder model. The network generates the restored image by down-sampling or the input image is encoded into the bottle-neck layer and then up-sampling or the output image is decoded from the bottle-neck layer. The skip connections are introduced in the middle of encoder and the decoder layers, which gives a 'U' shape to the architecture.

IOP Conf. Series: Materials Science and Engineering 1187 (2021) 012014 doi:10.1088/1757-899X/1187/1/012014

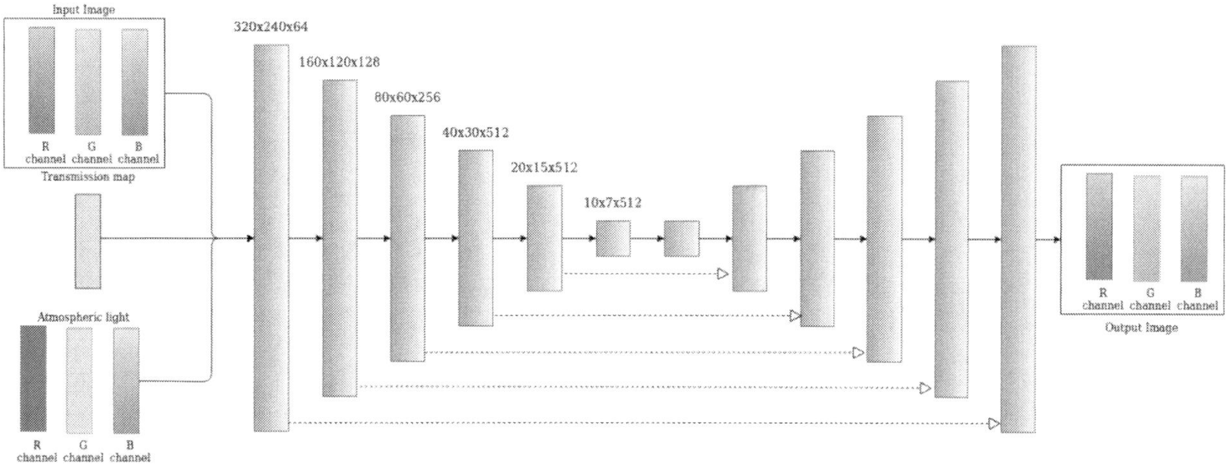

Figure 4. Generator architecture for Underwater restoration model

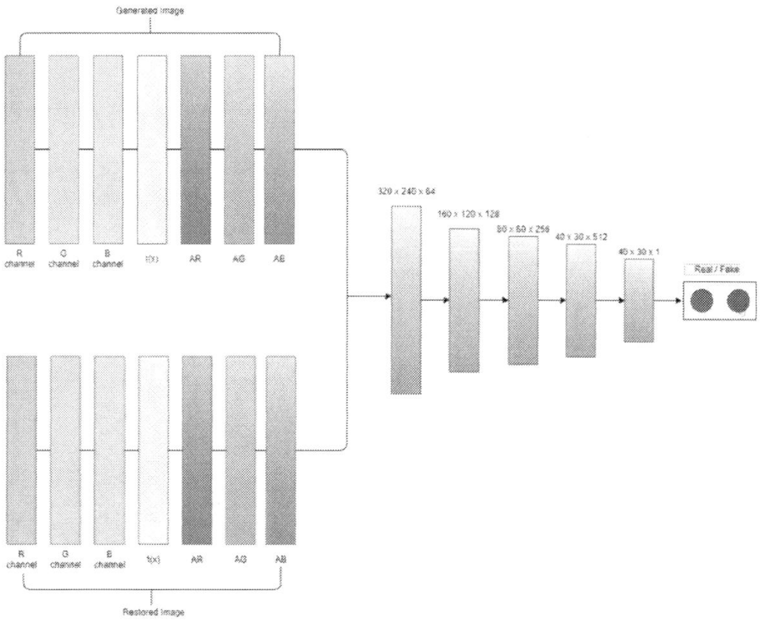

Figure 5. Discriminator architecture for Underwater restoration model

Discriminator :

The discriminator model takes a generated input and a real image from the data-set and along with transmission map and atmospheric light as a hint and predicts whether the image is a real or generated image. The Patch-GAN will be implemented in this proposed discriminator architecture. Here the discriminator architecture will generally try to categorise whether each of the MxM patch is fake or real for a given image. In this model a Patch-GAN of size 16×16 is applied, thus says that the output image will be mapped with the input image of 16x16 size. As a result, the 16x16 Patch-GAN will categorise 16x16 patches as fake or real in an input image.

IOP Conf. Series: Materials Science and Engineering 1187 (2021) 012014 doi:10.1088/1757-899X/1187/1/012014

4. Result Analysis

4.1. Results on atmospheric light

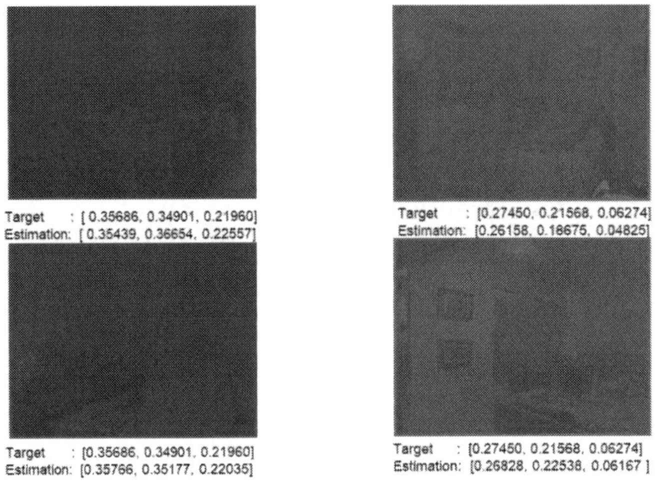

Figure 6. Target and estimation of testing images

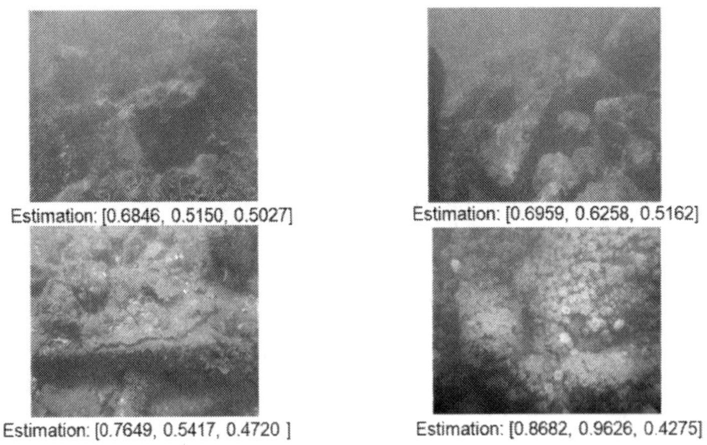

Figure 7. Atmospheric light estimation on UIEBD dataset

In Fig.6, we have estimated the atmospheric light on the testing images from the synthetic data-set. We infer from the Fig.9, that the atmospheric light estimated from the proposed Residual-Dense network works well. We can conclude from the test images that the prediction and target values are almost near. In the Fig.7, we have estimated the atmospheric light values on the Jerlov water type 7C and 9C of UIEBD data-set.

In Fig.8, we have estimated atmospheric light for the Jerlov water type - 1C of UIEBD data-set.

In Fig.9, We have tested our model on Jerlov water type - 1B. We can see that the predicted values are different from the actual values. Thereby, we can conclude that for different Jerlov water types we need to train different model for the estimation of atmospheric light.

Figure 8. Atmospheric light estimation on UIEBD data-set(Jerlov type - 1C)

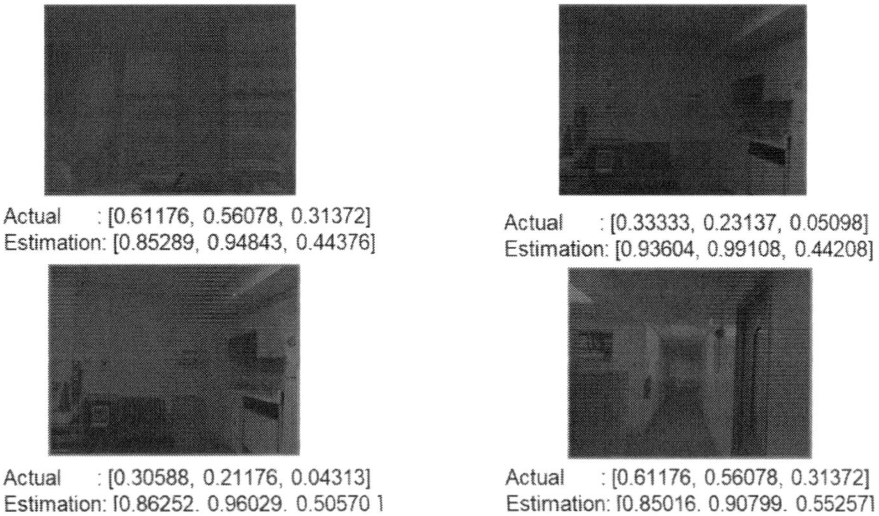

Figure 9. Atmospheric light prediction on Synthetic data(Jerlov type- IB)

4.2. Results of Transmission map

In Fig.10, The estimated transmission map for the β values 1, 1.2 and 1.6 respectively.

Here we have estimated transmission map using the CNN architecture mentioned in the Fig.5

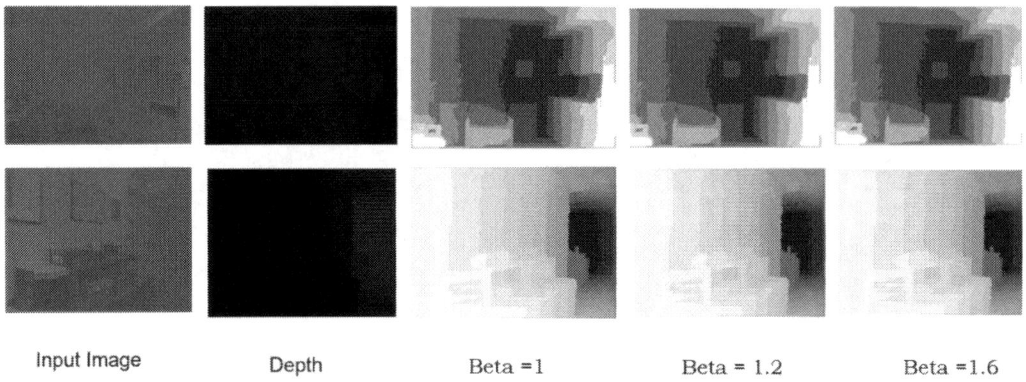

Figure 10. Generated transmission map for different β values along with depth

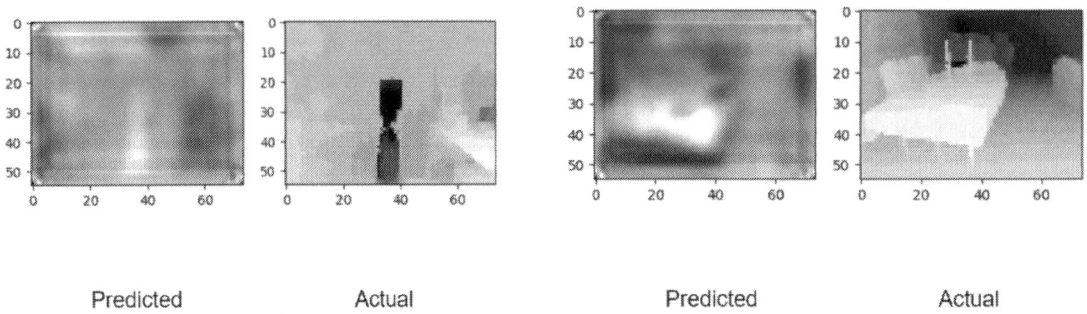

Figure 11. Transmission map estimated using CNN

4.3. Results on restoration

IOP Conf. Series: Materials Science and Engineering 1187 (2021) 012014 doi:10.1088/1757-899X/1187/1/012014

Figure 12. Results : Restoration on Synthetic data

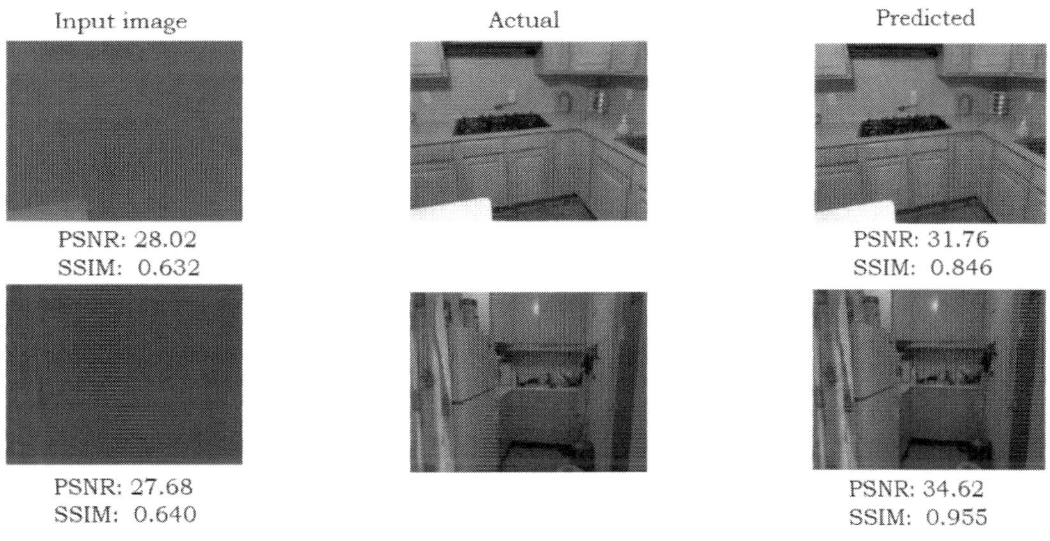

Figure 13. Results : Reference based quantitative metric - PSNR and SSIM

- Higher the UCIQUE score better the image quality.

IOP Conf. Series: Materials Science and Engineering 1187 (2021) 012014 doi:10.1088/1757-899X/1187/1/012014

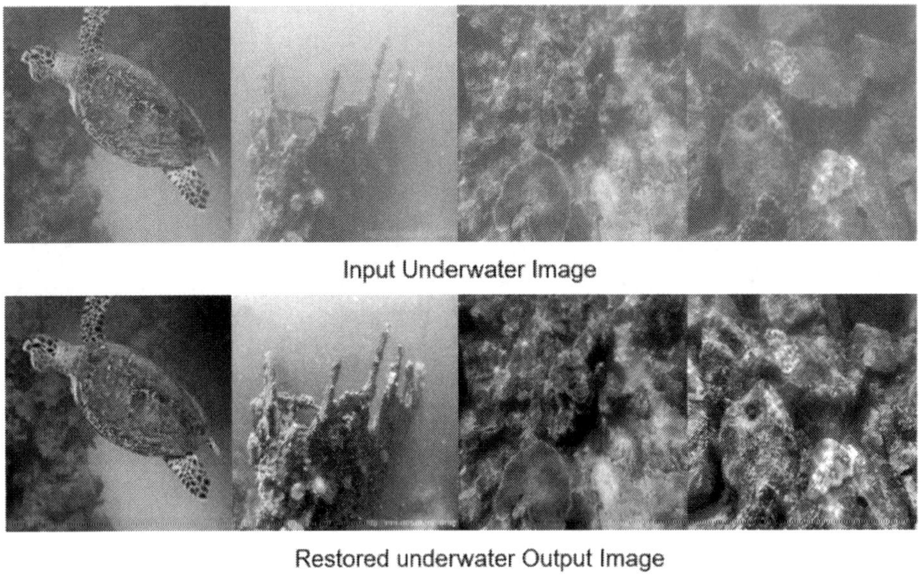

Figure 14. Results: Real underwater image

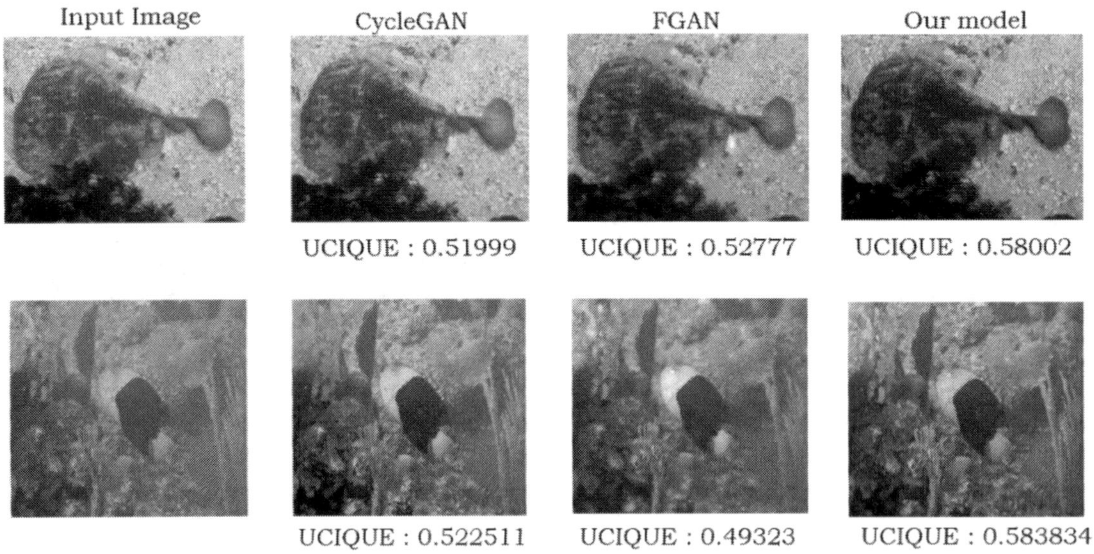

Figure 15. Result : Comparison with State-of-the-art methods(No reference quantitative metric)

5. Conclusion

In this project, the GAN framework is proposed which uses the attenuation coefficients for restoring an underwater image. The attenuation coefficients are estimated individually for the same. Here, a learning based estimation of atmospheric light using residual-dense

Figure 16. Result : Detailed view of the output

network is proposed and shows preferably a good result. The transmission map estimated using the underwater image formation model works better than the CNN based estimation. Further demonstrated the results on the benchmark data set and compared with state-of-the-art methods. The proposed GAN restoration framework performs well on the degraded images and the results obtained are comparatively better.

6. Refrences

[1] Derya Akkaynak, Tali Treibitz, Tom Shlesinger, Raz Tamir, Yossi Loya, and David Iluz "What Is the Space of Attenuation Coefficients in Underwater Computer Vision?", 2018.

[2] Phillip Isola, Jun-Yan Zhu Tinghui, Zhou Alexei, A. Efros "Image-to-Image Translation withConditional Adversarial Networks", 2018.

[3] Chongyi li 1,2, Jichang Guo1, Fatih Porikli3, Huazhu Fu 4, and Yanwei Pangi "A Cascaded Convolutional Neural Network for Single Image Dehazing", 2018.

[4] Ayush Singh, Ajay Bhave, Dilip K. Prasad "Single image dehazing for a variety of haze scenarios using back projected pyramid network"15 Aug 2020

[5] Jinjiang Li, Guihui Li, Hui Fan "Image Dehazing using Residual-Based Deep CNN"2018.

[7] Yincui Xu, Shouyi Yang "Single Image Haze Removal with Improved Atmospheric Light Estimation"2019

[8] Huimin Lua, Yujie Lia, Shota Nakashimad, Seiichi Serikawaa "Single Image Dehazing through Improved Atmospheric Light Estimation"2015

[9] Derya Akkaynak, Tali Treibitz "A Revised Underwater Image Formation Model" CVPR paper 2018.

[10] David Eigen, Christian Puhrsch, Rob Fergus "Depth Map Prediction from a Single Image using a Multi-Scale Deep Network"2019.

Design and evaluation of hybrid SHE+STT-MTJ/CMOS full adder based on LIM architecture

PRASHANTH BARLA[1], VINOD KUMAR JOSHI*, SOMASHEKARA BHAT[3]

Dept. of Electronics and Communication Engineering, Manipal Institute of Technology, MAHE, Manipal,Karnataka, India

E-mail: prashanthbarla@gmail.com[1], vinodkumar.joshi@manipal.edu[2], soma.bhat@manipal.edu[3]

Abstract. This work aimed at developing a full adder using hybrid magnetic tunnel junction/complementary metal oxide semiconductor (MTJ/CMOS) based on the logic-in-memory architecture (LIM). LIM has emerged as the most promising alternative to the standard von-Neumann architecture in the impeding post-CMOS era. Performance of the hybrid full adder is evaluated in terms of power, delay, power delay product (PDP), and device count. These results are compared with the existing double pass transistor logic-based clocked CMOS (DPTL-C^{2}MOS) full adder. Further, Monte-Carlo simulations on both variants of full adders were conducted to study their performance. Simulation results reveal that the hybrid full adder is superior to the DPTL-C^{2}MOS full adder and can be used in low power and high throughput computing systems in the near future.

1. Introduction

Exponential growth in the field of artificial intelligence, big data, and 5G technology demand high throughput and low power computing system. Though standard von-Neumann architecture has been the workhorse in the computing system for the last few decades, it cannot meet future demands as it poses memory wall problems [1]. Simultaneously, scaling down the technology node to obtain high packing density in the integrated circuits (ICs) poses its own challenges, such as an increase in power dissipation. An increase in the power dissipation, especially in the standby mode, is due to the secondary effects that kick in the standard CMOS technology below 45nm technology [2]. Although the higher clock frequency has increased the operating speed, at the same time, it also increases the dynamic power dissipation of the CMOS ICs. As a combined effect, the heat production in the ICs has been reaching its thermal limit [3]. To tackle this worsening condition, many efforts have been put forth at the device as well as architectural level. Among all the possible solutions, logic-in-memory (LIM) at the architectural level and magnetic tunnel junction (MTJ) at the device level are the most prominent ones [4]. A computational capability is instilled into the memory of LIM. Here the memory stores the information as well as takes part in the computational process [5, 6]. By doing so, the memory

*Corresponding author's E-mail: vinodkumar.joshi@manipal.edu

IOP Conf. Series: Materials Science and Engineering 1187 (2021) 012015 doi:10.1088/1757-899X/1187/1/012015

wall problem that is being seen in the standard von-Neumann architecture is resolved. MTJ is the best suited for LIM, because to its supremacy such as infinite endurance, fast reading capability, non-volatility, high density, 3D integration with the existing CMOS technology [7]. Figure 1 shows the difference between standard von-Neumann architecture and hybrid MTJ/CMOS LIM structure. The hybrid MTJ/CMOS LIM not-only avoids the unnecessary flow of data but-also reduces the area due to the feasibility of 3D integration on top of the silicon. Most importantly, in LIM, due to non-volatile memory, we can completely switch off the un-utilized blocks without losing the information, and there is no need for a backup and restoration process either. The information stored in the non-volatile memory is readily available for processing as soon as the power is resorted. This saves a significant amount of power in the standby mode [8, 9]. The process of storing the information into the MTJ is called MTJ switching/writing. There are several mechanisms used for the MTJ writing, such as field-induced magnetic switching (FIMS), thermal assisted switching (TAS), spin transfer torque (STT), spin-Hall effect (SHE), voltage assisted switching (VAS) [2]. Though the STT switching mechanism is commercialized in many applications, it suffers from incubation delay due to stochasticity [10]. This lowers the operating speed of the STT-MTJ devices. One can overcome this issue by increasing the write current, but this affects the endurance of the STT-MTJ device by causing the dielectric breakdown over a period of time. These limitations are strongly overcome with the recently introduced SHE-assisted STT (SHE+STT) switching mechanism in a three-terminal MTJ [11]. Here the SHE assists in overcoming the incubation delay and not-only speeds up the switching process but also increases the endurance of the MTJ device. Considering all these, in this paper a hybrid SHE+STT-MTJ/CMOS full adder (FA) based on LIM architecture has been developed. Simulations are carried out to study the performance of these hybrid circuits in-terms of key performance indicators (KPI) such as output response, power dissipation, power delay product (PDP), and the number of devices utilized. Comparison of these KPIs with the CMOS counterpart, i.e., double pass transistor logic-based clocked CMOS (DPTL-C^2MOS), is conducted. Monte-Carlo (MC) simulations on hybrid FA is also performed here, to study the power variations by incorporating process and mismatch variations in CMOS and extracted parameters of MTJ that would arise during the fabrication process.

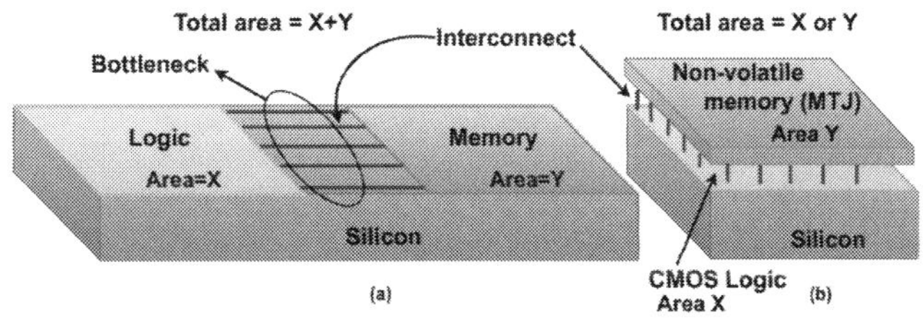

Figure 1. (a) Schematic of von-Neumann architecture with logic and memory blocks. Here logic as well as memory blocks are placed separately, and their combined area is X+Y. Interconnects aids the communication between them. (b) 3D integration of MTJ with CMOS in LIM with their combined area as either X or Y [6].

2. Background

2.1. MTJ structure and its switching with SHE+STT mechanism

Figure 2 shows the three-terminal (T1, T2, and T3) structure of SHE+STT-MTJ. T1 is connected to the pinned layer (PL), whereas T2 and T3 are connected to the opposite end of heavy metal (HM). The free layer (FL) is separated from the PL by a thin oxide barrier layer (BL) and is placed in contact with the HM. The HM material possesses a high atomic number

IOP Conf. Series: Materials Science and Engineering 1187 (2021) 012015 doi:10.1088/1757-899X/1187/1/012015

and converts a charge current into spin current due to high spin-orbit interaction (SOI), i.e., when an unpolarized charge current passes through the HM, due to SOI, electrons with opposite spins move in opposite directions thus generating a spin current in the HM. This effect is called SHE [12]. The relationship between spin current density ($\vec{J_S}$), charge current density ($\vec{J_C}$) and spin polarization ($\vec{\sigma}$) is given in Equation 1 as,

$$\vec{J_S} = \theta_{\mathrm{SH}} \left(\vec{\sigma} \times \vec{J_C} \right). \tag{1}$$

Where θ_{SH} is the spin-Hall angle. The $\vec{\sigma}$, $\vec{J_S}$ and $\vec{J_C}$ are all mutually perpendicular to each other. Thus, the SHE generated assists the STT to switch the MTJ from anti-parallel (AP) to parallel (P) state or vice versa. When the MTJ is in AP state, it is considered to be at a high resistance state (R_{AP}); on the contrary, when it is in P state, the MTJ is assumed to be of low resistance (R_P). Figure 2(b) shows switching of MTJ from AP to P state, where a charge current, with density (J_{SHE}) flow-through HM in the Y-direction and gets converted into the spin current, with density ($\vec{J_S}$) in Z-direction. This $\vec{J_S}$ exerts a torque known as SOT onto the FL's magnetic orientation. Due to SOT, the orientation of the FL will be tilted from -Z-direction to X-direction. At this point, the MTJ is in a meta-stable state for a brief moment. Now the spin polarized electrons entering from the PL to FL in -Z-direction will exert an STT onto the FL's magnetic orientation to change it from the X-direction to Z-direction. Hence MTJ switches from AP to P, with the combined effort of SHE and STT currents, and thus we call this switching as SHE-assisted STT switching mechanism. Similarly, switching the p-MTJ from P to AP (Figure 2(c)) can be understood.

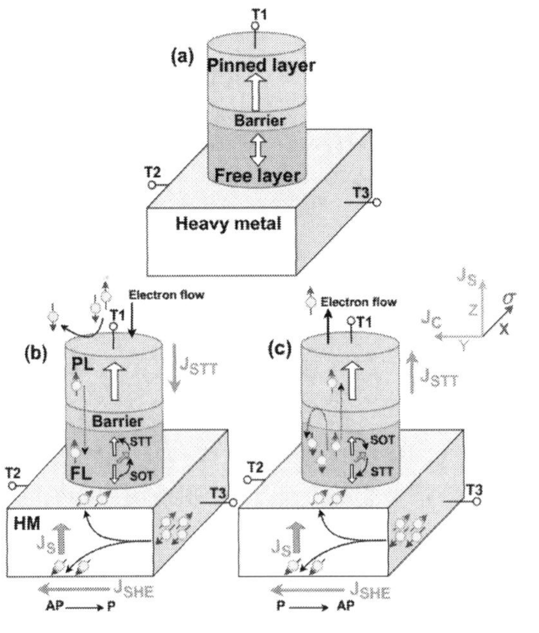

Figure 2. (a) Three terminal SHE p-MTJ device structure. When PL and FL are pointing in the same direction then MTJ is in P state or else when PL and FL are pointing in opposite direction MTJ is in AP stare. Using SHE+STT, MTJ switching take place from (b) AP to P with J_{SHE} in Y- and J_{STT} in -Z-direction. Whereas MTJ (c) P to AP switching take place with J_{SHE} in Y- and J_{STT} in Z-direction [2].

2.2. Logic-in-memory architecture

Block diagram of the hybrid MTJ/CMOS LIM architecture is shown in Figure 3. It consists of three main blocks: (a) sense amplifier (SA) [13], (b) logic network (LN) and (c) MTJ switching circuit [14]. SA detects the logic operations carried-out by the LN and produce OUT and $\overline{\text{OUT}}$. The LN is a combination of the MOS logic structure and MTJs. Here the volatile inputs are applied to the MOS logic structure, on the contrary non-volatile data is stored in MTJs. MTJ switching/writing block will alter the MTJ resistance from R_P to R_{AP} (thereby its state) or vice versa using the SHE+STT switching mechanism. The amalgamation of MTJs with the MOS logic structure is possible due to their resistance compatibility, ON resistance of the MOS (R_{ON}) is less than R_P; ($R_{ON} < R_P$) and the OFF resistance of the MOS (R_{OFF}) is more than R_{AP}; ($R_{OFF} > R_{AP}$).

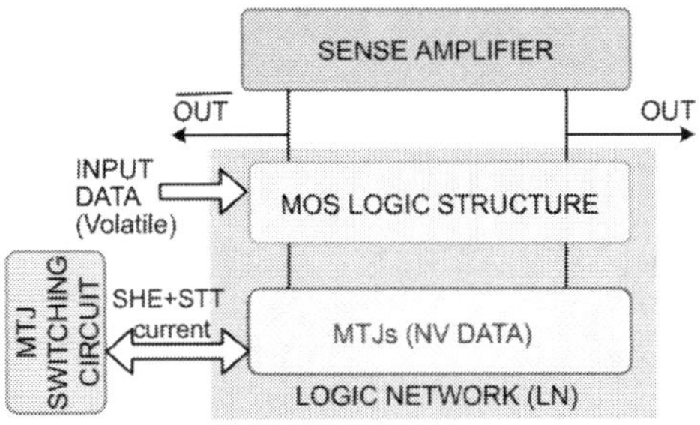

Figure 3. LIM architecture with sense amplifier, logic network and MTJ writing circuit blocks [7].

3. Design of hybrid full adder based on LIM

FA circuit is based on hybrid LIM architecture, and it consists of SUM (Figure 5(a)) and CARRY sub-circuit (Figure 5(b)). It produces output and its complement by accessing the differential current flow between the left and right arm. The inputs A and B are given to the MOS logic tree, whereas C_{in} is stored in the MTJ pair (MTJ0-MTJ1). When the MTJ pair is in AP-P configuration, we assume bit "0" is stored; on the contrary, when the MTJ pair is in P-AP configuration, we assume bit "1" is stored. Hybrid FA circuit operates in two phases viz, pre-charge and evaluation phase. In the pre-charge phase, clock (CLK) = "0" and inputs A, B, and C_{in} are applied to the circuit. In the evaluation phase, CLK = "1", and during this period, processing of these inputs are conducted in the LN to produce the corresponding output and its complement. As discussed earlier, writing the information into the MTJ pair is conducted using the SHE+STT switching mechanism, and this takes place in the pre-charge phase so that this information is available for processing at the LN in the next evaluation phase.

SHE+STT-MTJ writing driver is divided into two parts; control circuitry (Figure 4(a)) and writing core (Figure 4(b)). Table 1 shows the various combinations of input, intermediate signals, and corresponding status of the MTJ pair. When EnW is at logic "0", writing core is disabled by the intermediate signals (STTP, STTN, SHEP, SHEN = "1010") of the control circuit, and no writing takes place. Information will be written only when EnW = "1". When EnW, EnSHE, and Data are "111" respectively, intermediate signals STTP, STTN, SHEP, SHEN will be "1101" respectively. This will turn ON the transistors MP1, MP2, MP3, MN0, MN1, MN3. There are two types of current flowing in the circuit, i.e., SHE and STT current. SHE current path is; Vdda-MP1-MTJ0-MN1-gnd and Vdda-MP3-MTJ1-MN3-gnd. STT current path is; Vdda-MP2-MTJ1-MN3-gnd and Vdda-MP1-MTJ0-MN0-gnd. At this point, the MTJ pair is in a metastable state. After a brief period of time (200ps), EnSHE is made to "0", forcing

Table 1. Various combinations of input as well as intermediate signals and the corresponding states of MTJs during SHE+STT writing mechanism.

Input signals			Intermediate signals				MTJ status	
EnW	EnSHE	Data	STTP	STTN	SHEP	SHEN	MTJ0	MTJ1
0	X	X	1	0	1	0	X	X
1	1	1	1	1	0	1	Metastable	
1	0	1	1	1	0	0	AP-P	P-AP
1	1	0	0	0	1	1	Metastable	
1	0	0	0	0	1	0	P-AP	AP-P

*X represents don't care condition

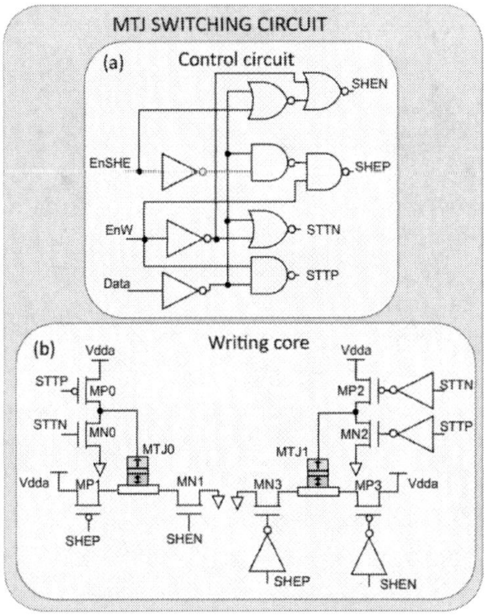

Figure 4. Schematic of SHE+STT-MTJ switching circuit consisting of (a) control circuit and (b) writing core [14].

STTP, STTN, SHEP, SHEN = "1100". This will turn OFF the transistors MP3 and MN1 to stop the flow of SHE current. But STT current continues to flow in the circuit to change the MTJ0-MTJ1 configuration to P-AP, respectively. Similarly, Data "0" can be written into the MTJ pair by suitably changing the input signals, as shown in Table 1.

Working of the hybrid, FA is explained with an example, where ABC_{in} is set as "010". In the pre-charge phase, when CLK= "0", transistors P3 is ON and pull-down (PD) transistor, N3 is OFF. The OUT (SUM/CARRY) and $\overline{\text{OUT}}(\overline{\text{SUM}}/\overline{\text{CARRY}})$ node voltages will be shared through P3 as a result that both OUT and $\overline{\text{OUT}}$ will be pre-charged to Vdd-Vth. In the evaluation phase, CLK = "1", transistor N3 is ON, providing a path for both outputs OUT and $\overline{\text{OUT}}$ to gnd through LN. In the MOS logic tree of SUM sub-circuit, inputs A and B will turn ON the transistors M2, M4, M5, M7, whereas the rest of them are OFF. Input C_{in} will configure the MTJ0-MTJ1 as AP-P. Now there are two paths for the outputs (SUM and $\overline{\text{SUM}}$) to discharge to ground, i.e., Path1 and Path2. The left-arm path (Path1) has a total resistance of two ON transistors (M5, M7) and one MTJ (MTJ1), and the right arm path (Path2) also has two ON transistors (M2, M4) and an MTJ (MTJ0). Since the MTJ0-MTJ1 are in AP-P

IOP Conf. Series: Materials Science and Engineering 1187 (2021) 012015 doi:10.1088/1757-899X/1187/1/012015

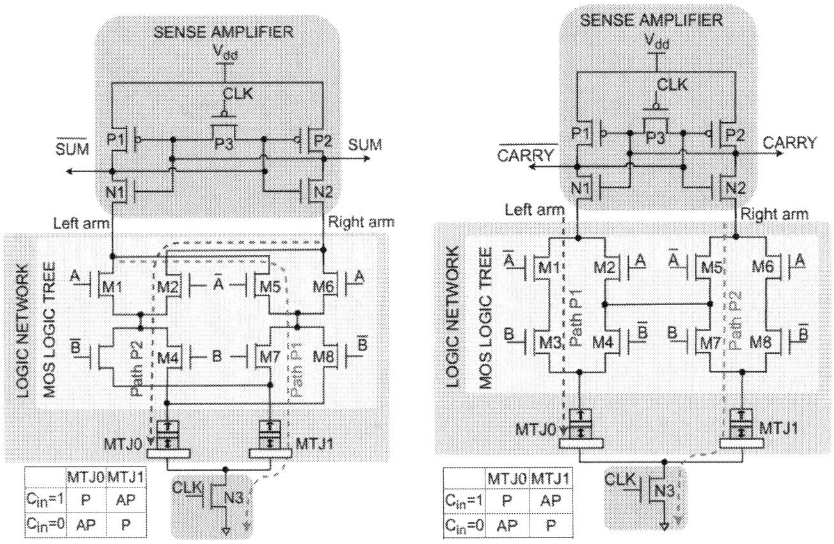

(a) (b)

Figure 5. (a) SUM and (b) CARRY sub-circuit of hybrid full adder based on LIM architecture.

configuration, the total resistance of Path1 ($2R_{ON} + R_P$) is less than the total resistance of Path2 ($2R_{ON} + R_{AP}$). Hence, though both output nodes start to discharge simultaneously, the discharge current in Path1 finds a less resistive path and discharges to gnd quicker than Path2. As a result, $\overline{\text{SUM}}$ is pulled down to the threshold voltage of P2 hence SUM raised to logic "1". Whereas $\overline{\text{SUM}}$ continues to discharge and attains logic "0". This produces SUM as "1" and $\overline{\text{SUM}}$ as "0". Similarly working of the CARRY sub-circuit can be understood. Figure 6 shows the simulated hybrid FA waveform.

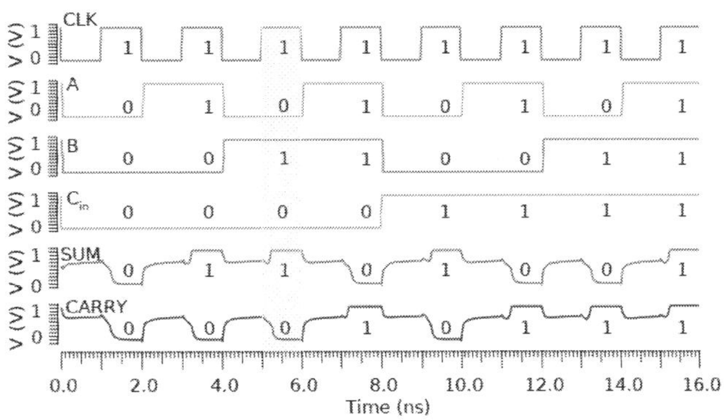

Figure 6. Waveforms of hybrid FA with SUM and CARRY outputs. The highlighted part shows a particular case, where input ABC_{in} are "010" and their corresponding outputs SUM and CARRY are "10".

4. Results and discussion

Simulations are carried out with Cadence (IC6.1.7-64b.500.19) 45nm GPDK design kit with the SHE+STT model developed using Verilog-A language [15]. Table 2 shows the MTJ device

IOP Conf. Series: Materials Science and Engineering 1187 (2021) 012015 doi:10.1088/1757-899X/1187/1/012015

parameters set during simulation. Other parameters are maintained as the default values as mentioned in [15]. The writing core transistors are set with W=480nm, whereas the rest of the transistors are set with default W=120nm and L=45nm. A power supply of Vdda=1.2V is used for writing core to drive write current, whereas rest of the circuit operates with the power supply of Vdd=1V.

Table 2. SHE+STT-MTJ parameters set during the simulation [15].

Parameter	Description	Unit	Default value
t_{sl}	Free layer thickness	nm	0.7
t_{ox}	MgO barrier thickness	nm	0.85
TMR	TMR ratio under zero bias voltage	%	200
Shape	MTJ Surface shape	–	circle
r	MTJ Surface radius	nm	16
w	Heavy-metal width	nm	32
d	Heavy-metal thickness	nm	3
l	Heavy-metal length	nm	60
σ_{TMR}	Standard deviation of TMR	%	3% of TMR
$\sigma_{t_{sl}}$	Standard deviation of t_{sl}	nm	3% of t_{sl}
$\sigma_{t_{ox}}$	Standard deviation of t_{ox}	nm	3% of t_{ox}

Comparison between hybrid FA and DPTL-C^2MOS FA in terms of power dissipation, delay, PDP, and device count is shown in Table 3. The dynamic and total power dissipation of hybrid FA is 97.86% and 47.11% less than DPTL-C^2MOS FA respectively. The static power dissipation of the hybrid FA is almost zero when compared to its DPTL-C^2MOS FA. This is because input (C_{in}) is stored in the MTJ pair, imparting non-volatility to the hybrid FA. So, in the standby mode, the power supply to the hybrid FA can be completely cut-off without a backup process for the C_{in} information. C_{in} information is readily available in the subsequent active mode, and there is no restore process either. However, there is steady-state power dissipation for hybrid FA, which is 2.81% lesser than DPTL-C^2MOS FA.

Table 3. Comparison of hybrid ALU with CMOS-ALU arithmetic operation @500 MHz.

Particulars	DPTL-C^2MOS FA	Hybrid FA
Static power(nW)	712^a	0^b (692^c)
Dynamic power(nW)	621.5	13.28
Total power(nW)	1333.5^d	705.3^d
Worst case delay(ps)	82.17	94.94
PDP(aJ)	109.57	66.96
Device count	48 MOS	32MOS+4MTJ

Note: Write circuit is excluded while obtaining the above values.

[a] Power supply can not be turned off in DPTL-C^2MOS FA, as it is volatile in nature.

[b] In standby mode power is switched off in hybrid FA, as it is non-volatile in nature.

[a,c] In the steady state condition, static power dissipation is observed.

[d] Total power dissipation = dynamic + static power, in active mode.

A higher delay of 13.45% is observed in hybrid FA than DPTL-C^2MOS FA. The delay of hybrid design is comparatively more than its CMOS counterpart; it is because of the dependency of discharge current on the width of N3 transistor (Figure 5). A large N3 transistor would assist to quickly discharge the differential current in the hybrid FA circuit and thereby alleviate the worst-case delay. This can be understood by looking at Figure 7, where the size of the N3 transistor is varied and it is observed that there is a decrease in delay for the increase in the N3 transistor size. However, as the N3 transistor size increases, there is an increase in the total power dissipation also. Hence delay versus power trade-off needs to be balanced in the hybrid FA circuit.

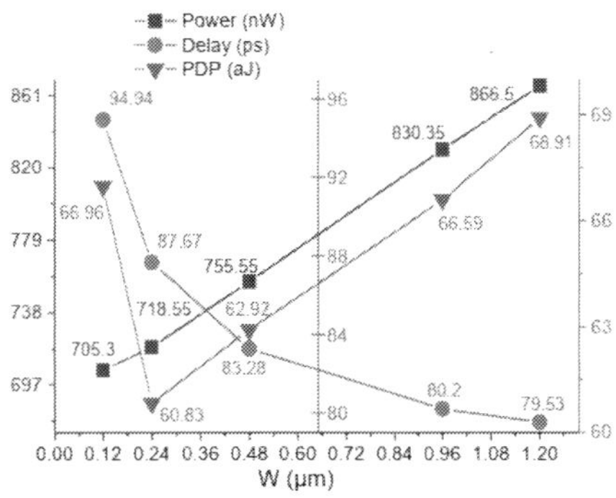

Figure 7. Power dissipation, delay and PDP dependency on N3 transistor's size. As N3 transistor's width increase, delay reduces, but the power dissipation increases, while PDP decreases initially but later on increases.

In terms of transistor count, hybrid FA needs 32 MOS, whereasDPTL-C^{2}MOS FA requires 48 MOS. That means in hybrid FA, the transistor count is less by 33.33% than its CMOS counterpart and helps to save the silicon area. Moreover, the MTJs of the hybrid FA can be built on top of the silicon layer, thanks to the 3D integration capability of MTJs. Thus, the die area for the hybrid FA is lesser as compared to the DPTL-C^{2}MOS FA.

Table 4. Total power dissipation of DPTL-C^{2}MOS FA versus hybrid FA with MC simulation of 200 runs.

Design type	Min (nW)	Max (nW)	Mean (nW)	Standard divination(nW)
DPTL-C^{2}MOS FA	1195	1576	1315	74.85
Hybrid FA	668.75	763.3	712.25	15.62

Further, MC simulation on DPTL-C^{2}MOS FA and hybrid FA circuits are conducted for 200 runs to study the process and mismatch variations that may affect its performance during the nano-scale fabrication, and the total power variation is shown in Table 4. A 3% variations in TMR, t_{sl} and t_{ox} which follow Gaussian distribution is included for MTJs during MC simulations. It is observed that hybrid FA dissipates lesser power than DPTL-C^{2}MOS FA in all three cases, i.e., min, max, and mean.

5. Conclusion

In this paper, a design and analysis of hybrid FA based on LIM architecture is presented. Due to the use of SHE+STT-MTJs in the hybrid FA, the challenges posed by the STT-MTJs such as lower writing speed, high write energy consumption, and risk of dielectric break down are easily overcome in this design. Compared to the DPTL-C^{2}MOS FA, the hybrid FA dissipates significantly lower power and PDP by using a lesser number of transistors. Lower power dissipation in the hybrid FA is also supported by MC simulation. Moreover, due to the non-volatility nature of hybrid FA, its static power dissipation in standby mode is almost zero.

So, this hybrid FA based in LIM architecture can be considered as an alternative to standard von-Neumann architecture in computing systems for various applications.

Appendix A. CMOS full adder using double pass transistor logic based clocked circuit

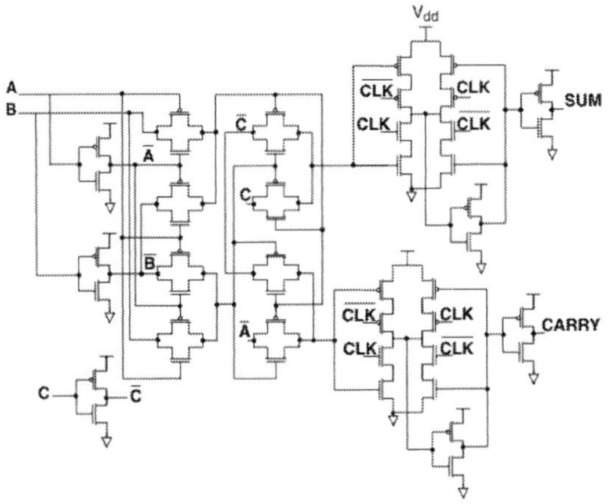

Figure A1. CMOS full adder based on double pass transistor logic based clocked circuit [16,17].

References

[1] Ankit A, Chakraborty I, Agrawal A, Ali M and Roy K 2020 *IEEE Micro* **40** 8–22 ISSN 1937-4143
[2] Barla P, Joshi V K and Bhat S 2021 *J. Comput. Electron.* 1–33 ISSN 1572-8137
[3] Kang W, Deng E, Wang Z and Zhao W 2020 *Springer Series in Advanced Microelectronics* **63** 215–229
[4] Joshi V K, Barla P, Bhat S and Kaushik B K 2020 *IEEE Access* **8** 194105–194146 ISSN 2169-3536
[5] Joshi V K 2016 *Eng. Sci. Technol.* **19** 1503–1513 ISSN 2215-0986
[6] Barla P, Shet D, Joshi V K and Bhat S 2020 *2020 5th International Conference on Devices, Circuits and Systems (ICDCS)* pp 41–45
[7] Zhao W and Prenat G 2015 *Spintronics-based Computing* (Berlin, Germany: Springer International Publishing) ISBN 978-3-319-15179-3.
[8] Barla P, Joshi V K and Bhat S 2020 *IEEE Access* **8** 6876–6889 ISSN 2169-3536
[9] Barla P, Joshi V K and Bhat S 2020 *Ain Shams Eng. J.* ISSN 2090-4479
[10] Devolder T, Hayakawa J, Ito K, Takahashi H, Ikeda S, Crozat P, Zerounian N, Kim J V, Chappert C and Ohno H 2008 *Phys. Rev. Lett.* **100** 057206 ISSN 1079-7114
[11] van den Brink A, Cosemans S, Cornelissen S, Manfrini M, Vaysset A, Van Roy W, Min T, Swagten H J M and Koopmans B 2014 *Appl. Phys. Lett.* **104** 012403 ISSN 0003-6951
[12] Hirsch J E 1999 *Phys. Rev. Lett.* **83** 1834–1837 ISSN 1079-7114
[13] Thapliyal H, Sharifi F and Kumar S D 2018 *IEEE Trans. Magn.* **54** 1–8 ISSN 1941-0069
[14] Wang Z, Zhao W, Deng E, Zhang Y and Klein J O 2015 *Phys. Status Solidi RRL* **9** 375–378 ISSN 1862-6254
[15] Wang Z, Zhao W, Deng E, Klein J O and Chappert C 2015 *J. Phys. D: Appl. Phys.* **48** 065001 ISSN 0022-3727
[16] STMicroelectronics, Manual of design kit for CMOS 65nm, 2009
[17] Gang Y, Zhao W, Klein J O, Chappert C and Mazoyer P 2011 *IEEE Trans. Magn.* **47** 4611–4616 ISSN 0018-9464

Design & Simulation of S Band Planar Microwave Filter for Wireless Application

Bhumikaba Gohil[1] and Khyati Chavda[2]

[1] Department of electronic and communications, / Shantilal Shah Engineering College, Bhavnagar, Gujarat, India, [1]bhumika.g.0901@gmail.com.

[2] Department of electronic and communications, / Shantilal Shah Engineering College, Bhavnagar, Gujarat, India, [2]khyati.chavda@gmail.com

Abstract. Reducing the size of today's wireless communication systems has become an important criterion in filter design with minimal possible value of filter parameters. To achieve this goal in wireless communication system, filter design is the most important part of the signal transmission process. Conventional filters are very important for modern wireless communication systems as they do not provide the desired miniaturization and results. Recently, using metamaterials and dielectric resonators, the filter size has been drastically reduced to a very small size for practical applications. This filter is flat and lightweight. It consumes very low power consumption. In this proposed work, a compact S-band microwave band stop filter was designed using a simple technique. These filters are small in size and light in weight. The design covers the S-band and is suitable for the desired frequency for Wi – Fi and wireless LAN application.

Key words – Planar microwave filter, Metamaterial, Square split ring resonator (SSRR), Wi - Fi range, Wireless LAN application.

1. Introduction

Now, in recent year wireless communication requires compact, less complex and easy to fabricated devices which necessitated the use of microstrip structures. Microstrip filters are inevitable part of communication system. Various filters are design according to this structure. Now a day's many people have focused on filter design with square split ring resonator (SSRR) or improved SRR structures. The split-ring resonator presented by Pendry can have a negative permeability and is used for the realization of left-handed materials (LHM) with a negative refractive index. The SRRs are made up of concentric rings of non-magnetic metal such as copper. When they are generated by a magnetic field, their real part of permeability can be very positive to negative near their resonance frequency. This resonance frequency is a narrow band that can be explained by analogy with the LC resonator, where the inductance L is the length of the ring and the capacitance C is the size of the cut. Because of their properties, the SRRs are used to reduce the size of the microwave filter. And these filters are also narrow-band. [8] The position and the width of the stop band filters are depend on the size, number and position of SRRs. Here SRRs with different size and position gives the dual frequency response with reduced structure of SRRs in S – band. By varying the size of SRRs we can varies the frequencies for desired frequency of response. And also by varying the distance between SRRs we can achieve good frequency of response.

2. Design of stop band filter

The Design of stop band filter with different geometry of SRRs and single feed line is proposed for different variation in results. Here in this proposed work by using reduced structure of SRRs designed stop band filter with dual frequency for wireless LAN application in S – band. Basic ideology behind this design is come from one of the reference paper which covers design for stopband filter with single frequency [1]. Here designed the stop band filter for dual frequency by changing size of ring and some of the parameters with compact size of filter. Stop band filter with compact size is the main research motive for today's technology. For this here chose single microstrip line with two parallel sides of n-cells of SRRs. By which filter size is drastically reduced to very small size and getting good frequency of response.

2.1 Design for first iteration of stop band filter

Here for first iteration of stop band filter there is only 1 pair of SRRs arranged parallel to microstrip line. Which used to exhibits negative permeability and give good frequency of response. The gap between strip line and resonators are varies from design to others but it is constant between cells in one design. This design gives one frequency of response at Wi - Fi range in WLAN channel.

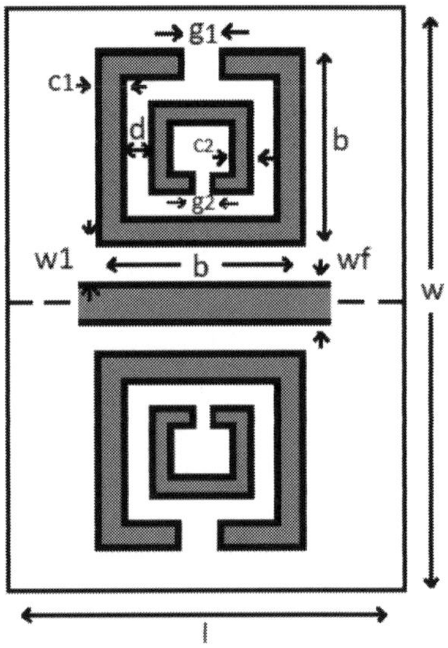

Parameters	b	c1	c2	d	g1	g2	wf	w_1	w	l
Values(mm)	13.268	1	0.5	2	1	0.5	4.8	0.4	40	22

2.2 Design for second iteration of stop band filter

Here for second iteration of stop band filter there is two pairs of SRRs are arranged parallel to microstrip line. These pair of SRRs consist same parametric loop. These loops of SRRs arranged at some distance from each other to get good frequency of response. This design gives one cut of frequency with higher insertion loss at Wi – Fi range in WLAN channel.

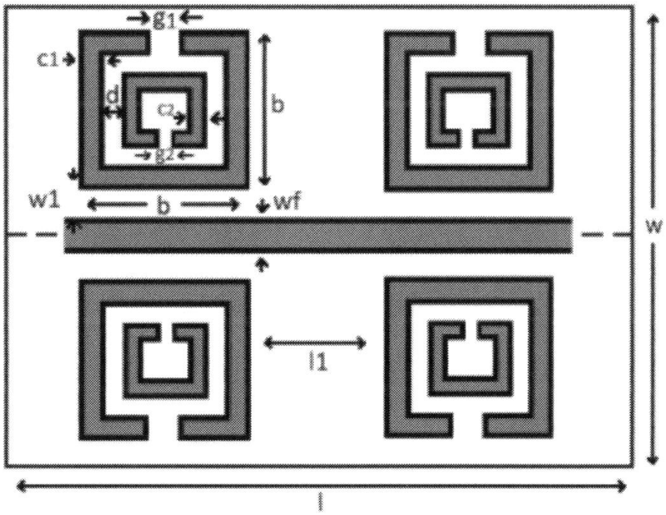

Parameters	b	c1	c2	d	g1	g2	wf	w_1	l_1	w	l
Values(mm)	13.2281	1	0.5	2	1	0.5	4.8	0.4	11	40	52

2.3 Design for final iteration of stop band filter

Here for final iteration of stop band filter two different parametric pair of SRRs are arranged parallel to microstrip line. To get dual frequency of response here the size variation is caused in second pair of SRR. So, by doing this variation we get dual frequency of response at reduced structure of SRRs. In this design we get required range of insertion loss and return loss at dual frequency for Wi – Fi and WLAN application.

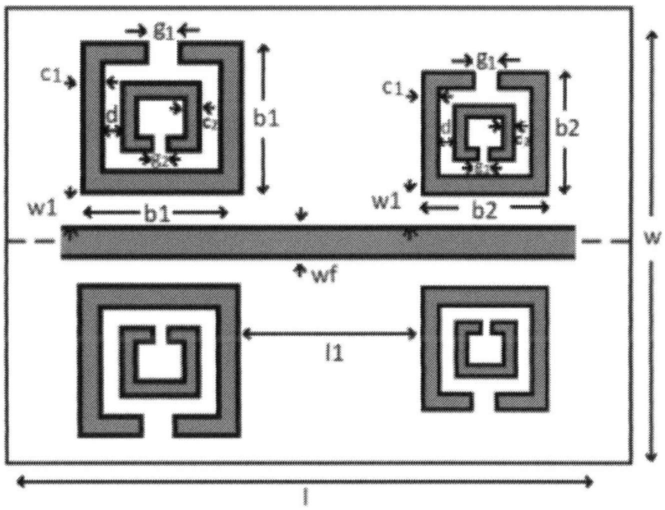

Parameters	b1	b2	c1	c2	d	g1	g2	Wf	w_1	l_1	w	l
Values(mm)	13.2281	9.384	1	0.5	2	1	0.5	4.8	0.4	11	36	48

3. Simulation and result

The different geometry of band stop filter gives different aspects of output. Here are some simulated designs and their results which gives varieties of responses. For simulation of these designs I used CST studio suit software. Which is basic software for filter design and easy to use. Here the iteration of band stop filter is chosen according to their geometry and compact size. Here to design this filter ground and upper patch of SRRs and feed line material is copper and substrate material is a RogerRT5880. These filters give response at 2.4 GHz and 3.6 GHz frequencies. Among all three iteration of this design first two design of band stop filters gives response at only 2.4 GHz Wi – Fi frequency range. And last and final iteration of these designs gives both frequency of response for Wi – Fi and WLAN application.

Here the Rogers RT5880 (lossy) substrate was used with a thickness of 1.57 mm and a relative permittivity of 2.2. The copper thickness is 0.035 mm. The simulation of final design shows good performance at 2.4 GHz (2.36 to 2.44 GHz) and 3.6 GHz (3.55 to 3.66 GHz). There are two waveguide port given to both the side of feed line for simulation. Overall thickness of this filter is 1.64 mm.

Here I choose Roger RT5880 (lossy) substrate in place of most common material FR – 4. Because Roger gives better compactness compare to FR – 4. And by using this material insertion loss also decreases at the reduced structure of SRRs. By using FR – 4 material size of filter will increases for better response. And tangent loss of FR-4 material also high compare to Roger material which is results in a slow substrate. So, as tangent loss of substrate material is decreases the substrate becomes fast substrate.

3.1 Simulated design and result of first iteration

For first iteration of band stop filter we get insertion loss up to -22.465 dB and return loss up to 0.68 dB. Here B.W. of this filter is up to 85 MHz and one rejection is observed and located at 2.4 GHz.

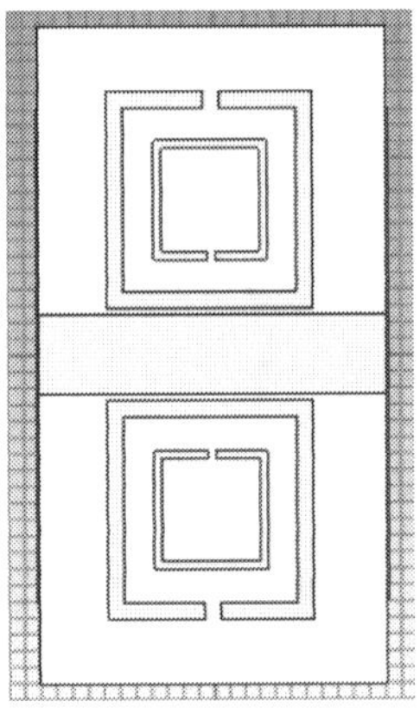

Fig.-1(a) Top-view of design-1

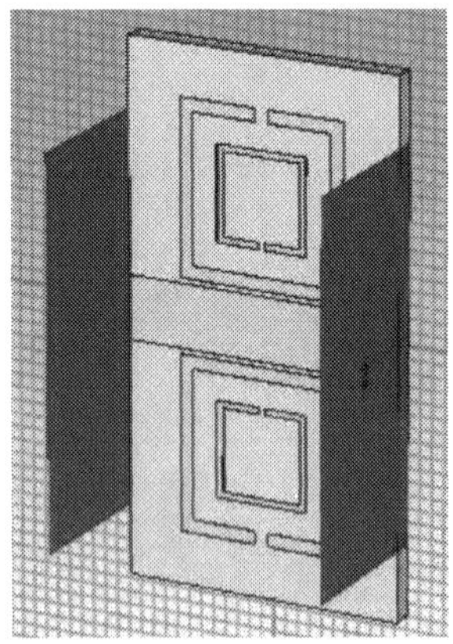

Fig.-1(b) Cross-view of design-1

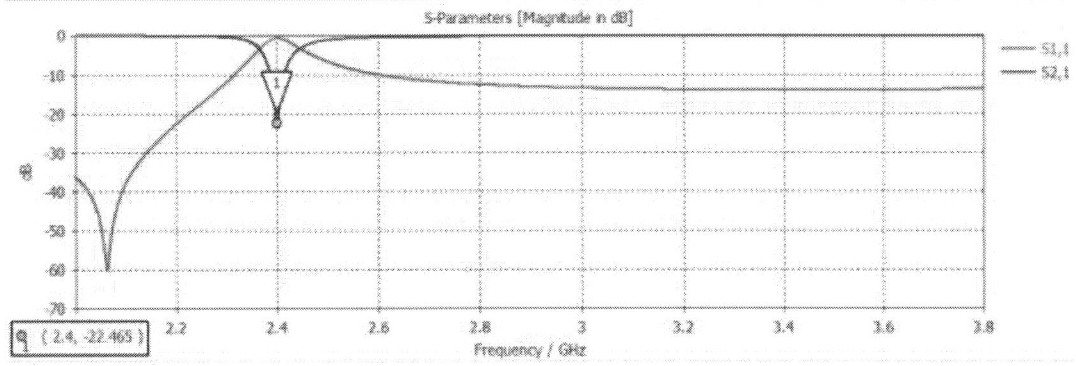

Fig.-1(c) S – parameters of result-1

3.2 Simulated design and result of second iteration

For second iteration of band stop filter we get insertion loss up to -37.915 dB and return loss up to 0.56 dB. Here B.W. of this filter is up to 150 MHz and one rejection is observed and located at 2.4 GHz.

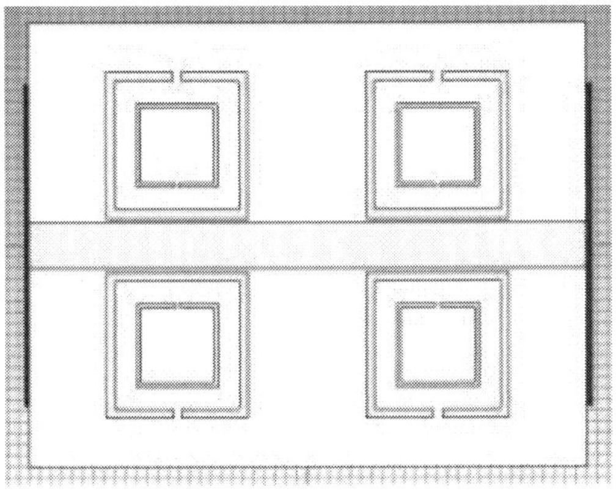

Fig.-2(a) Top-view of design-2

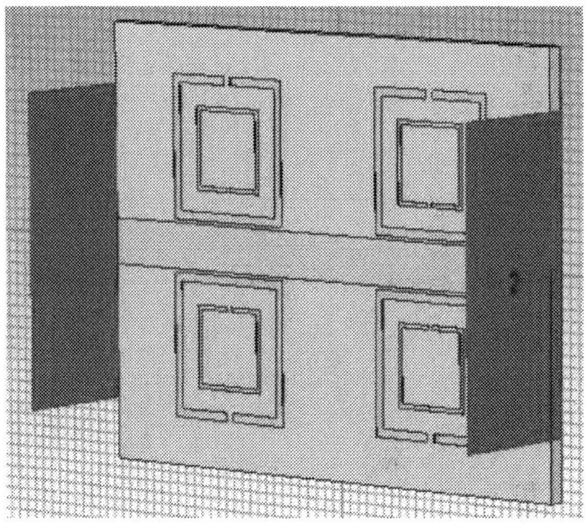

Fig.-2(b) Cross-view of design-2

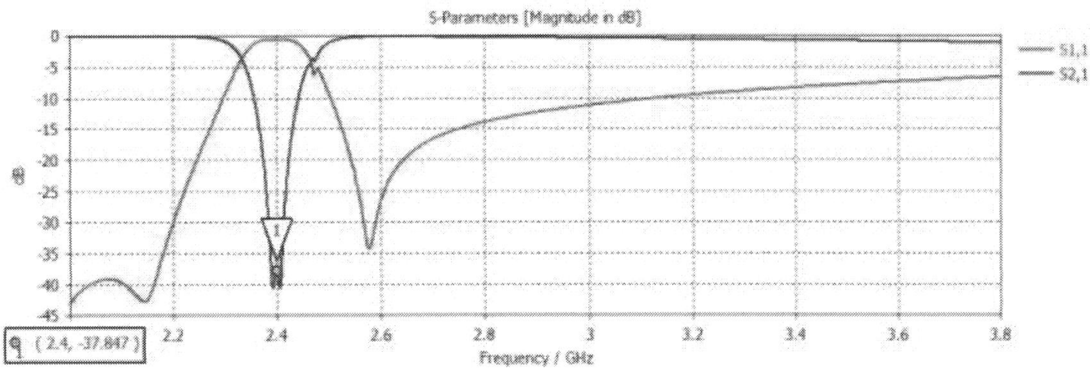

Fig.-2(c) S – parameters of result-2

3.3 Simulated design and result of final iteration

For final iteration of band stop filter we get insertion loss at 2.4 GHz frequency up to -22.566 dB and at 3.6 GHz frequency insertion loss is -21.568 dB. Here B.W. of this filter is 85 MHz and 100 MHz at 2.4

IOP Conf. Series: Materials Science and Engineering 1187 (2021) 012016 doi:10.1088/1757-899X/1187/1/012016

GHz and 3.6 GHz frequencies. The rejection observed and located at 2.4 GHz and 3.6 GHz frequencies.

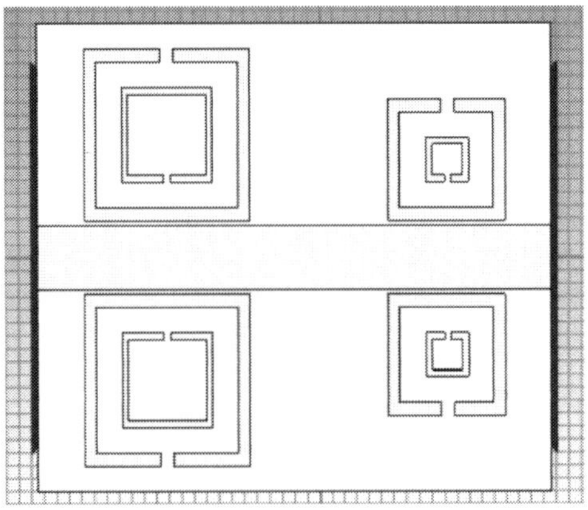

Fig.-3(a) Top-view of final design

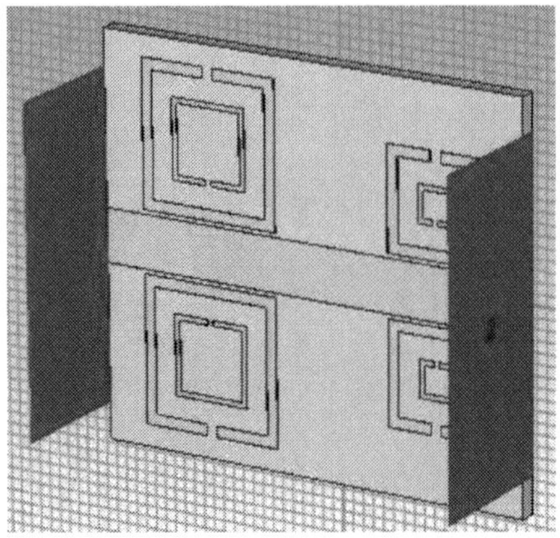

Fig.-3(b) Cross-view of final design

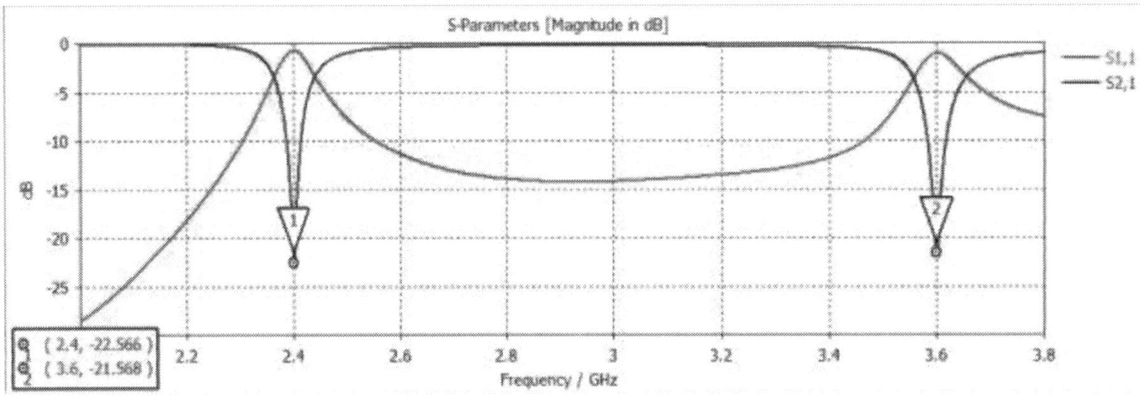

Fig.-3(c) S – parameters of final design

Insertion loss:-

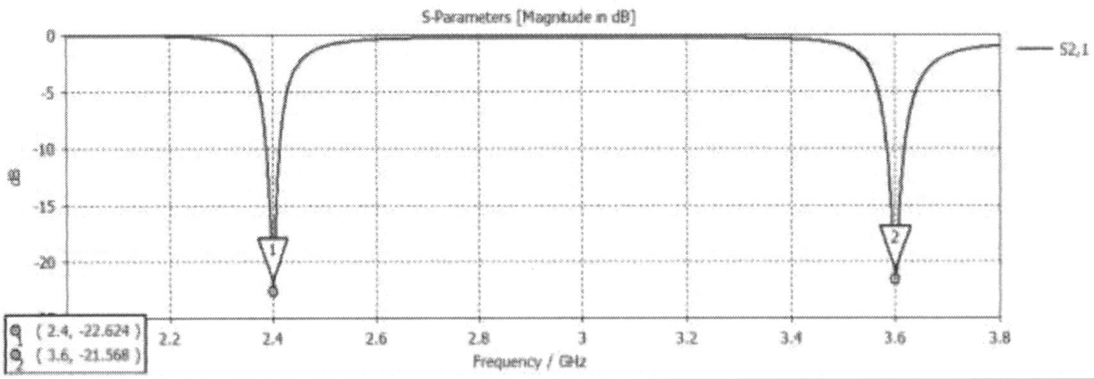

Return loss:-

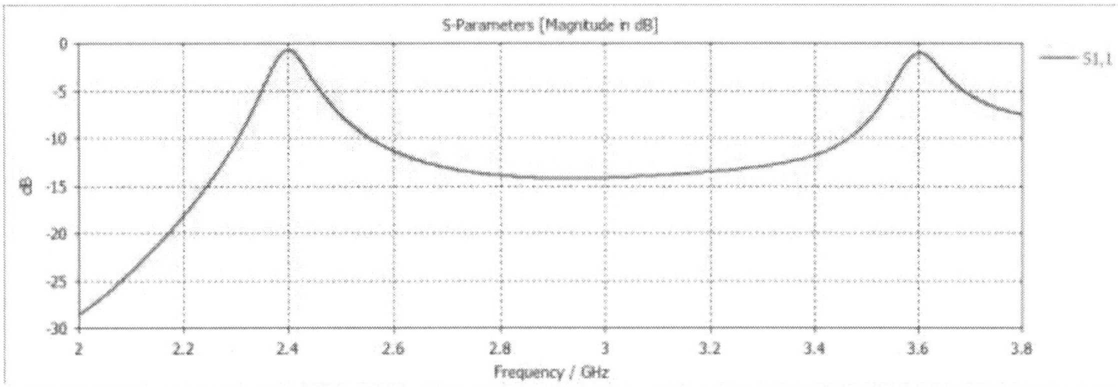

Fig.-3(d) Return loss of final design

4. Fabrication work and result

Here fig.-4(a) shows the fabricated microwave band stop filter and fig.–4(b) shows the tested result of S-parameters. This result is measured in VNA of key sight at GEC Bhavnagar. Fabrication shows some deviation in frequency response because of the fabrication loss and loss. The overall result of fabricated filter was good and exists in the S-band. Here fabricated result shows result at 2.48 GHz and 3.69 GHz frequencies. The insertion loss at these frequencies is around 16.83dB and 21dB. The overall size of filter is around 4 cm.

Fig.-4(a) Fabricated filter

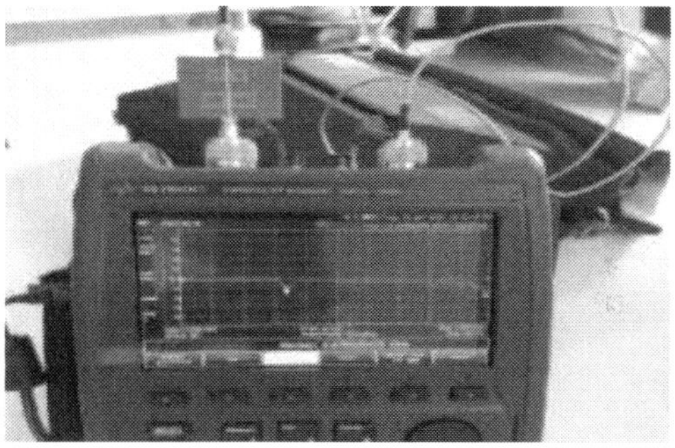

Fig.-4(b) Fabricated S-parameters

5. Conclusion

This work is based on effect of metamaterial component on stop band filter using SRRs. The filters are loaded with two parallel sides of split ring resonators (SRRs). In which numbers and size of resonators varies in every design to get different frequency of response. Here by improving the same sized filter with some parameter changes in their geometry of design getting good frequency of response for Wi-Fi and WLAN application in S – band. By doing this, filter size becomes extremely compact. And these filters are becomes more suitable for new technology in wireless application. The fractional B.W. of proposed work is 3.54% (2.36 to 2.44 GHz) and 3.05% (3.55 to 3.66 GHz). Here you can see the fractional B.W. is lower than 5%. This causes sharp response at lower frequencies. Finally narrow band stop filter is designed and simulated for Wi-Fi and WLAN applications.

6. References

[1]. A. H. Reja and S. N. Ahmad, "Studying the Effects of Metamaterial Components on Microwave Filters," IEEE publication, 2015.

[2]. S. Jarchi, M. Khalily and R. Tafazolli, "Effects of Metamaterial Loading on Miniaturization of Loop and Open Loop Microstrip Filters," 2020 International Conference on UK-China Emerging Technologies (UCET), 2020, pp. 1-4, doi: 10.1109/UCET51115.2020.9205325.

[3]. X. Cao, B. Luo, Y. Zhu, Z. Xia and Q. Cai, "Research on the Defected Ground Structure With Von Koch Snowflake Fractals," in IEEE Access, vol. 8, pp. 32404-32411, 2020, doi: 10.1109/ACCESS.2020.2973295.

[4]. Khani, Shiva & Danaie, Mohammad & Rezaei, Pejman & Shahzadi, Ali. (2019). Compact Ultra-Wide Upper Stopband Microstrip Dual-Band BPF Using Tapered and Octagonal Loop Resonators. Frequenz. 74. 10.1515/freq-2019-0060.

[5]. R. Chen, S. Wong, J. Lin and Y. He, "Miniaturized Microwave Filter Using Circular Spiral Resonators in a Single Metal Cavity," 2019 IEEE MTT-S International Microwave Symposium (IMS), 2019, pp. 1347-1350, doi: 10.1109/MWSYM.2019.8700992.

[6]. G. Vasudev, A. S., A. M. and S. K. Menon, "Direct Coupled Spiral Resonator for Band Stop Filter Applications," 2020 7th International Conference on Signal Processing and Integrated Networks (SPIN), 2020, pp. 520-523, doi: 10.1109/SPIN48934.2020.9070325.

[7]. A. Marzah and J. S. Aziz, "Design and Implementation of Compact Dual-Mode BPF Using Moore Fractal Geometry," 2020 Fourth World Conference on Smart Trends in Systems, Security and Sustainability (WorldS4), 2020, pp. 341-345, doi : 10.1109/WorldS450073.2020.9210378.

[8]. Carver, J., Reignault, V. & Gadot, F. Engineering of the metamaterial-based cut-band filter. *Appl. Phys. A* **117,** 513–516 (2014).

[9]. M. K. T. Al-Nuaimi and W. G. Whittow, "Compact microstrip band stop filter using SRR and CSSR: Design, simulation and results," Proceedings of the Fourth European Conference on Antennas and Propagation, 2010, pp. 1-5.

[10]. I. Gil, J. Bonache, J. Garcia-Garcia, F. Falcone and F. Martin, "Metamaterials in microstrip technology for filter applications," 2005 IEEE Antennas and Propagation Society International Symposium, 2005, pp. 668-671 Vol. 1A, doi: 10.1109/APS.2005.1551409.

[11]. G. Vasudev, A. S., A. M. and S. K. Menon, "Direct Coupled Spiral Resonator for Band Stop Filter Applications," 2020 7th International Conference on Signal Processing and Integrated Networks (SPIN), 2020, pp. 520-523, doi: 10.1109/SPIN48934.2020.9070325.

[12]. K. L. Chung, R. Liu, Y. Li and B. Feng, "A Study of Composite Microstrip/Coplanar-Waveguide Transmission Line for Miniaturization of Microwave Planar Circuits," 2019 Cross Strait Quad-Regional Radio Science and Wireless Technology Conference (CSQRWC), 2019, pp. 1-3, doi: 10.1109/CSQRWC.2019.8799255.

[13]. Jindal, Swati & Sharma, Jigyasa. (2012). Review of Metamaterials in Microstrip Technology for Filter Applications. International Journal of Computer Applications. 54. 48-54. 10.5120/8548-2108.

[14]. I. Gil, J. Bonache, J. Garcia-Garcia, F. Falcone and F. Martin, "Metamaterials in microstrip technology for filter applications," 2005 IEEE Antennas and Propagation Society International Symposium, 2005, pp. 668-671 Vol. 1A, doi: 10.1109/APS.2005.1551409.

[15]. Microstrip Lines and Slotlines, Third Edition (Artech House Microwave Library (Hardcover))

[16]. Microwave integrated circuit component design using MATLAB (S. Raghavan).

[17]. Engheta Metamaterials-Physics and Engineering Explorations_0471761028. (John Wiley & Sons, Inc.)

IOP Conf. Series: Materials Science and Engineering 1187 (2021) 012017 doi:10.1088/1757-899X/1187/1/012017

A Different Shaped Radiating Element Wide Band Multi-Band Massive MIMO Antenna for 5G/WLAN applications with Enhanced Correlation Coefficient

Shrenik Suresh Sarade[1], **S. D. Ruikar**[2]

[12]Electronics Engineering, WCE, Sangli, India

[1]shreniks2k7@rediffmail.com

Abstract. In this paper, a different shaped 8-element multiband multiple input multiple output (MIMO) massive antenna is investigated for 5G and various applications. It can be operated on various wireless communication application such as WLAN (5.15GHz to 5.30GHz and 5.73GHz to 5.85GHs) and 5G band & C-band (4 to 8GHz) for satellite communication with wideband. The 8-element Massive MIMO antenna is designed for 6GHz. This antenna is working on multiband with wideband bandwidth. The defected ground structure (DGS) systems are used for enhancement of return loss, isolation, total active reflection coefficients (TARC), correlation coefficients (CC) and envelope correlation coefficients (ECC). The rectangular shaped DGS is used in the ground plane for improvement of various parameter of antenna. This antenna provides return loss (RL) and isolation is less than -10dB.
Keywords. MIMO, DGS, TARC, CC.

1. Introduction

In recent years, wireless technology is rapidly developed with increasing tremendously of its users. Therefore, there is high demands of data rate and reliability to handle the growth of users. MIMO antenna system has bring the attention of researchers for the data rate, capacity of channel and reliability. For good communication, there is a requirement of higher bandwidth. Therefore, it is required to design the antenna which operates on wideband bandwidth. The MIMO technology is mostly useful for upcoming 5G mobile systems, as well various application of wireless system. Meanwhile, the today's wireless systems has required the more antennas, which operates on multiple frequencies. But, more antenna on single substrate which increases the isolation. Hence, that effects on the radiation, efficiency and overall performance of antenna. So, for improvement of performance of the antenna, the isolation is required to improve. Therefore, enhancement of isolation, CC and ECC values of MIMO antenna are critical task. The different techniques are used for improvement of isolation of MIMO antenna. The decoupling method is used for enhancement of isolation [1-3]. In [4-7], a different shaped DGS is used on the ground to change the characteristics of the microstrip line to reduce isolation between the radiating elements. In [8], electromagnetics bandgap (EBG) was designed for improvement of isolation. The mutual coupling is enhanced using the metamaterial [9] and the neutralization line [10-11]. The isolation is reduced using the parasitic elements between the radiating element [12-13] and the using the traditional methods [14].

To summarize previous paper, in this paper the multi-band wideband massive MIMO antenna that consists of different shaped radiating element is designed for various application. The different shaped 8-element Massive MIMO antenna are investigated for mm-wave. The DGS is used for improvement

IOP Conf. Series: Materials Science and Engineering 1187 (2021) 012017 doi:10.1088/1757-899X/1187/1/012017

of isolation between the radiating elements. The rectangular shaped DGS is used on the ground plane. This antennas are operated multiband, hence covers the different application in single antenna. Also, this antennas are operated on wideband bandwidth. These antennas works at multi- band with less than -10dB RL and -10dB isolation with wideband operation.

2. Antenna design procedure

The transmission line model (TLM) is used for calculation of geometric structure of radiating element and ground plane of antenna. This antenna is designed using the following formula and analysed using antenna parameter.

2.1. TLM technique

The $length\,(L_p)$ and $width\,(w_p)$ radiating patch are computed using the following equation 1 to equation 5,

$$Width\,(W_p) = \frac{1}{2f_0\sqrt{\mu_0\epsilon_0}}\sqrt{\frac{2}{\epsilon_{0r}+1}} \tag{1}$$

$$L_\nabla = \frac{4.12}{10}h\,[\frac{\left(\epsilon_{eff}+\frac{3}{10}\right)\left(\frac{W_p}{h}+\frac{2.64}{10}\right)}{\left(\epsilon_{eff}-0.258\right)\left(\frac{W_p}{h}+0.8\right)}] \tag{2}$$

$$L_{eff} = L_p + 2L_\nabla \tag{3}$$

$$f_0 = \frac{1}{2L_p\sqrt{\epsilon_{0r}}\sqrt{\mu_0\epsilon_0}} \tag{4}$$

$$\epsilon_{eff} = \frac{\epsilon_{0r}+1}{2} + \frac{\epsilon_{0r}-1}{2}\,[1+12\,\frac{h}{W_p}]^{-1/2} \tag{5}$$

Where,

$W_p = Width\ of\ Patch$

$L_p = Length\ of\ Patch$

The $length\,(L_g)$ and $width\,(W_g)$ of ground plane are computed using the following equation 6 to equation 7,

$$L_g = L_p + 6h \tag{6}$$

$$W_g = W_p + 6h \tag{7}$$

Where,

$L_g = Length\ of\ Ground$

$W_g = Width\ of\ Ground$

The $length\,(L_f)$ and $width\,(W_f)$ for inset feed line (IFL) are computed using the equation 8 and 10 to excite the radiating patch of antenna.

$$Feed\ line\ length\,(L_f) = \frac{\lambda_g}{4} \tag{8}$$

$$Guided\ wavelength\,(\lambda_g) = \frac{\lambda_0}{4} \tag{9}$$

The $width\,(W_f)$ of the IFL required to excite the radiating patch is given by,

$$W_f = 0.636h \times \left[\frac{592.19}{Z_0\sqrt{\epsilon_{0r}}} - 0.7 - \ln\left(\frac{2\times 377\pi}{2Z_0\sqrt{\epsilon_{0r}}}+1\right)\right] + \frac{2h}{\pi}\times[\frac{(\epsilon_{0r}-1)}{2\epsilon_{0r}}\left(\ln\left(\frac{592.19}{Z_0\sqrt{\epsilon_{0r}}}-1\right)-\frac{0.6}{\varepsilon_{0r}}\right)] \tag{10}$$

Where,

IOP Conf. Series: Materials Science and Engineering 1187 (2021) 012017 doi:10.1088/1757-899X/1187/1/012017

$Characteristics\ Impedance = Z_0 = 50\Omega$

2.2. MIMO antenna parameter

1) *TARC*

This is calculated using equation 11.

$$TARC\ (\tau) = \frac{\sqrt{[(S_{11}+S_{12}e^{\theta})^2 + (S_{21}+S_{22}e^{\theta})^2}}{\sqrt{2}} \qquad (11)$$

.Where:

$\theta = Phase\ of\ the\ input\ feeding$

$S_{11}\ and\ S_{22} = Reflection\ coefficient$

$S_{12}\ and\ S_{21} = Isolation$

2) $CC\ (\rho_{xy})and\ ECC(\rho_{ecc_{xy}})$

These are calculated by using the equation-12 for 2-element.

$$\rho_{xy}{}^2 = \rho_{ecc_{xy}} = [\frac{(S_{11}S_{12}+S_{21}S_{22}}{\sqrt{(1-S_{11}{}^2-S_{21}{}^2)(1-S_{22}{}^2-S_{12}{}^2)}}]^2 \qquad (12)$$

Where:

$\rho_{xy} = CC\ value\ between\ the\ 1^{st}(x)and\ 2^{th}\ (y)\ element$

$\rho_{ecc_{xy}} = ECC\ value\ between\ the\ 1^{st}(x)and\ 2^{th}\ (y)element$

3) Diversity Gain (DG)

DG of the antenna is find out using equation 13.

$$DG = 10\sqrt{1-(0.99ECC)^2} \qquad (13)$$

4) Bandwidth

Bandwidth of antenna is calculated by using equation 14.

$$\%BW = \{\left(\frac{F_h - F_l}{F_0}\right) \times 100\} \qquad (14)$$

$$F_0 = \{F_l + \left(\frac{F_h - F_l}{2}\right)\} \qquad (15)$$

Where:

$BW = Bandwidth$

$F_h = Higher\ frequency$

$F_l = Lower\ frequency$

$F_0 = Center\ frequency$

3. Antenna Design and analysis

In this paper, 8-element massive MIMO antennas are designed.

3.1. Design of Different Shaped 8-radiating element massive MIMO antenna.

The 8-element Massive MIMO antenna using DGS is designed on FR4 substrate as shown in figure 1. The characteristics of FR4 substrate is shown in table-1. This antenna consists of 8-radiating element (P1 to P8), patch-1 to patch-8 is designed for 6GHz frequency (mm-wave) as shown in figure 1 and figure 2. The geometrical structure of radiating element for 6GHz frequency is as shown in figure 2 and table 1. Every radiating element is energized using the inset feed techniques having dimensions are shown in table 1. The antenna dimensions are 32.58mm × 74mm × 1.6mm. The rectangular shape structure has used as DGS (G1 to G8) in ground plane as shown in figure 3 having dimensions are shown in figure 4 and table 2. The spacing between the radiating elements is less than $\frac{\lambda}{2}$ with respected to 6GHz frequency.

Table 1. Dimensions of Massive MIMO antenna

Parameter	Value of Parameter
Frequency	6 GHz
Dielectric Constant (ϵ_{0r})	4.4
Height of Substrate	1.6mm
Patch Width	W_p = 15.21mm
Length of patch	L_p = 11.34 mm
Width (W_f) of inset feed line	M = 2.942 mm
Length (L_f) of inset line	N = 6.387 mm
Length "K"	K= 12.69mm

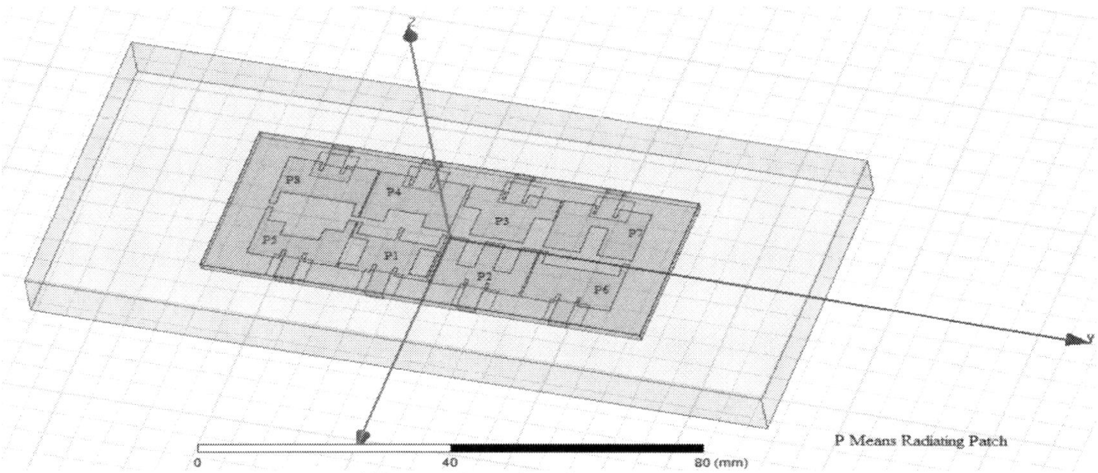

Figure 1. Different Shaped 8-element Massive MIMO antenna

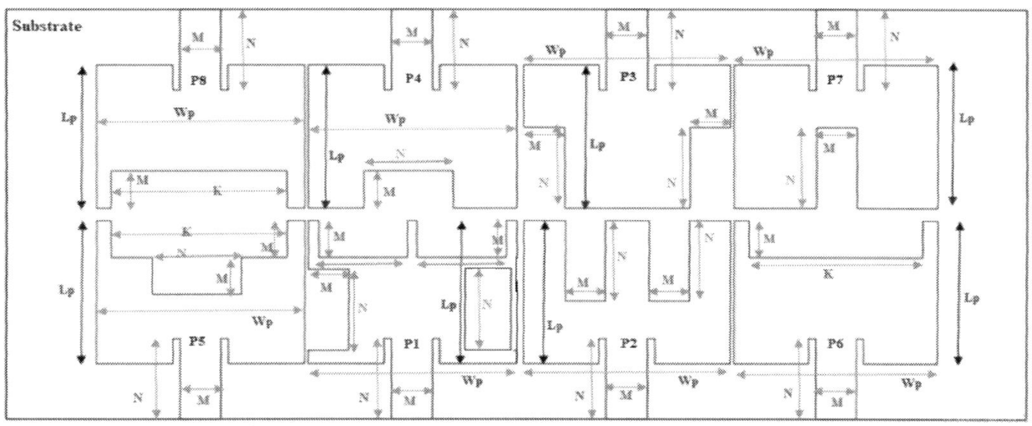

P1 to P8 stands for Radiating Patch

Figure 2. Top side of antenna geometry

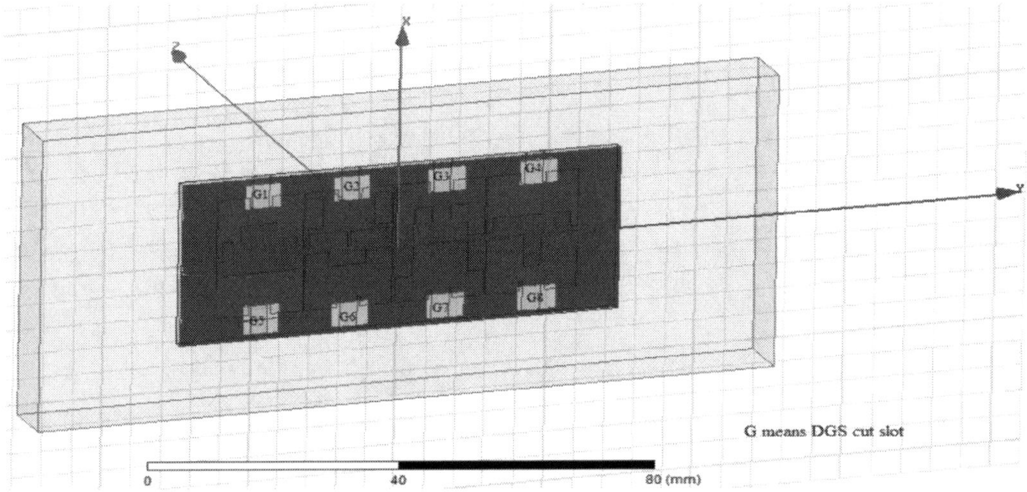

Figure 3. Ground Plane with DGS cut slot

Table 2. Dimensions of Ground plane with DGS

Parameter	Values in mm
L_g	32.58
W_g	74
Length of DGS (G1 to G8) cut slot	A= 3.884
Width of DGS (G1 to G8) cut slot	B= 6.38
'C' and 'E'	8.84
'D'	9.64
'G' and 'F'	10.52
'Z'	1.058

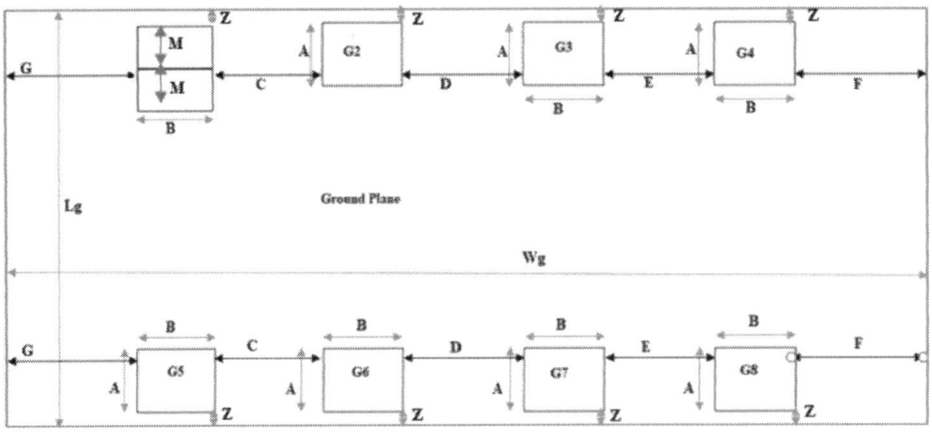

Figure 4. Bottom side (Ground plane) with DGS geometry

After analysis of antenna using HFSS software, the analysis of RL and isolation are obtained as shown in following figures. This antenna is operated on multiband for the 5.0GHz, 5.5GHz, 6.0GHz, 6.1GHz, 6.3GHz, 6.4GHz, 6.7GHz and 6.9GHz frequencies.

This antenna is consists of 8 different shaped radiating element and all 8 different shape are placed on the substrate. Which forms the 8-element Massive MIMO antenna. This antenna is operated on multiband with wideband application. The parameter RL and isolation is enhanced using DGS structure. Also, the all resonating band are operated on wide band (Bandwidth is in between 1% to 20%) with respect to 6GHz frequency.

The RL S_{11}, S_{22}, S_{33}, S_{44}, S_{55}, S_{66}, S_{77} *and* S_{88} and relative isolation is as shown in following figure 5 to figure 12 with in the table 3 to table 10 respectively (m1 and m2 are show points on graph of RL and isolation). The bandwidth (% bandwidth) of antenna is calculated using the equation 14 and equation 15.

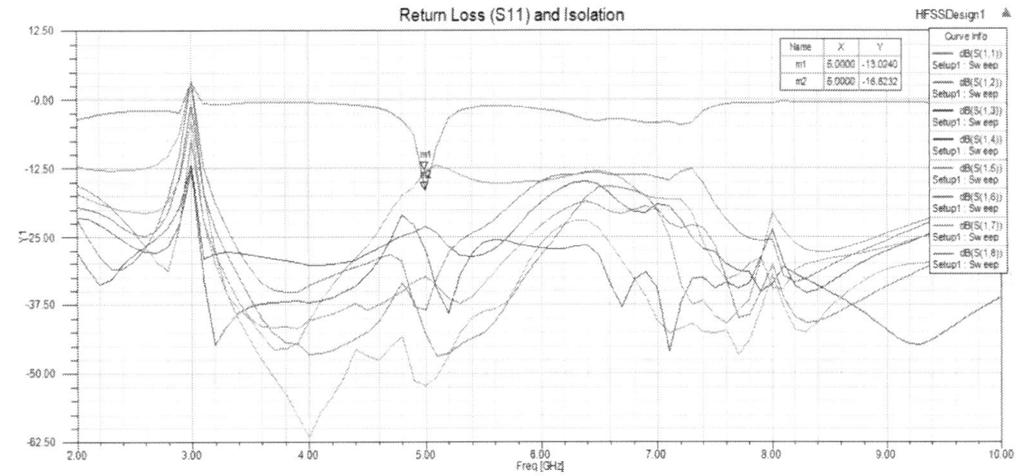

Figure 5. RL (S_{11}) and related isolation

Table 3. The results of radiating patch-1

Parameter	Operating Frequency (in GHz)	Values
S_{11} (m2)		-16.62 dB
S_{12}, S_{13}, S_{14}, S_{15} (m1), S_{16}, S_{17} and S_{18}	5	Less than -13.02 dB
Bandwidth		132MHz (2.19%)

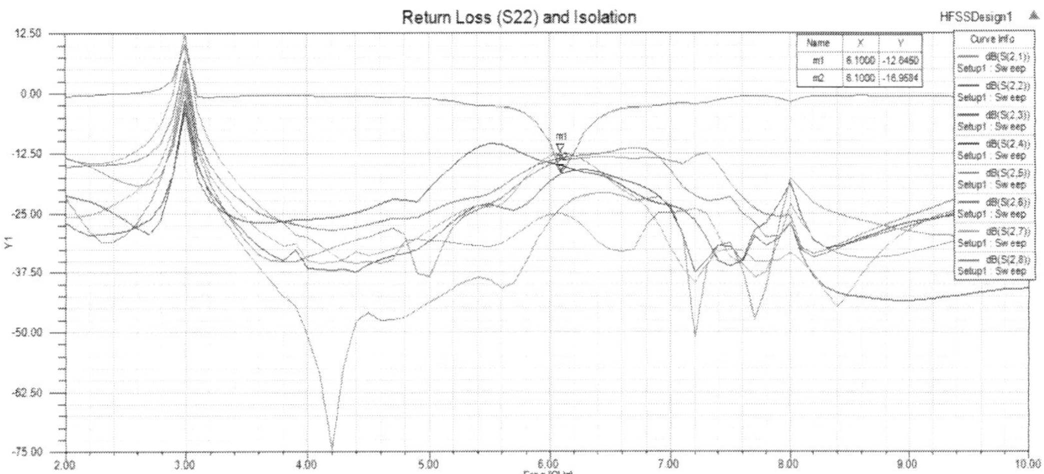

Figure 6. RL (S_{22}) and related isolation

Table 4. The results of radiating patch-2

Parameter	Operating Frequency (in GHz)	Values
S_{22} (m2)		-16.96 dB
S_{21}, S_{23}, S_{24}, S_{25}, S_{26}, S_{27} (m1) and S_{28}	6.1	Less than -12.64 dB
Bandwidth		294MHz (4.9%)

Table 5. The results of radiating patch-3

Parameter	Operating Frequency (in GHz)	Values
S_{33} (m2)		-11.41 dB
S_{31}, S_{32} (m1), S_{34}, S_{35}, S_{36}, S_{37} and S_{38}	5.5	Less than -10.55 dB
Bandwidth		152MHz (2.53%)

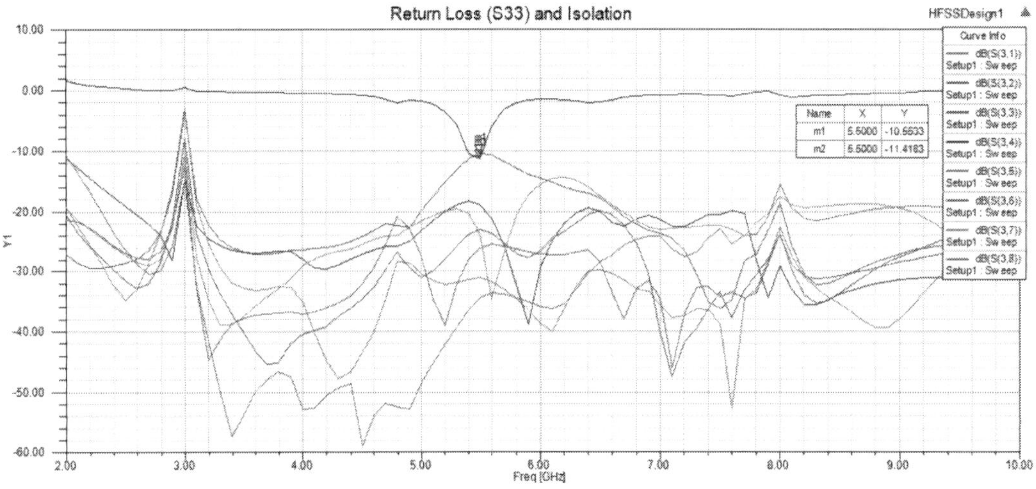

Figure 7. RL (S_{33}) and related isolation

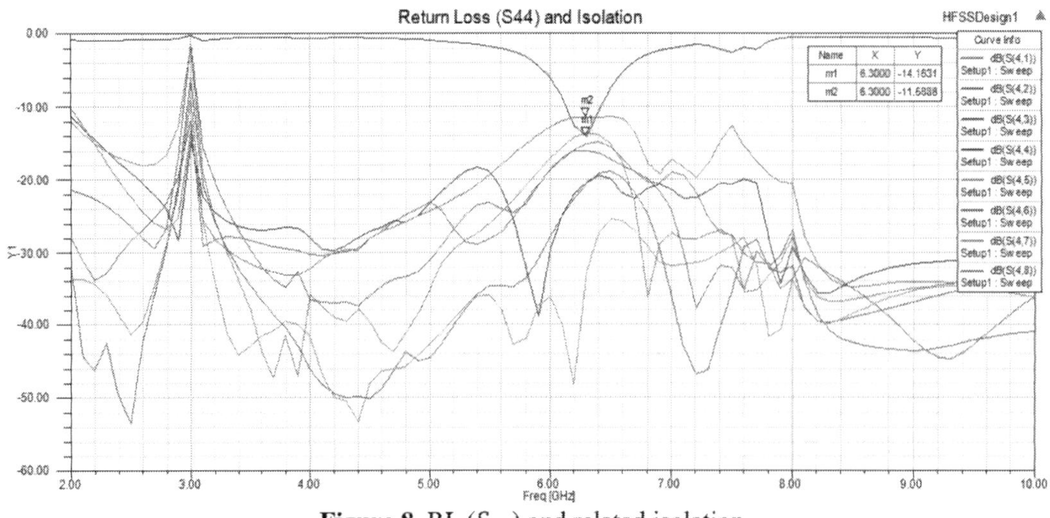

Figure 8. RL (S_{44}) and related isolation

Table 6. The results of radiating patch-4

Parameter	Operating Frequency (in GHz)	Values
S_{44} (m1)		-14.16 dB
S_{41}, S_{42}, S_{43}, S_{45}, S_{46}, S_{47} and S_{48} (m2)	6.3 GHz	Less than -11.58 dB
Bandwidth		98 MHz (1.63%)

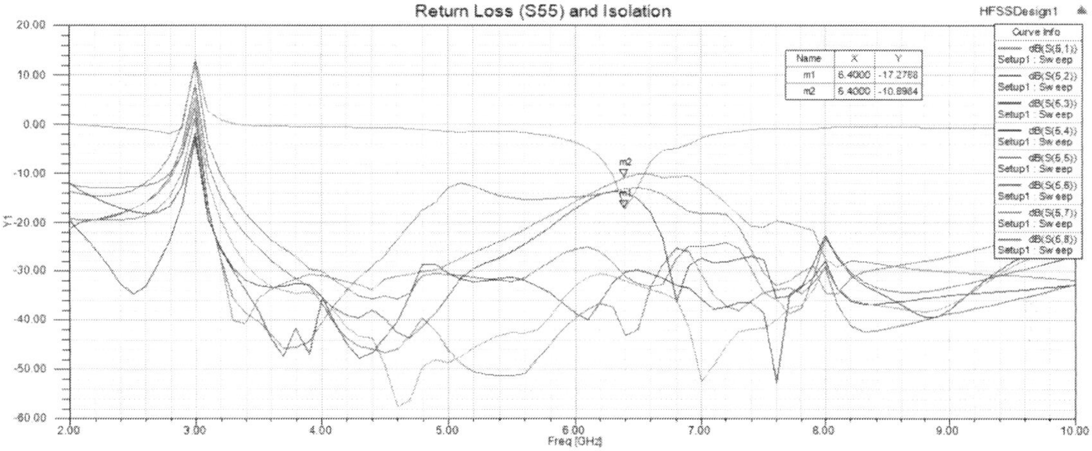

Figure 9. RL (S_{55}) and related isolation

Table 7. The results of radiating patch-5

Parameter	Operating Frequency (in GHz)	Values
S_{55} (m1)		-17.27 dB
S_{51}, S_{52}, S_{53}, S_{54}, S_{56}, S_{57} and S_{58} (m2)	6.4 GHz	Less than -10.89 dB
Bandwidth		242 MHz (4.04%)

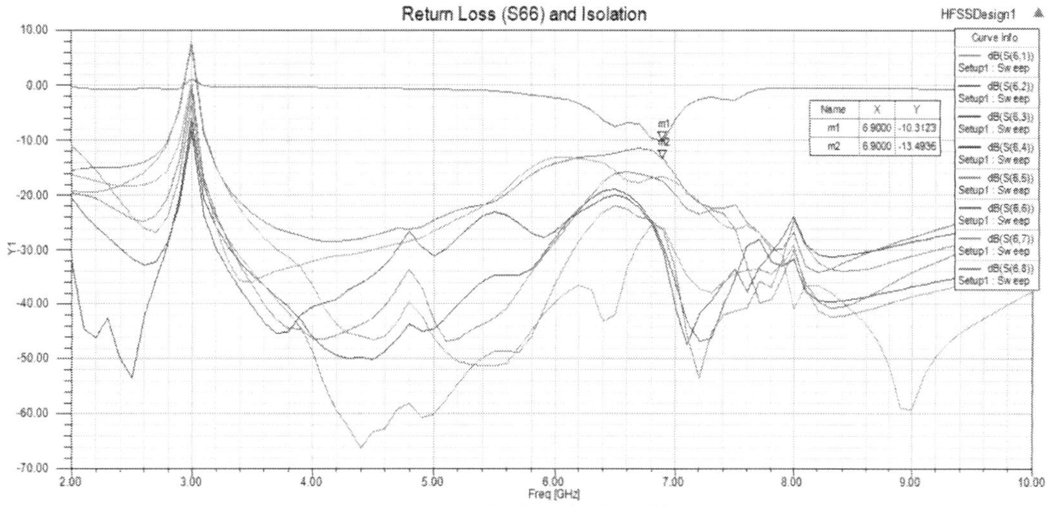

Figure 10. RL (S_{66}) and related isolation

Table 8. The results of radiating patch-6

Parameter	Operating Frequency (in GHz)	Values
S_{66} (m1)		-10.31 dB
S_{61} , S_{62} (m2), S_{63}, S_{64}, S_{65}, S_{67} and S_{68}	6.9	Less than -13.49 dB
Bandwidth		43 MHz (0.72%)

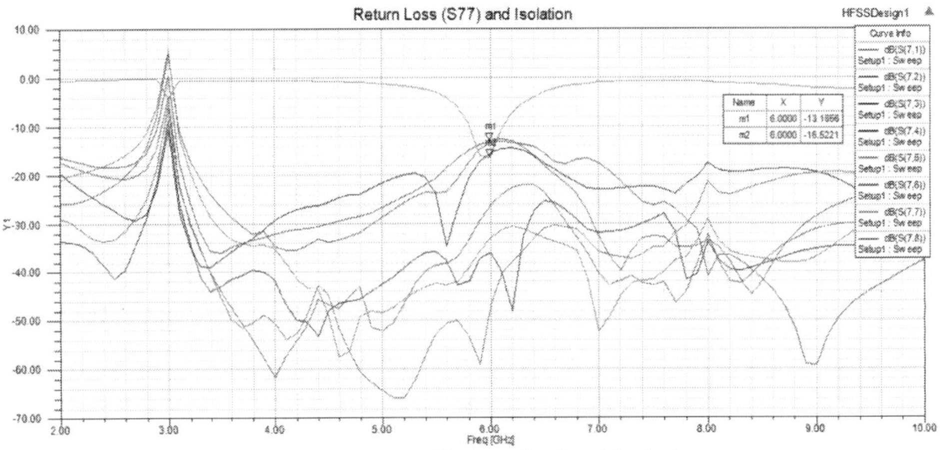

Figure 11. RL (S_{77}) and related isolation

Table 9. The results of radiating patch-7

Parameter	Operating Frequency (in GHz)	Values
S_{77} (m2)		-16.52 dB
S_{71} , S_{72}, S_{73}, S_{74}, S_{75}, S_{76} (m1) and S_{78}	6.0 GHz	Less than -13.18 dB
Bandwidth		311 MHz (5.19%)

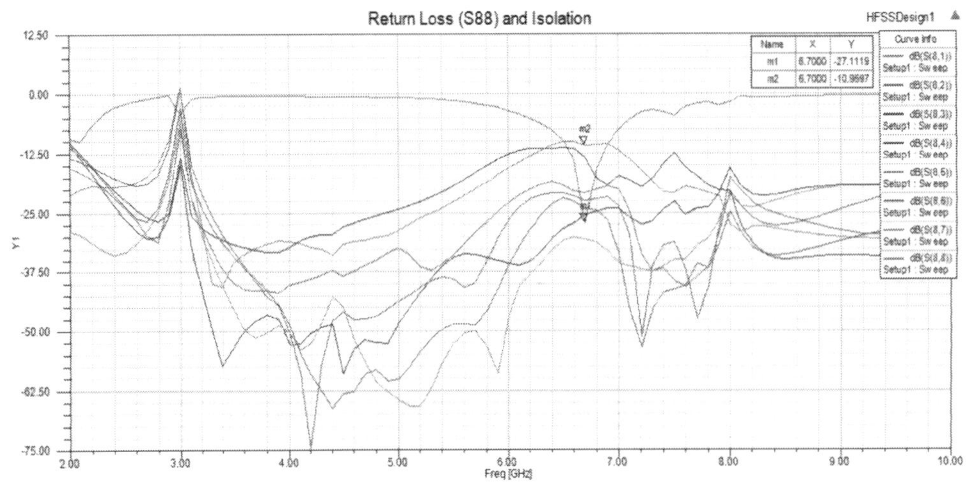

Figure 12. RL (S_{88}) and related isolation

Table 10. The results of radiating patch-8

Parameter	Operating Frequency (in GHz)	Values
S_{88} (m1)		-27.11 dB
S_{81}, S_{82}, S_{83}, S_{84}, S_{85} (m2), S_{86} and S_{87}	6.7 GHz	Less than -10.95 dB
Bandwidth		358 MHz (5.96%)

The VSWR of radiating patch is as shown in figure 13 and table 11. The gain (G_A) and directivity (D_A) of antenna is as shown in figure 14 and values shown in table 12. The radiation pattern of antenna is as shown in figure 15. The antenna parameters are as shown in table 13 which is calculated using equation 11 to equation 15.

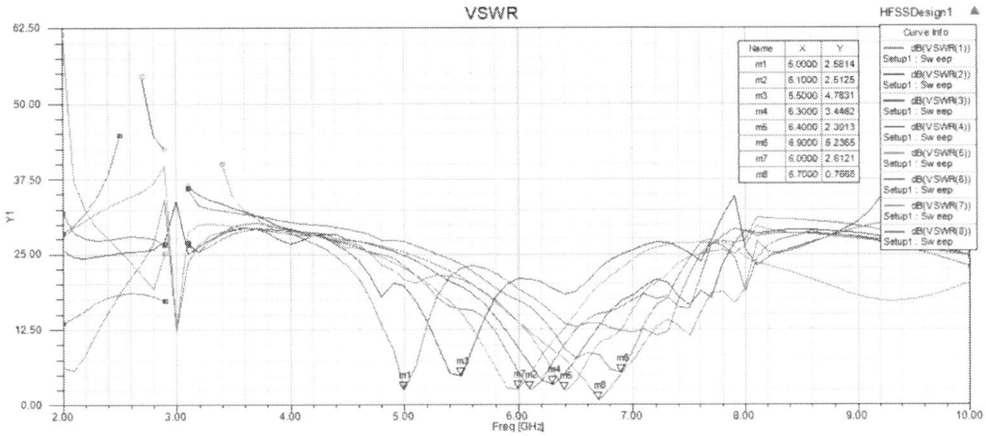

Figure 13. VSWR

Table 11. VSWR

Parameter	Frequency (GHz)	Values (dB)	Absolute Values
Patch-1 $VSWR$ (m1)	5.0	2.58	1.81
Patch-2 $VSWR$ (m2)	6.1	2.51	1.78
Patch-3 $VSWR$ (m3)	5.5	4.7	2.95
Patch-4 $VSWR$ (m4)	6.3	3.4	2.18
Patch-5 $VSWR$ (m5)	6.4	2.39	1.73
Patch-6 $VSWR$ (m6)	6.9	5.2	3.31
Patch-7 $VSWR$ (m7)	6.0	2.61	1.82
Patch-8 $VSWR$ (m8)	6.7	0.67	1.16

IOP Conf. Series: Materials Science and Engineering 1187 (2021) 012017 doi:10.1088/1757-899X/1187/1/012017

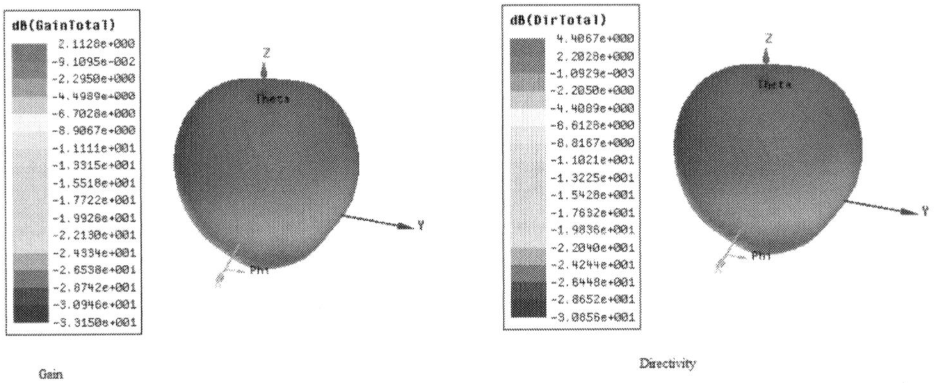

Figure 14. Gain G_A and Directivity D_A

Table 12. Gain (G_A) and Directivity (D_A) of antenna

Parameter	Values in dB
Gain	2.11
Directivity	4.40

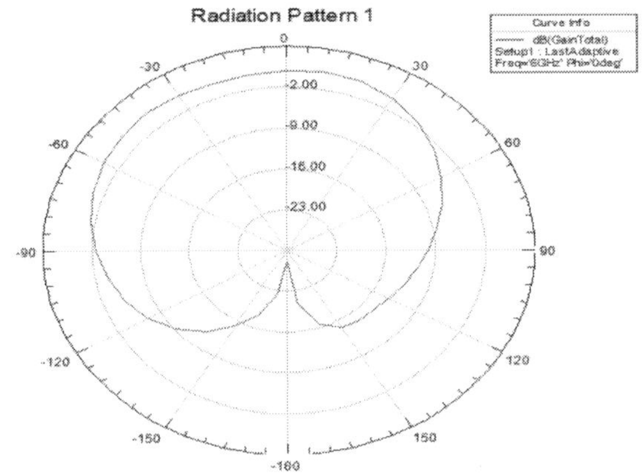

Figure 15. Radiation Pattern

Table 13. Parameter of MIMO antenna

Parameter	Values
TARC	Less than 0.159 (less than -7.98dB)
CC	Less than 0.12
ECC	Less than 0.0148
DG	Close to 10dB

4. Conclusion

The 8-different shaped element massive MIMO antenna is investigated for millimetre wave. The analysis of antenna shows that, it is operated on multi-band for 5.0GHz, 5.5GHz, 6.0GHz, 6.1GHz, 6.3GHz, 6.4GHz, 6.7GHz and 6.9GHz frequencies. This antenna is operated on wideband bandwidth. The analysis of antenna shows that less than -10dB RL and less than -10dB isolation with 2.11 dB gain, 4.40 dB directivity and bandwidth is in between 43MHz to 358MHz. The absolute VSWR values are in between

1.16 to 3.31. The analysis of antenna shows less than 0.12CC and less than 0.0148ECC. The diversity gain of antenna is close to 10dB and less than -7.98dB TARC. This antenna is used for the WLAN (5.15GHz to 5.30GHz and 5.73GHz to 5.85GHs) and 5G band & C-band (4 to 8GHz) for satellite communication.

References

[1] W. Jiang, et al., "A Dual-Band MIMO Antenna With Enhanced Isolation for 5G Smartphone Applications," in IEEE Access, vol. 7, pp. 112554-112563, 2019.
[2] L. Cui, et al., "An 8-Element Dual-Band MIMO Antenna with Decoupling Stub for 5G Smartphone Applications," in IEEE Antennas and Wireless Propagation Letters, vol. 18, no. 10, pp. 2095-2099, Oct. 2019.
[3] Z. Niu, et al.,"Isolation Enhancement in Closely Coupled Dual-Band MIMO Patch Antennas," in IEEE Antennas and Wireless Propagation Letters, vol. 18, no. 8, pp. 1686-1690, Aug. 2019.
[4] H. T. Chattha, et al., "4-Port 2-Element MIMO Antenna for 5G Portable Applications," in IEEE Access, vol. 7, pp. 96516-96520, 2019.
[5] Y. Liu, et al., "A Low Correlation and Mutual Coupling MIMO Antenna," in IEEE Access, vol. 7, pp. 127384-127392, 2019.
[6] Y. Li, et al., "High-Isolation 3.5 GHz 8-Antenna MIMO Array Using Balanced Open-Slot Antenna Element for 5G Smartphones," in IEEE Transactions on Antennas and Propagation, vol. 67, no. 6, pp. 3820-3830, June 2019.
[7] K.-L. Wong, et al.,"2-asymmetrically mirrored gap-coupled loop antennas as a compact building block for 8-antenna MIMO array in the future smartphone," IEEE Trans. Antennas Propag., vol. 65, no. 4, pp. 1765-1778, Apr. 2017.
[8] S. Xiumei, et al., "A Miniaturized Microstrip Antenna Array at 5G Millimeter Wave Band," in IEEE Antennas and Wireless Propagation Letters.
[9] P. Garg and P. Jain, "Isolation Improvement of MIMO Antenna using a Novel Flower Shaped Metamaterial Absorber at 5.5 GHz WiMAX band," in IEEE Transactions on Circuits and Systems II.
[10] J. Guo, et al., "Side-edge frame printed eight-port dual-band antenna array for 5G smartphone applications," IEEE Trans. Antennas Propag., vol. 66, no. 12, pp. 7412-7417, Dec. 2018.
[11] X. Shi, et al., "Dual-band 8-element MIMO antenna with short neutral line for 5G mobile handset, " in Proc. 11th Eur. Conf. Antennas Propag., May 2017, pp. 3140-3142.
[12] N. O. Parchin et al., "8-Element Dual-Polarized MIMO Slot Antenna System for 5G Smartphone Applications," in IEEE Access, vol. 7, pp. 15612-15622, 2019.
[13] Q. Chen et al., "Single ring slot-based antennas for metal-rimmed 4G/5G smart- phones," IEEE Trans. Antennas Propag., vol. 67, no. 3, pp. 1476-1487, Mar. 2019.
[14] J. Li et al., "Dual-Band 8-Antenna Array Design for MIMO Applications in 5G Mobile Terminals," in IEEE Access, vol. 7, pp. 71636-71644, 2019.

Influence of pH and salinity on the photocatalytic dye degradation and heavy metal ion reduction with cobalt ferrite photocatalysts.

Anna Heba Chakkunny [1], Bintu Thomas [2] and Libu K Alexander[2,*]

[1]Department of Physics, University of Calicut, India; annahebachakkunny@gmail.com
[2]Department of Physics, University of Calicut, India; bintuthomas16@gmail.com
*Correspondence: Department of Physics, University of Calicut, India; LKA@uoc.ac.in

Abstract. The magnetic nanoparticle assisted photocatalytic oxidation-reduction is an effective method to remove and detoxify the organic dyes and the heavy metal ions present in the industrial wastewater. In the present study, the photocatalytic degradation of Methyl Orange(MO) dye and Cr(VI) ions in the presence of cobalt ferrite ($CoFe_2O_4$) superparamagnetic photocatalysts under visible light illumination are investigated as a function of wide range of pH and salinity conditions. The effect of pH on the photodegradation rates and the dual effect observed in the presence of Cl^- ions were discussed. The photocatalytic degradation rates in single and binary impurity systems were also examined and the synergetic effect of MO oxidation and Cr(VI) reduction were established at different salinity and pH conditions.

Keywords: Photocatalysis; cobalt ferrite catalysts; dye degradation; heavy metal ion reduction; pH; salinity; synergetic catalysis.

1. Introduction

One of the significant threats recently confronted by most developing countries is the scarcity of clean drinking water free of chemicals and pathogens[1]. Heavy metal ions and organic dyes are the chemicals usually found in industrial effluents that cause a high degree of water contamination. The bio-accumulative nature of heavy metals in biotic systems substantiates the consequences of its hazardous nature[2]. The hexavalent chromium Cr(VI) ion, which is highly toxic and potentially harmful to living organisms, is introduced in the drinking water through the effluents from electroplating activities, paper, rubber, paint, nylon and plastic factories[3]. Methyl Orange(MO) is a hazardous anionic organic dye commonly found in industrial discharges which persists together with the heavy metal ions in the surface water sources[4].

The conventional adsorption methods by activated carbon[5],[6], biomass[7] etc. are widely studied water treatment methods. Even though the pollutants are removed from the bulk liquid onto the adsorbent surface, in most cases, the detoxification of these hazardous contaminants hardly occurs[3]. Photocatalytic oxidation-reduction is an alternative that can remove and detoxify the heavy metal ions and the organic dye simultaneously in a mixed impurity system[8],[9]. A wide variety of photocatalysts has been studied so far, incredibly magnetic nanoparticle photocatalysts, one amongst is the cobalt ferrite ($CoFe_2O_4$) which is proved to have higher adsorption, photocatalytic and magnetic separable efficiency[4]. The superparamagnetic property and its sensitivity to visible light make cobalt ferrite an attractive catalyst for water treatment application.

IOP Conf. Series: Materials Science and Engineering 1187 (2021) 012018 doi:10.1088/1757-899X/1187/1/012018

To enhance the adsorption efficiency and the photocatalytic degradation rates, it is necessary to determine the optimal process parameters like solution pH, reaction temperature, the concentration of the catalysts and the concentration of the pollutants and the salts present. The pH is a critical operational variable in actual wastewater treatment. It is a parameter that plays a key role in determining the adsorption capacity of a photocatalyst since its surface properties explicitly depend on the reaction pH. It also influences the reaction mechanisms of the photocatalytic oxidation of organic dyes and the reduction of heavy metal ions [5],[7]. Sodium chloride being a primary constituent in the dye effluents coming with the sectional water of textile mills, the study of its influence on the photo mineralization efficiency is much relevant[10]. Considering the necessity for optimization of reaction parameters, this study investigates the effect of solution pH and salinity in the photocatalytic degradation of MO dye and Cr(VI) ions. This study also demonstrates the effect of the synergism of MO oxidation and Cr(VI) reduction in the photocatalytic degradation rates in acidic, neutral and alkaline medium and in the presence of Cl^- ions.

2. Materials and Methods

2.1 Materials

The Cobalt ferrite nanoparticles are synthesized from the nitrates of Co and Fe, $Co(NO_3)_2.6H_2O$ (97%) and $Fe(NO_3)_3.9H_2O$(98%), respectively, in Ethylene Glycol solvent using the solvothermal route. Poly ethylene glycol -PEG, Sodium acetate and Urea were used as surfactants. Methyl Orange (MO: $C_{14}H_{14}N_3NaO_3S$) and potassium dichromate ($K_2Cr_2O_7$) are used as the organic dye and the source of heavy metal ion respectively, to study the photocatalytic experiments. The pH of the test solution is adjusted by the addition of sodium hydroxide and hydrochloric acid in required concentrations. The salinity of the test solution is varied by adding the right concentration of sodium chloride (NaCl) to it. All the chemicals are used as obtained without further purification and double distilled water is employed here in all the studies.

2.2 Methods

The salinity dependence of the photocatalytic degradation of MO and Cr(VI) was studied by varying the concentration of NaCl from 0 mM to 900 mM. The pH dependence of the same is studied in the acidic, neutral and basic pH conditions with pH values ranging from 2 to 12. The photocatalytic degradation were investigated in both single impurity systems, i.e. (i) with MO only and (ii) with $K_2Cr_2O_7$ only and in binary impurity system, ie, a mixture of MO and $K_2Cr_2O_7$ (10 ppm each). In each experiment, 50 mg of cobalt ferrite photocatalysts were suspended in 100 ml of the impurity solution. The photocatalytic experiments were carried out under visible light illumination using a 500 W halogen lamp with an illuminance of 70,000 lux. The detailed experimental procedure is cited elsewhere[4]. For each experiment, the absorbance spectra of methyl orange and chromium ions are studied using a UV-Vis Spectrophotometer to determine the photocatalytic reaction rate constants.

3. Results

3.1. *Effect of pH*

The effect of pH on the photocatalytic degradation of MO dye and chromium ion in the presence of cobalt ferrite photocatalysts are studied in the solution pH ranging from pH 2 to 12 and is shown in figure 1.

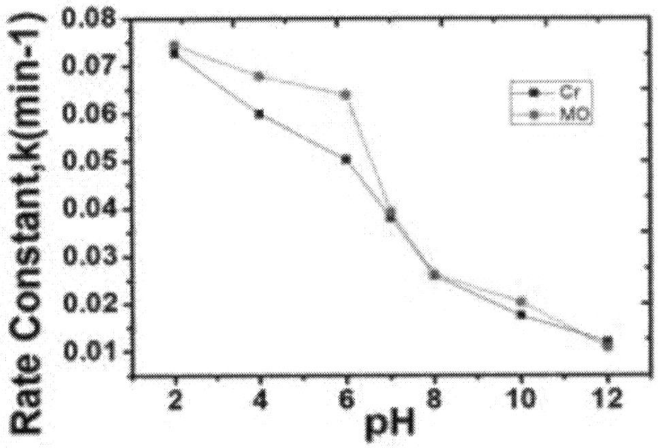

Figure 1 Effect of pH on the photocatalytic degradation of MO dye (Red Line) and Cr(VI) (Black line) in the presence of cobalt ferrite photocatalyst.

The photocatalytic dye degradation rate enhancement at strong acidic conditions may be attributed to the improved adsorption of the anionic MO dye molecules onto the protonated catalyst surface. Point of Zero charge(PZC) of a catalyst is the pH at which its surface becomes neutral. The PZC of cobalt ferrite photocatalyst is reported to be around pH ~ 6.6 [11][12]. At pH < PZC, the surface of the catalyst will be positively charged. The catalyst then strongly adsorbs the anionic dye molecules onto its surface by coulombic attraction, increasing the degradation efficiency. As pH of the solution increases below PZC, the positive charges on the catalyst surface decreases which causes a decrease in the degradation rate. When the solution pH becomes greater than PZC, ie, pH > 6.6, the degradation rate decreases markedly due to the electrostatic repulsion of the dye molecules with the surface of the photocatalyst which is negatively charged and also with the hydroxyl ions present in the solution.
The different forms of hexavalent chromium that exist in aqueous solution are $HCrO_4^-$, $Cr_2O_7^{2-}$, CrO_4^{2-} and the stability of these forms mainly depends on the pH of the system[7]. $HCrO_4^-$, $Cr_2O_7^{2-}$, $Cr_3O_{10}^{2-}$,$Cr_4O_{13}^{2-}$ are the different forms of chromium ions that co-exists in the pH range of 1.0 -6.0 of which $HCrO_4^-$ predominates. As the pH of the solution increases, the predominant species are then $Cr_2O_7^{2-}$ and CrO_4^{2-}. The greater adsorption at acidic pH is due to the strong electrostatic attraction between the H^+ ions at the highly protonated adsorbent surface and the chromate ions. The adsorption of Cr(VI) at pH values greater than 6.0 decreases due to the dual competition of both the anions , CrO_4^{2-} and OH^-, to get adsorbed on the surface of the adsorbent, of which OH^- predominates[5].

3.2. Effect of salinity

The effect of salinity on the photocatalytic oxidation of methyl orange and the reduction of chromium ions were studied in the chloride ion concentrations ranging from 0-900 mM keeping all other parameters a constant. It is noted that the photocatalytic degradation rate of both MO and Cr ion increases initially till the Cl^- ion concentration reaches 100 mM and decreases on further addition of Cl^- ions (100-900 mM).

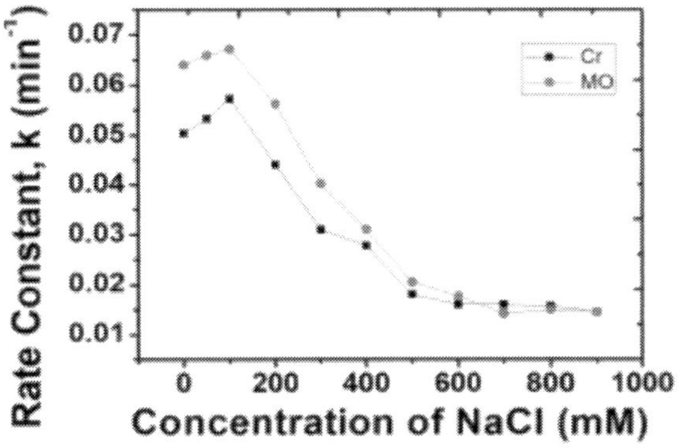

Figure 2 Effect of salinity on the photocatalytic degradation of MO dye (Red Line) and Cr(VI) (Black Line) in the presence of cobalt ferrite photocatalyst.

The enhancement of dye degradation rate at lower Cl^- ion concentration can be attributed to a mechanism involving chloride($Cl^{\cdot}$) and dichloride radicals($Cl_2^{\cdot-}$). The chloride and dichloride radicals are efficient in inhibiting the e^- - h^+ recombination which enhances the photodegardation. The chlorine radicals involved in the degradation mechanism are formed through the following reactions[13],Eqs. (1) to (4):

$$Cl^- + OH^{\cdot} \longrightarrow HOCl^{\cdot -} \quad (1)$$

$$HOCl^{\cdot -} + H^+ \longrightarrow Cl^{\cdot} + H_2O \quad (2)$$

$$Cl^- + OH^{\cdot} \longrightarrow Cl^{\cdot} + OH^- \quad (3)$$

$$Cl^{\cdot} + Cl^- \longrightarrow Cl_2^{\cdot -} \quad (4)$$

The photocatalytic degradation mechanism of MO dye in the presence of Cl^- ions is illustrated in figure 3 [13-15].

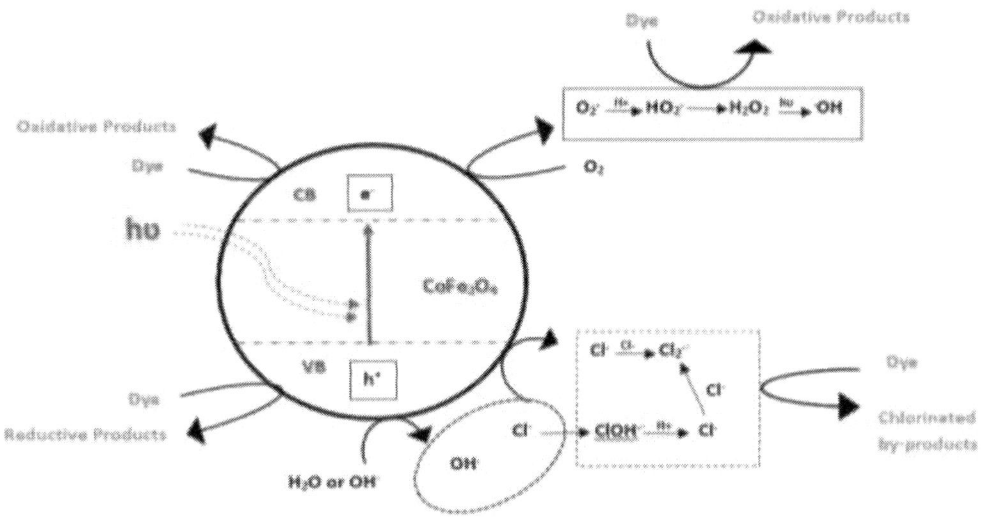

Figure 3. The proposed chlorine radical-mediated mechanism for the photocatalytic degradation of MO in the presence of Cl^- ions.

The degradation rate enhancement at lower Cl^- ion concentration can also be explained based on the PZC of the photocatalyst[16]. The photocatalytic experiments for the salinity dependence studies were carried out at the natural solution pH 5.75(axcidic) which is less than the PZC of the cobalt ferrite. Thus the positively charged photocatalyst surface at the reaction pH, easily adsorbs the anionic MO dye molecules through an electrostatic interaction at low Cl^- ion concentration which in turn improves the photocatalytic performance. Similar is the case of chromium ion where $HCrO_4^-$, the predominant form of Cr ion in acidic pH condition get easily adsorbed onto the protonated catalyst surface which undergoes further reduction.
The considerable reduction in the degradation rates observed at higher salt concentrations can be ascribed to the inhibitory effect of chloride ions. At higher salinity, there is a competitive adsorption between the chloride ions and the anionic dye molecules/ $HCrO_4^-$ on to the protonated catalyst surface[17]. A higher number of Cl^- anions adsorbing onto the catalyst surface, masks and effectively block the active sites, thus decreasing the photodegardation rates notably[15].

3.3. Effect of Salinity and pH on the synergetic photocatalysis of Cr(VI) reduction and MO oxidation

The salinity and pH dependence of the photocatalytic experiments were studied both in single impurity systems containing either Cr(VI) ions or MO dye alone and the binary impurity systems containing both the impurities together. In both cases the concentration of Cr(VI) and MO are 10 ppm each. It is observed that the value of the rate constant, k is higher in the binary impurity system than in the single one. This enhancement can be attributed to a mutual promotion in the degradation mechanisms through a simultaneous Cr(VI) reduction and methyl orange oxidation. The hole-scavenging property of the MO dye prevents the electron-hole recombination which facilitates the utilization of the conduction band electrons by the Cr(VI) ions for undergoing reduction. In addition, the electrons are available for the Cr(VI) ions, which are transferred to them through the conduction band of $CoFe_2O_4$ by the photoexcited MO molecules[4].

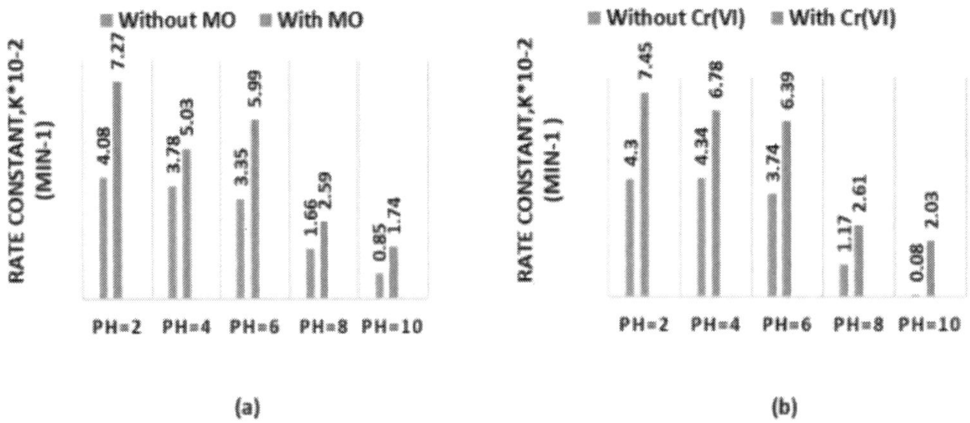

Figure 4. (a) The photocatalytic reduction of Cr(VI) in the absence (Green bar) and presence (Blue bar) of MO under different pH conditions, **(b)** The photocatalytic oxidation of MO in the absence (Green bar) and presence (Blue bar) of Cr(VI) under different pH conditions

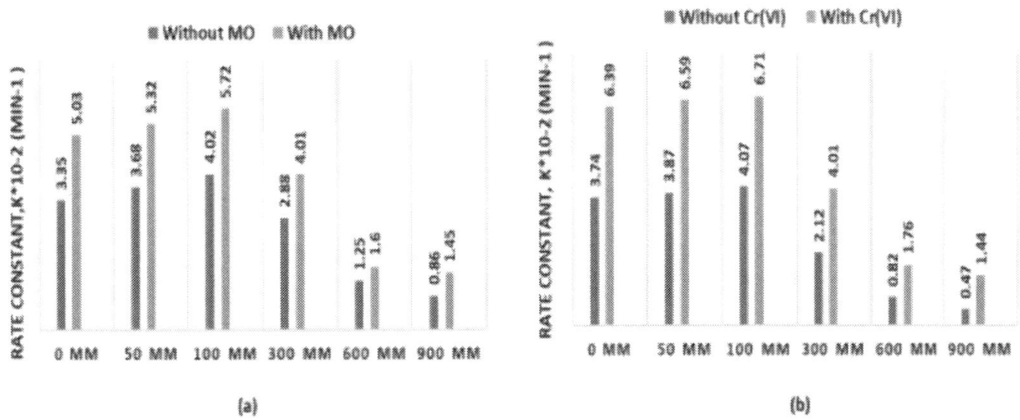

Figure 5. (a) The photocatalytic reduction of Cr(VI) in the absence (Blue bar) and presence (Red bar) of MO under different concentration of NaCl (mM), **(b)** The photocatalytic oxidation of MO in the absence (Blue bar) and presence (Red bar) of Cr(VI) under different concentration of NaCl (mM).

4. Conclusions

The effect of salinity and pH on the photocatalytic oxidation of Methyl Orange(MO) and reduction of Chromium(VI) ion in the presence of $CoFe_2O_4$ photocatalyst under visible light illumination was investigated. The adsorption of both the MO dye and Cr(VI) on the photocatalyst is highly dependent on the pH and the concentration of Cl^- ions present in the waste water to be treated. The photodegardation rates of both MO dye and Cr(VI) ions are observed to be the highest at acidic pH conditions. The salinity of the test solutioSn presented a double effect on the photodegradtion rates; a promotion effect at lower salt concentrations and an inhibitory effect at higher salt concentrations. This work also demonstrated the enhancement of photocatalytic degradation rates in binary impurity system that arises from the synergistic MO oxidation and Cr(VI) reduction and proves to hold in acidic, neutral and alkaline conditions and also in the presence of salt.

References

[1] Tiwari, D.; Behari, J.; Sen, P. 2008,Application of nanoparticles in waste water treatment. *World Applied Sciences Journal, 3.*

[2] Gautam, R.; Sharma, S.; Mahiya, S.; Chattopadhyaya, M. 2014, *Heavy metals in water : Presence, Removal and Safety.*

[3] Wahyuni, E.; Aprilita, N.; Hatimah, H.; Wulandari, A.; Mudasir, M. 2015, Removal of toxic metal ions in water by photocatalytic method. *American Chemical Science Journal*, **5**, 194.

[4] Thomas, B.; Alexander, L. K. 2018, Enhanced synergetic effect of Cr(VI) ion removal and anionic dye degradation with superparamagnetic cobalt ferrite meso–macroporous nanospheres. *Appl Nanosci*, ***8*** (1), 125–135.

[5] Attia, A. A.; Khedr, S. A.; Elkholy, S. A. 2010, Adsorption of chromium ion (VI) by acid activated carbon. *Brazilian Journal of Chemical Engineering*, **27** (1), 183–193.

[6] Taiwo, A.; Amadi, S.; Deinkuro, N. 2010, Adsorption and treatment of organic contaminants using activated carbon from waste nigerian bamboo. *Journal of Applied Sciences and Environmental Management, 13.*

[7] Demiral, Y.; Öztürk, N. 2013, Cr(VI) Removal from aqueous solution by adsorption onto sugar beet bagasse. *The Journal of Selcuk University Natural and Applied Science*, **2,** 246–255.

[8] Chen, D.; K. Ray, 2001, A. Removal of toxic metal ions from wastewater by semiconductor photocatalysis. *Chemical Engineering Science*, **56** (4), 1561–1570.

[9] Barakat, M. A. 2011,New trends in removing heavy metals from industrial wastewater. *Arabian Journal of Chemistry*, **4** (4), 361–377.

[10] Rupa, V.; Manikandan, D.; Duraiswami, D.; Revathi, S.; Esther, M.; Preethi, M.; Shanthi, K.; Sivakumar, T.2007, Photocatalytic degradation of tatrazine dye using TiO~ 2 catalyst: salt effect and kinetic studies. *Indian Journal of Chemical Technology*, **14**, 71–78.

[11] Kosmulski, M. 2009, pH-dependent surface charging and points of zero charge. IV. update and new approach. *Journal of Colloid and Interface Science*, **337** (2), 439–448.

[12] Barale, M.; Mansour, C.; Carrette, F.; Pavageau, E. M.; Catalette, H.; Lefèvre, G.; Fedoroff, M.; Cote, G. 2008, Characterization of the surface charge of oxide particles of PWR primary water circuits from 5 to 320°C. *Journal of Nuclear Materials*, **381** (3), 302–308.

[13] Luk, M. K. Photocatalytic degradation and chlorination of azo dye in saline wastewater, UTAR, 2016.

[14] Yuan, R.; Ramjaun, S. N.; Wang, Z.; Liu, J.2012, Photocatalytic degradation and chlorination of azo dye in saline wastewater: kinetics and AOX formation. *Chemical Engineering Journal*, **192**, 171–178.

[15] Wang, Y.; Lu, K.; Feng, C. 2013, Influence of inorganic anions and organic additives on photocatalytic degradation of methyl orange with supported polyoxometalates as photocatalyst. *Journal of Rare Earths*, **31** (4), 360–365.

[16] Muthukumaran, S.; Song, L.; Zhu, B.; Myat, D.; Chen, J.-Y.; Gray, S.; Duke, M. 2014, UV/TiO2 photocatalytic oxidation of recalcitrant organic matter: effect of salinity and pH. *Water science and technology : a journal of the International Association on Water Pollution Research*, **70**, 437–443.

[17] Andreozzi, M.; Álvarez, M. G.; Contreras, S.; Medina, F.; Clarizia, L.; Vitiello, G.; Llorca, J.; Marotta, R. 2018, Treatment of saline produced water through photocatalysis using RGO-TiO2 nanocomposites. *Catalysis Today*, **315**, 194–204.

IOP Conf. Series: Materials Science and Engineering 1187 (2021) 012019 doi:10.1088/1757-899X/1187/1/012019

Mathematical model and analysis of PV Converter - Inverter System

K R Pushpa[1] and R S Geetha[2]

[1] Department of EEE, PES University, Bangalore, India
[2] Departments of EEE, BMSCE, Bangalore, India

Abstract— Photo Voltaic (PV) system is commonly preferred renewable energy source for most of the residential, commercial as well as for utility applications. Photo Voltaic systems are extensively used for standalone and grid tied applications. To utilize the PV generated power to the best possible extent, control and conversion of power in PV systems plays an important role. In this paper, mathematical model of the complete PV system that comprises of solar panel, DC-DC Boost converter and a H bridge inverter is developed. The developed model is validated for constant load condition at varying irradiation levels and at fixed duty cycle. The performance analysis of the PV system in terms of efficiency is carried out for 200W resistive load.

Keywords: Photo Voltaic (PV), converter-Inverter, DC-DC converter, Inverter, modeling

1. Introduction

PV system used in most of grid tied and standalone application consists of a PV module, DC-DC converter and Inverter as shown in Figure 1. The variable DC from PV panel is initially boosted to a higher voltage using DC-DC converter and then converted to AC using inverter that can be fed into the grid or used by standalone load [1].The performance of PV converter–Inverter system depends on irradiation, temperature and conversion efficiency. The current generated from PV panel varies with irradiation and output obtained is unregulated DC. To obtain steady and regulated power, DC-DC converters are used in PV system. DC-DC converter output power is controlled by varying duty ratio. To reduce the voltage ripple a suitable DC capacitor is connected across PV module [2].

PV converter-Inverter system design includes PV panel identification and power converter topology selection which is based on cost, efficiency and reliability. Various control strategies are used in grid tied and standalone applications. Inverters used in PV system can have bipolar or unipolar PWM switching.PWM technique is generally used in controlling inverter output. PWM technique compares high frequency triangular carrier signal with low frequency sinusoidal signal. Power injected to grid or supplied to a standalone load can be controlled using voltage or current controlled PWM technique. With proper filter design pure sine wave can be obtained from PWM Inverter [3][5][6][7].

The performance of PV converter-Inverter system can be analyzed using mathematical modeling approach. The mathematical modeling describes physical system by a set of equations that relates a set of variables between input and output of a system. These mathematical equations provide an insight of theoretical and numerical data of PV converter-Inverter system variables which helps in analyzing the PV system precisely and apply various computation methods for better results for a particular application. The PV system is represented as a set of differential equations that describe the dynamic behavior of the PV converter-Inverter system. Solution of these differential equations are analyzed and redesigned as per the desired output [4][6][7].In this paper, the mathematical modeling of PV system that includes modeling of PV module, DC-DC converter, and Inverter is carried out. The PV system performance and conversion efficiency is analyzed for a resistive load. The performance of the developed system is validated for variation in irradiation condition. The effect of irradiation on efficiency of the PV system is analyzed.

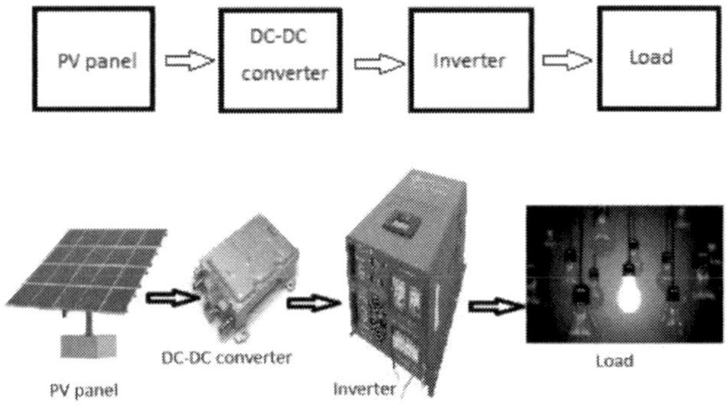

Figure 1. PV converter-Inverter system

2. Design and Simulation

The simulation model for the PV converter-Inverter system represented in Figure 1 is shown in Figure 2.The selected PV module generates maximum power of 213W at 1000W/m^2 at constant temperature of 25^{0}C at maximum load condition. PV module output voltage of 29V DC is boosted to 120V DC using boost converter. To obtain 120V DC output, the boost converter is operated with 75.8% duty ratio. With this duty ratio 120V DC output from converter is converted to AC of 120V peak at 50Hz frequency using single phase H Bridge Inverter with amplitude modulation 1 and frequency modulation ratio 20. Inverter output is supplied to a standalone resistive load.

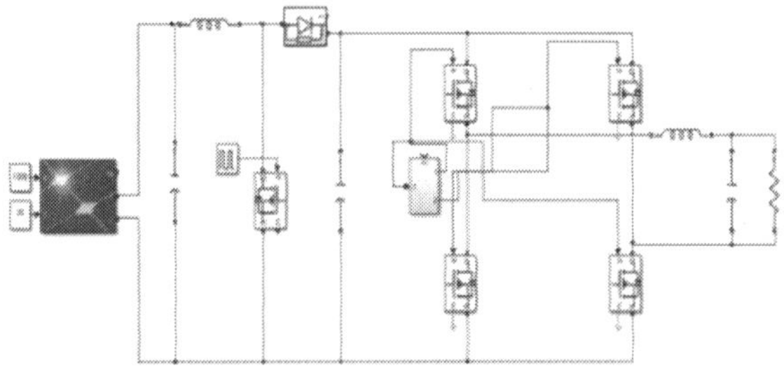

Figure 2. Physical Model of the PV Converter- Inverter system

The analysis of PV converter –Inverter system includes

- Modeling and analysis of effect of irradiation on 213 W PV panel.
- Modeling and control of 200W DC-DC Boost converter.
- Modeling and control of 200VA single phase H Bridge Inverter.

IOP Conf. Series: Materials Science and Engineering 1187 (2021) 012019 doi:10.1088/1757-899X/1187/1/012019

3. Modeling of PV module

PV module parameters used for modeling in Simulink is shown in Table 1.The solar irradiance G is set to 1 representing 1000W/m^2 as a standard test condition.

Table 1. PV module parameters

Short Circuit Current(I_{sc})	7.84A
Open Circuit Current(V_{oc})	36.3V
Current at maximum power (I_{mp})	7.35A
Voltage at maximum power (V_{mp})	29V
Number of cells per module(N)	60
Diode saturation current(I_s)	$2.9259*10^{-10}$
Ideality factor(A)	0.98117
Series Resistance(R_s)	0.39383Ω
Shunt Resistance(R_{sh})	313.3991 Ω
Electron charge (q)	$1.6*10^{-19}$C
Boltzmann constant (k)	$1.38*10^{-23}$

The equation that represents the photovoltaic model is shown in equation (1). The current I_{ph} generated from photo voltaic module is proportional to the solar irradiance G and is shown in equation (2) .The equation used for calculating thermal voltage vt of diode is shown in equation (3) [4].Mathematical model of PV module implemented in Simulink using equations (1-3) is shown in Figure 3.

$$I = I_{ph} - I_S\left(e^{\frac{(V+I.R_S)}{vt}} - 1\right) - \frac{V+I.R_S}{R_{sh}} \quad (1)$$

$$I_{ph} = I_{sc}.G \quad (2)$$

$$vt = \frac{(A.K.t.N)}{q} \quad (3)$$

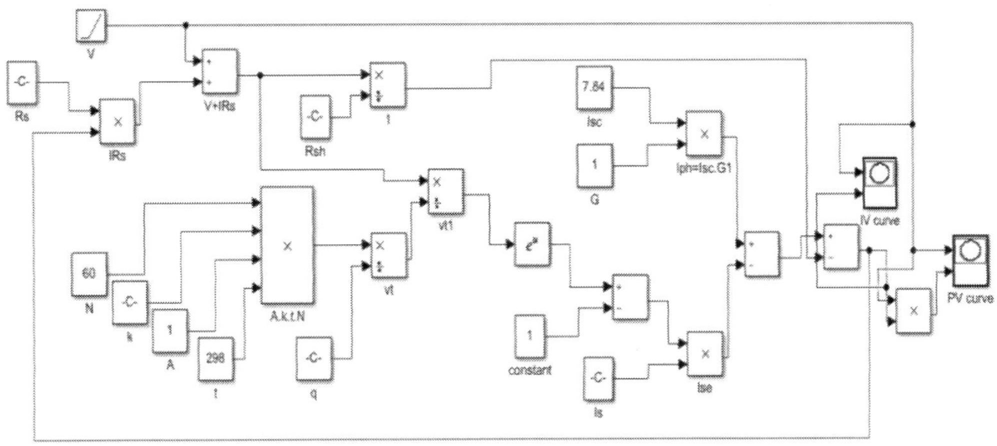

Figure 3. Mathematical model of PV module

As per the selected PV module parameters maximum power will be generated from PV module at 1000W/m^2 at maximum load condition. With maximum voltage from PV module as 29V and with maximum current 7.35A, maximum power of 213W will be obtained when load resistance is nearly 3.945Ω.The simulated I-V and P-V curves of 213W panel are shown in Figure 4.

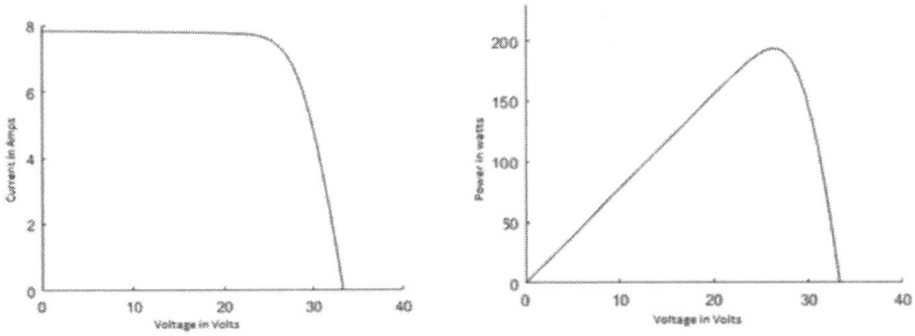

Figure 4. I-V and PV curve of PV module

4. Modeling of Boost converter

In Boost converter, the output voltage is always greater than the input voltage. When switch is 'on' input supplies energy to the inductor. When the switch is 'off' output stage receives energy from the inductor as well as from input [6] [7].The physical model of Boost converter is shown in Figure 5 which uses MOSFET as the switch. Table 2 shows Boost converter specifications for 200W output power. Input to boost converter is supplied from PV module.

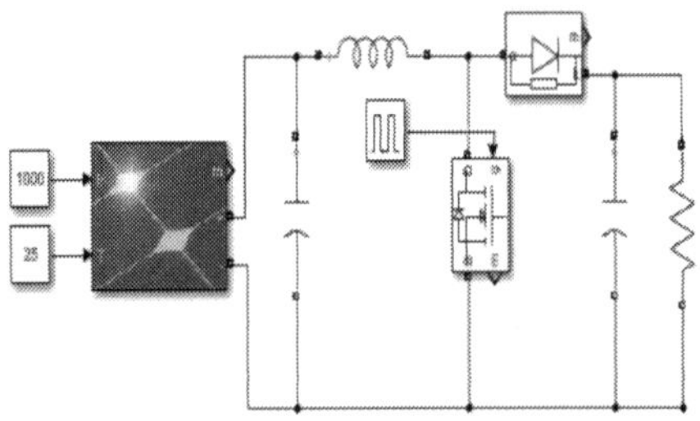

Figure 5. Physical Model of Boost converter

Table 2. Boost Converter specification

Input voltage(V)	Output voltage(V)	Duty cycle(%)	Switching frequency(kHz)
29	120	75.8	30

The relation between duty cycle, input and output voltage in Boost converter is shown in equation (4).

$$\frac{V_0}{V_{IN}} = \frac{1}{1-D} \quad (4)$$

With input voltage V_{IN}=29V and duty ratio D=0.758 output voltage of Boost converter V_0 = 120V .The dynamic equations are obtained by applying KVL to the circuit during on and off state of the switch [6] [7].

IOP Conf. Series: Materials Science and Engineering 1187 (2021) 012019 doi:10.1088/1757-899X/1187/1/012019

When switch is 'off', by applying KVL to the resultant circuit, the voltage and current equations as shown in equations (5-7) are obtained for inductor.

$$V_L = V_{IN} - V_0 \tag{5}$$

$$L\frac{di_L}{dt} = V_{IN} - V_0 \tag{6}$$

$$i_L = \frac{1}{L}\int (V_{IN} - V_0)dt \tag{7}$$

When switch is 'on', by applying KVL to the resultant circuit, the voltage and current equations as shown in equations (8-10) are obtained for inductor.

$$V_0 = -V_L \tag{8}$$

$$-V_0 = L\frac{di_L}{dt} \tag{9}$$

$$di_L = \frac{1}{L}\int -V_0 dt \tag{10}$$

Voltage and current equations of capacitor are obtained as shown in equations (11-14)

$$V_C = V_0 \tag{11}$$

$$i_C = C\frac{dv_c}{dt} \tag{12}$$

$$(i_L - i_0)dt = Cdv_c \tag{13}$$

$$V_C = \frac{1}{C}\int (i_L - i_0)dt \tag{14}$$

Using the dynamic equations of Boost converter mathematical model is developed as shown in Figure 6 .

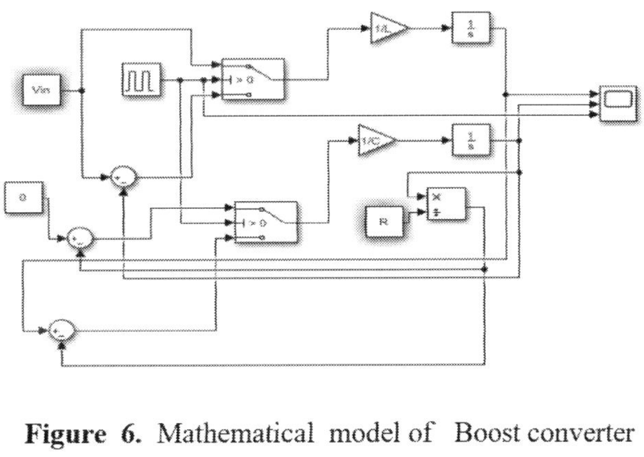

Figure 6. Mathematical model of Boost converter

The required value of inductor and capacitor when operated at different frequency and with 10% ripple is calculated using equations(15-16)[6].Table 3 shows values of L and C calculated for 10% ripple voltage and ripple current..For 200W boost converter with 120V output voltage, output current is 1.6667A.

$$\Delta I = \frac{V_{IN}D}{f_s L} \quad (15)$$

$$\Delta V_c = \frac{I_0 D}{f_s C} \quad (16)$$

Table 3. Boost converter L and C design

Ripple(%)	f_s(kHz)	ΔI(A)	ΔV_c(V)	L(mH)	C(uF)
10	30	0.1667	12	4.349	3.472
10	15	0.1667	12	8.69	6.9

5. Modeling of single phase bridge inverter

A single phase H Bridge Voltage Source Inverter consists of four switches S1, S2, S3 and S4 as shown in Figure 7. When S1 and S2 are turned on simultaneously the input voltage appears across the load. When S3 and S4 are turned on simultaneously voltage across the load is reversed [6] [7]. Table 4 shows Single phase H Bridge Inverter specification for 200VA output power.

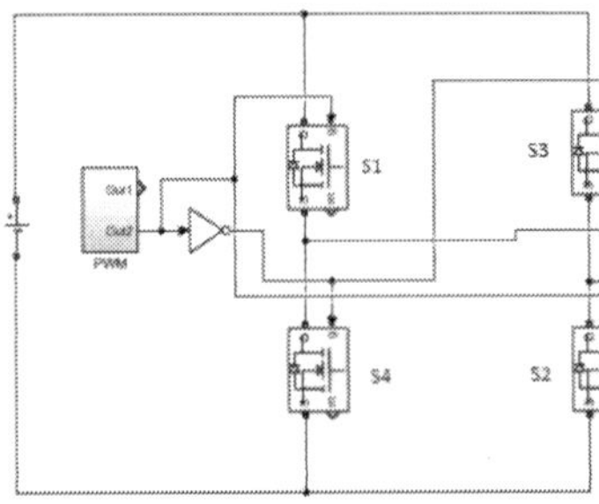

Figure 7. Physical model of Single phase H Bridge Inverter

Table 4. Inverter specification

Input voltage(V)	Output power(VA)	Output frequency(Hz)	Switching frequency(kHz)
120	200	50	1

When switches S1 and S2 are 'on' the voltage and current equations (17-19) are obtained for inductor.

$$V_L = V_{IN} - V_0 \quad (17)$$

$$L\frac{di_L}{dt} = V_{IN} - V_0 \quad (18)$$

$$i_L = \frac{1}{L}\int(V_{IN} - V_0)dt \quad (19)$$

When switches S3 and S4 are 'on' the voltage and current equations (20-22) are obtained for inductor.

$$V_L = -V_{IN} - V_0 \quad (20)$$

$$L\frac{di_L}{dt} = -V_{IN} - V_0 \quad (21)$$

$$i_L = \frac{1}{L}\int -(V_{IN} + V_0)dt \quad (22)$$

Voltage across capacitor is obtained as shown in equation (23).

$$V_C = \frac{1}{C}\int(i_L - i_0)dt \quad (23)$$

Using the dynamic equations of single phase H Bridge Inverter mathematical model is developed as shown in Figure 8.

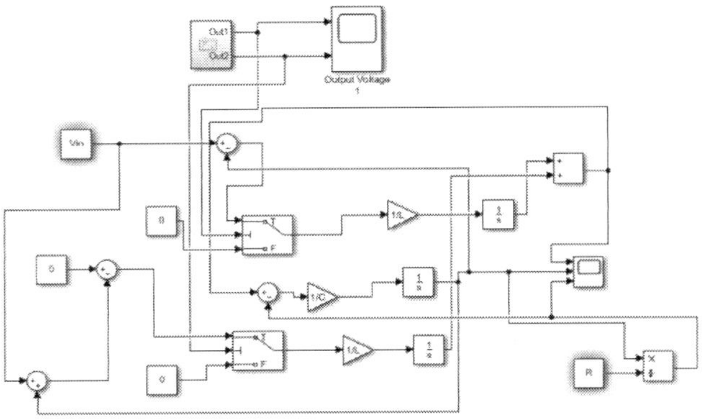

Figure 8.Mathematical model of Single phase H Bridge Inverter

The required values of filter components L and C of Inverter to obtain sine output is calculated using equations (24-25) [10][11] and is tabulated in Table 5.

$$L = \frac{V_{dc}}{4f_s\Delta I_0} \quad (24) \qquad C = \frac{\Delta I_0}{8f_s\Delta V_0} \quad (25)$$

Table 5. Inverter L and C designed values

Ripple(%)	f_s(kHz)	ΔI_0(A)	ΔV_0(V)	L(mH)	C(uF)
10	1	0.1667	12	179.9	1.736
10	5	0.1667	12	35.99	0.34

6. Modeling of PV converter inverter system

Implementation of the mathematical model of PV converter-Inverter system is shown in Figure 9. 213 W capacity PV module is connected to 200 W boost converter which in turn is connected to 200VA inverter. Generated power from PV module, power output from Boost converter and Inverter is calculated for ideal conditions neglecting switching losses of the power semiconductor devices. The simulation results obtained are tabulated in Table 6 for different modulation index.

Table 6. Voltage output at various stages in a PV converter-Inverter system

Modulation Index	Value	Panel output (V)	Boost Output(V)	Inverter output (V_{Peak})
Ma=0.6 M_f=20	Ideal	29	120	72
	Simulation	29	117	61.37
Ma=0.9 M_f=20	Ideal	29	120	108
	Simulation	29	118.4	103
Ma=1 M_f=20	Ideal	29	120	120
	Simulation	29	118.4	115.54

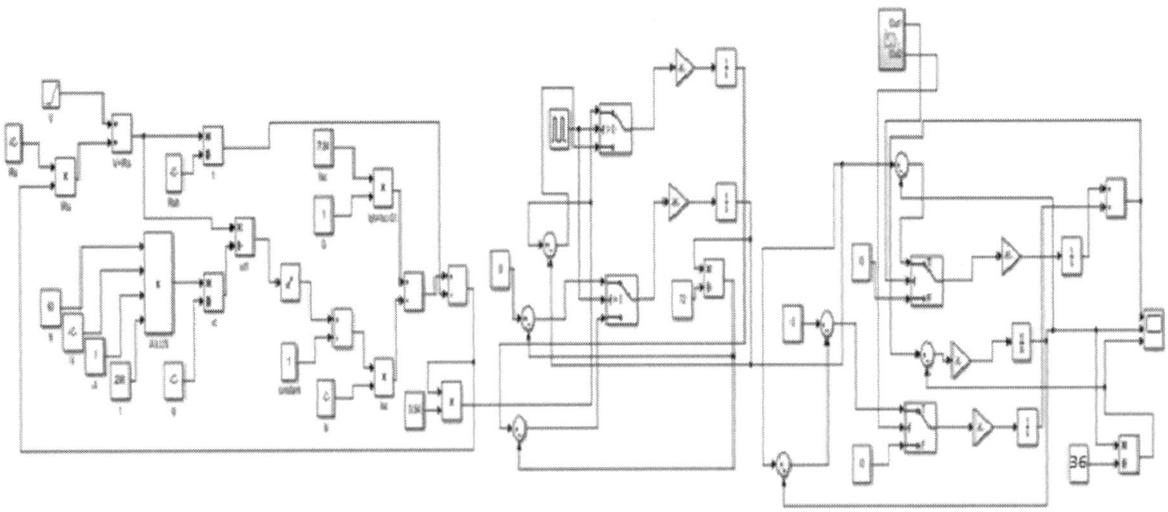

Figure 9. Mathematical model of PV converter-Inverter system

Implementation of the mathematical model of PV converter-Inverter system is shown in Figure 9. A 213 W capacity PV module is connected to 200 W boost converter which in turn is connected to 200VA inverter. The simulation results obtained are tabulated in Table 6 for different modulation index

IOP Conf. Series: Materials Science and Engineering 1187 (2021) 012019 doi:10.1088/1757-899X/1187/1/012019

7. Simulation results and discussion

The analysis of PV converter-Inverter system is carried out for a resistive load of 200W. Simulations are carried out for the following two cases.

- Constant load at different modulation index.
- Varying irradiation with fixed duty ratio.

By varying modulation index inverter output is controlled. Power output and conversion efficiency at various stages in PV system with varying irradiation is shown in Table7.For the designed values at constant load of 33ohm with constant duty ratio 75.8%, Boost converter, Inverter and overall PV system efficiency is around 94%, 82.6% and 77.8% at any irradiation level and at any modulation index. The variation in percentage of power output from PV module and power output at various stages of the PV converter–Inverter system for varying irradiation is tabulated in Table 8.The percentage panel output power is calculated with respect to the maximum power output capacity of the panel which is 213 W for an irradiance of 1000W/m^2.

Table 7. Power output & conversion efficiency of PV converter-Inverter system

Irradiation (W/m^2)	Ppanel (W)	Pdc (W)	Pac (W)	η_{Boost}(%)	$\eta_{Inverter}$(%)	$\eta_{Overall}$(%)
1	206.7	194.6	160.9	94.14	82.68	77.84
0.9	167.5	157.8	130.4	94.2	82.63	77.85
0.8	132.3	124.6	103	94.17	82.66	77.85
0.7	101.1	95.17	78.67	94.13	82.66	77.81
0.6	73.93	69.61	57.54	94.15	82.66	77.83
0.5	50.95	47.97	39.65	94.15	82.65	77.82

Table 8. Percentage of power output at different irradiation in PV converter-Inverter system

Irradiation	Ppanel (W)	Pdc (W)	Pac (W)	Ppanel(%)
1	206.7	194.6	160.9	97
0.9	167.5	157.8	130.4	78.63
0.8	132.3	124.6	103	62.11
0.7	101.1	95.17	78.67	47.46
0.6	73.93	69.61	57.54	34.7
0.5	50.95	47.97	39.65	23.92

7.1. PV output simulation results

PV panel output voltage, current and power generated at 1000W/m^2 is shown in Figure 10. 213W power is generated from PV panel with maximum voltage of 29V and with maximum current 7.35A with a load resistance of 3.945Ω.

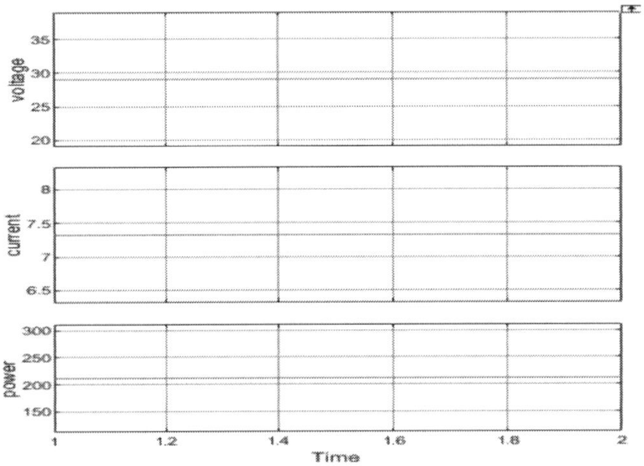

Figure 10. PV output voltage, current and power

7.2 .DC-DC Converter Simulation results

Figure 11.shows DC-DC converter output voltage, current and power at constant irradiation of 1000W/m^2.With DC input of 29V to Boost converter with 75.8% duty ratio an output of 120V DC is obtained. 200W output is obtained from Boost converter when connected to load resistance of 72Ω.

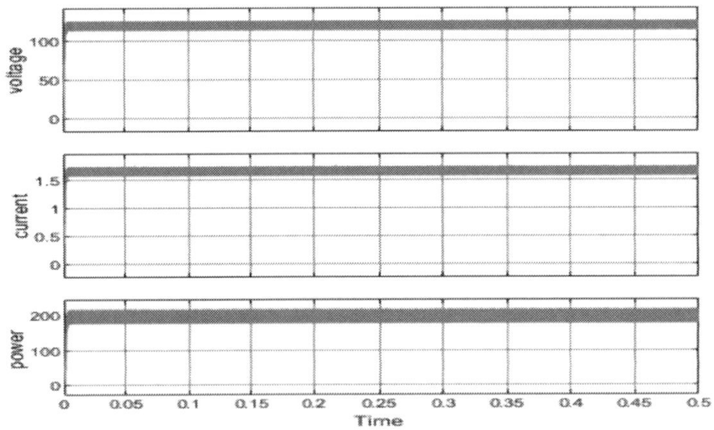

Figure 11. Boost converter output voltage, current and power

7.3. Single phase H Bridge Inverter Simulation results

The single phase H Bridge Inverter output with bipolar voltage switching is obtained as per the equation (26)

$$V_{01}= M_a*V_{dc}\ (M_a \leq 1.0) \qquad (26)$$

For amplitude modulation index 1 and frequency modulation index 20 output obtained is shown in Figure 12.It is observed that, the peak value of Inverter voltage increases from 61.37V to 115.54V when modulation index M_a is changed from 0.6 to 1.

IOP Conf. Series: Materials Science and Engineering 1187 (2021) 012019 doi:10.1088/1757-899X/1187/1/012019

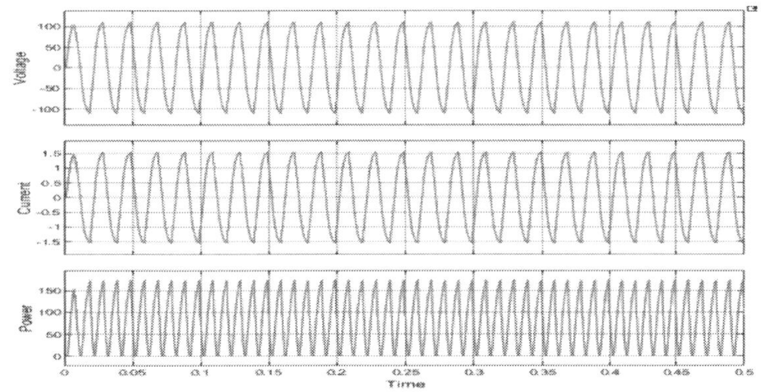

Figure 12. Single phase Inverter output waveform with M_a=1,M_f=20

7.4. PV converter –Inverter system Simulation results

For amplitude modulation index of 1 and frequency modulation index of 20 Figure 13 shows PV system output for fixed load resistance of 33ohms at 75.8% fixed duty cycle .Maximum output voltage from panel 29V is boosted to 120V DC and is converted to 120Vac peak which is shown in Figure.13.With lower value of modulation index it can be observed that the output voltage decreases as shown in Figure 14.With modulation index Ma as 1, maximum output is obtained from single phase inverter. As observed from the results obtained, irrespective of variation in modulation index and irradiation level, the power conversion efficiency of DC-DC converter, inverter and overall system remains constant.

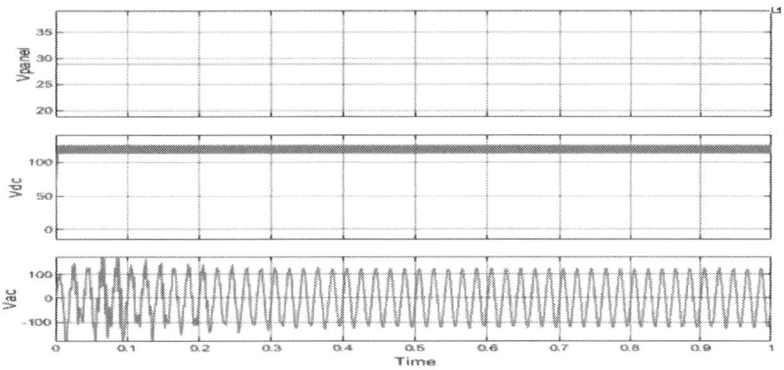

Figure 13. PV system output voltage with M_a=1,M_f=20

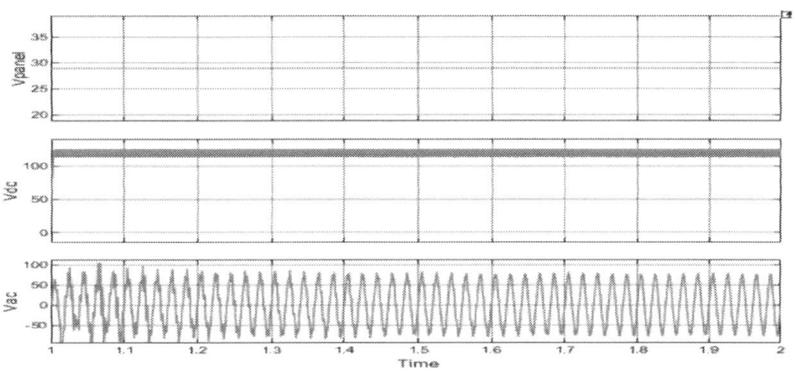

Figure 14. PV system output voltage with M_a=0.6,M_f=20.

8. Conclusion

In this paper, simulation and analysis of mathematical model of the PV converter – Inverter system is carried out. The simulation is carried out for constant resistive load at varying irradiation and also for constant load at varying modulation index. Power output at various stages of PV system and power conversion efficiency is discussed. It is observed that the power available from the panel is reduced with reduction in the irradiation. However, it is also observed that, the overall efficiency of the PV converter-inverter system remains to be a constant for different levels of irradiation as it is operated at fixed duty cycle at constant load condition.

9. References

[1] Oluwaseun M.Akeyo , Vandana Rallabandi, Nicholas Jewell, Dan M. Ionel , May/June 2020, *The Design and Analysis of Large Solar Farm Configurations with DC- Connected Battery Systems,* vol no **3** , IEEE Transactions on Industry Applications

[2] Sebastian Strache , Ralf Wunderlich and Stefan Heinen ,, July 2014, *A Comprehensive Quantitative Comparison of Inverter Architectures for Various PV Systems, PV cells .and Irradiance Profiles,* vol no **5**, IEEE Transactions on Sustainable Energy

[3] Michael A. Boost and Phoivos D. Ziogas , March/April 1988, *State -of-the-Art Carrier PWM Techniques: A Critical Evaluation,* vol no **24**, IEEE Transactions on Industry Applications

[4] Rodney H.G.Tan and Matthew Y.W.Teow, 2016, *A Comprehensive Modeling of Photovoltaic Module Characteristic Curve in MATLAB/Simulink* , 4th IET clean Energy and Technology conference

[5] Thandar Aung and Tun Lin Naing,2016, *Modeling and Simulation of DC- DC Boost Converter - Inverter System with Open –Source Software Scilab/Xcos* Science PG, Software Engineering ,6(2):27- 36

[6] Umanand L. (2009) , *Power Electronics: Essentials & Applications.* John Wiley & Sons.

[7] Mohan, Undeland, Robbins. (2003), *Power Electronics: Converters, Applications and Design,* John Wiley & Sons, New York

[8] Jeanette Lam Min Yi, R.T. Naayagi Thillainathan Logenthiran *Modelling and Implementation of Single Phase Dual Stage Grid- Tied Solar Power Inverter* 978-1-5090-2597- 8/16/$31.00 @2016 IEEE

[9] Subhash Chander, Pramod Agarwal and Indra Gupta *Design, Modeling and Simulation of DC-DC converter* 978-1-4244-7398-4/10/$26.00 ©2010 IEEE

[10] Pekik A. Dahono, Agus Purwadi and Qamaruzzaman , *An LC Filter Design Method for Single-phase PWM Inverters* , Catalogue No. 95TH8025 0-7803-2423-4/95/$4.00 & 31995 IEEE

IOP Conf. Series: Materials Science and Engineering 1187 (2021) 012020 doi:10.1088/1757-899X/1187/1/012020

Ring-Based Quantum Network with Quantum Key Distribution (QKD) and Quantum Digital Signature (QDS)

Osman Ceylan[1] & İhsan Yılmaz[2]

[1]Student of Department of Computer Engineering, Çanakkale 18 Mart University, Çanakkale, Turkey

[2]Department of Computer Engineering, Çanakkale 18 Mart University, Çanakkale, Turkey

E-mail: [1]`osman.semi.ceylan@gmail.com`

E-mail: [2]`iyilmaz@comu.edu.tr`

Abstract. Quantum technological developments offers a great promise for implementing secure quantum internet. In this study, we propose a ring topology-based quantum network with QKD and QDS for the quantum internet. The combination of QKD and QDS allows both the detection of the intruder and the increase in security in the network. The proposed network has been demonstrated to work by being implemented in the SQUANCH network simulator [1].

Keywords: quantum internet, ring-based quantum network, quantum key distribution

1. Introduction

Superior features of quantum technology such as entanglement and superposition enable teleportation and super-dense coding. In addition to these, when the feature that it is impossible to measure a feature of a system without changing its state, the superiority of quantum technologies over classical technologies becomes clear. These superior features not only offer many opportunities to humanity, but also offer secure communication. In recent years, scientists have developed quantum key generation devices that enable secure communication and have used them in practice. In this context, long-distance QKD networks with reliable nodes are also currently possible [2–5]. Recently, the distance of 500 km has been achieved on fibre based channels [6]. Similarly, the applications of entangled interaction between quantum memories in recent years show that the quantum internet can be applied in the very near future [7]. For example, countries such as China [8,9], European countries [10–12] and America [13] invest heavily in the quantum internet. Recently, satellite-based quantum internet applications have been developed by many countries such as China, America, Canada, Japan, Singapore and Europe to carry the quantum internet to longer distances [14–20]. In this study, we propose a quantum network based on a central network topology consisting of nested rings. Since this network uses both QKD and QDS, it offers a more secure quantum internet opportunity. The paper is organized as follows: In Section 2, the structure of Ring Based Quantum Network Topology is given. Also quantum communication on the suggested network is explained with example. In Section 3, implementation of Ring-Based Quantum Network with QKD and QDS is given. In Section 4 concludes the article.

2. Ring Based Quantum Network Topology

In this study, the quantum Internet is modelled as a ring-based quantum network consisting of quantum nodes, repeaters and channels . Since the quantum network will be long distance, there should be quantum repeaters at certain intervals between the nodes. Each nodes consist quantum devices and classic controlled agent devices. These quantum devices will be able to create qubits, apply some quantum transformations and have acceptable error rates. In addition to this, classic controlled agent devices can do classical work on all nodes. All this is shown in Figure 1 below.

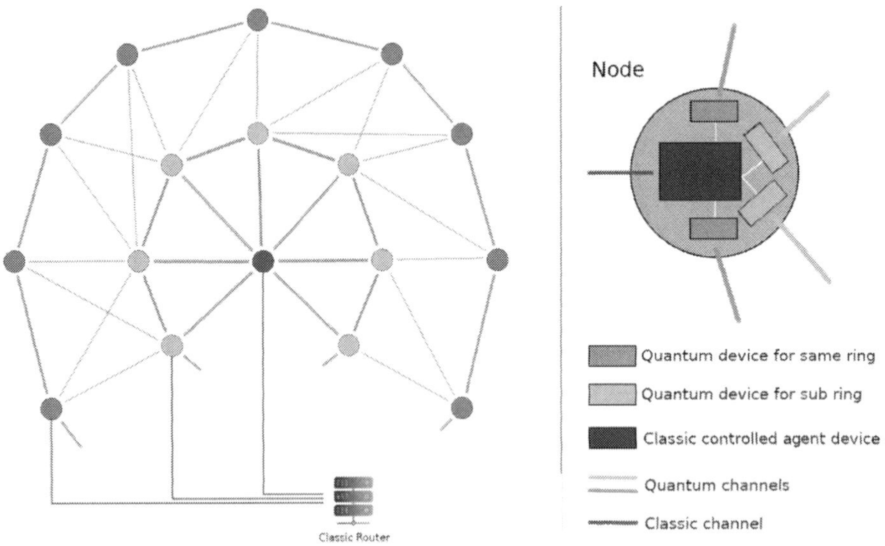

Figure 1: An example of nodes and channels in a ring-based network topology.

There is a central node, as can be seen in Figure 1. This is the zero ring. Other nodes are opened ring-by-ring on it. Each node in a ring is connected by quantum channels with the nearest nodes up to the degree of the ring. Classical channels between all nodes in the network can be arranged as desired. Let's examine how communication takes place on the proposed topology. For this purpose, let's consider a cross-section of the mesh as in Figure 2.

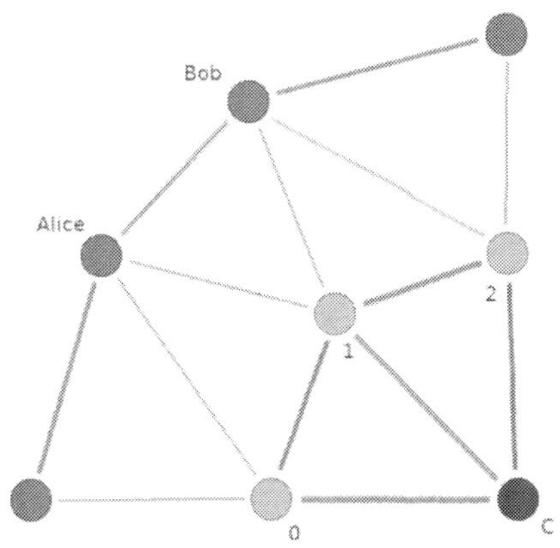

Figure 2: Portion of the network.

The communication between Alice and Bob over the proposed ring-based quantum network takes place as follows.

- Alice's Key Generation
- Generation of Sign as Key
- Alice Inherits Sign as Key
- Alice Prepares the Message
- Bob Handles Message
- Decryption of Message
- Return of Message Pieces

a-) Alice's Key Generation

The communication is start with generic quantum key distribution process. Alice and Bob runs an quantum key generation protocol between themselves. After a successful run of protocol both party generates the S_{AB} key. If this protocol ends with no success Alice prepares and sends an error report to center node. Table 1 is holds current knowledge of nodes.

Table 1: Knowledge table after step 1.

Alice	Bob	0	1	2	C
S_{AB}	S_{AB}	-	-	-	-

b-) Generation of Sign as Key

Alice now needs keys for encryption other than generated key. Alice request signs from the connected sub-level nodes. Sub-level nodes must holds an generated sign from center node. If these sub-level nodes grade low enough to hold original sign then they request a generated sign from center node. Else they inherit the signs from one of connected sub-level nodes. In this example "0", "1" and "2" named nodes grade is 1 which is low enough to hold original signs from center node. So they contact to center node for getting these sign. After successful quantum key distribution process between "0" to center and "1" to center S_{C0} and S_{C1} keys are generated. Center node generates unique M_0 and M_1 signs then encrypts them with related keys from QKD protocol. Center node sends $V_{C0} = S_{CO}(M_0)$ and $V_{C1} = S_{C1}(M_1)$ information to related nodes.

Table 2: Knowledge table after step 2.

Alice	Bob	0	1	2	C
S_{AB}	S_{AB}	-	-	-	-
-	-	V_{C0}	V_{C1}	-	M_0, M_1

c-) Alice Inherits Sign as Key

In this stage Alice inherit encrypted sign as keys from connected low-level nodes. These signs was generated in previous step. With the success of QKD protocol, inherit communication process happens.

Table 3: Knowledge table after step 3.

Alice	Bob	0	1	2	C
S_{AB}	S_{AB}	-	-	-	-
-	-	V_{C0}	V_{C1}	-	M_0, M_1
V_{C0}, V_{C1}	-	-	-	-	-

d-) Alice Prepares the Message

In this step Alice processes the K message she wants to send. Alice then selects an partition method B from the partition method pool which designed for the topology. Alice then divides the message as $K = K_0 + K_1 + ... + K_n$ with the selected method. n is equal to degree of the Alice in topology. Alice encrypts every piece of the message with encrypted sign as keys that she was got and translates them into $F = V(K)$. Alice then prepares package P containing partition method, connected low-level node identifications and message F. Alice encrypts the P package with S_{AB} and translate P into $P' = S_{AB}(P)$. Lastly Alice sends P' to Bob.

Table 4: Knowledge table after step 4.

Alice	Bob	0	1	2	C
S_{AB}	S_{AB}	-	-	-	-
-	-	V_{C0}	V_{C1}	-	M_0, M_1
V_{C0}, V_{C1}	-	-	-	-	-
P, K	P'	-	-	-	-

e-) Bob Handles Message

Bob runs successful QKD protocol with each of connected nodes from low-level ring after Bob decrypts P' package and extract P information. Bob sends them every piece of message F with using generated keys. If node identification in package P matches with one of node from Bob's connected nodes, Bob sends related piece of message to them. In this example Bob sends $F0$ piece to node "2" and $F1$ piece to node "1". Bob also sends them identification of nodes with related piece because target nodes needs to verify if they are bequeath of sign.

Table 5: Knowledge table after step 5.

Alice	Bob	0	1	2	C
S_{AB}	S_{AB}	-	-	-	-
-	-	V_{C0}	V_{C1}	-	M_0, M_1
V_{C0}, V_{C1}	-	-	-	-	-
P, K	P'	-	-	-	-
-	-	-	$F0$, node1	$F1$, node0	-

f-) Decryption of Message

The low-level nodes who got message from Bob tries decrypts the message. If related node identification matches with themselves they decrypts piece of message F_n with its encrypted sign as key V_n and extracts the piece of real message K_n. If match process returns false then sends them to low-level nodes until any identification match happens. At the worst case scenario piece of message reaches center node and center node decrypts the piece of massage. Since center node only one sign provider of network he knows all signs exits on whole network.

Table 6: Knowledge table after step 6.

Alice	Bob	0	1	2	C
S_{AB}	S_{AB}	-	-	-	-
-	-	V_{C0}	V_{C1}	-	M_0, M_1
V_{C0}, V_{C1}	-	-	-	-	-
P, K	P'	-	-	-	-
-	-	-	F_0, node1	F_1, node0	-
-	-	-	K_0	-	K_1

g-) Return of Message Pieces

In this final step every piece of message that decrypted returns to Bob. Bob easily gather all pieces he got since he know partition method B and reaches the K message. Alice and Bob agrees upon a successful communication and reports the whole process to center node. As we demonstrated an example communication; all protocol about is splitting and encrypting message while other party tries to recover the message with the helps from other nodes.

Table 7: Knowledge table after step 7.

Alice	Bob	0	1	2	C
S_{AB}	S_{AB}	-	-	-	-
-	-	V_{C0}	V_{C1}	-	M_0, M_1
V_{C0}, V_{C1}	-	-	-	-	-
P, K	P'	-	-	-	-
-	-	-	F_0, node1	F_1, node0	-
-	-	-	K_0	-	K_1
-	K	-	-	-	

3. Implementation of Ring-Based Quantum Network with QKD and QDS on SQUANCH

SQUANCH is simple quantum node and channel simulation framework written in Python. This framework allows us to simulate quantum networks. While SQUANCH unfortunately allows only few nodes for reasonable simulation, it sure runs fast among the other tools. We coded the same network in Figure 3 with SQUACNH and run the exact same example we gave previous section.

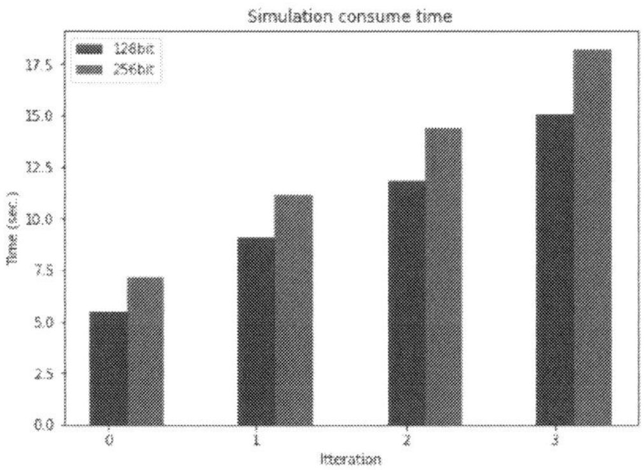

(a) Simulation consume time.

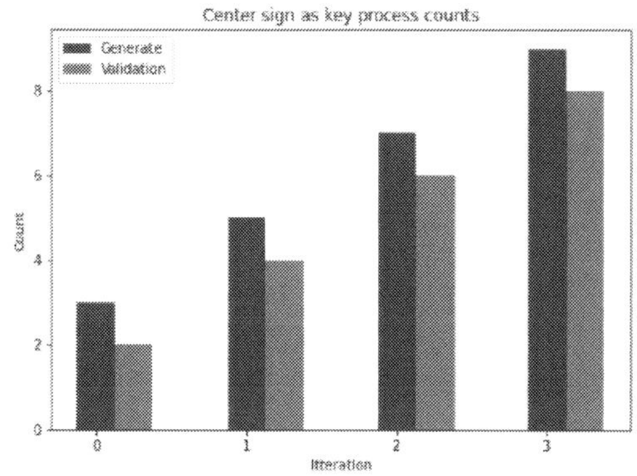

(b) Center node sign as key process counts.

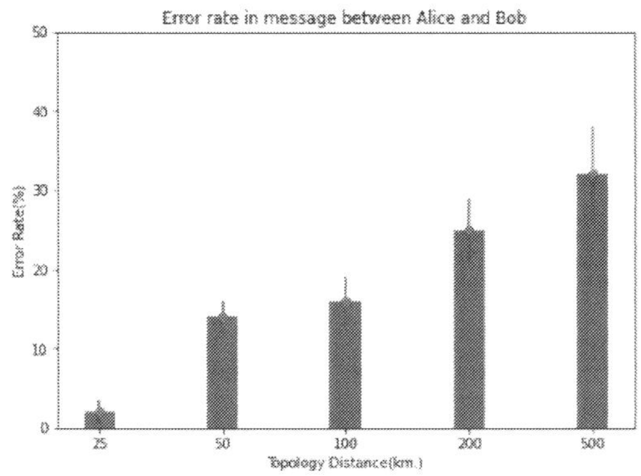

(c) Error rate in different total distances. A special noise pattern is used.

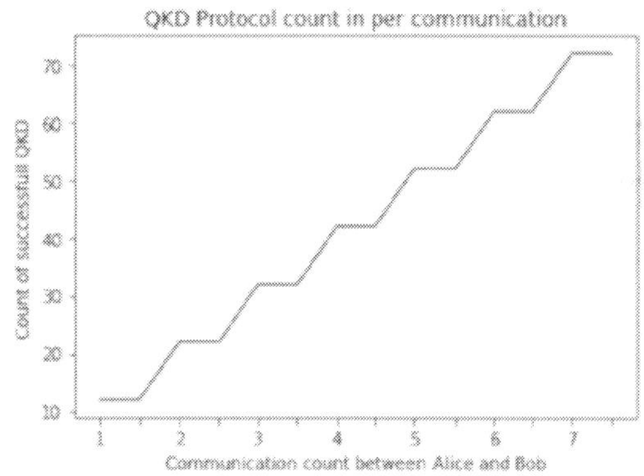

(d) QKD protocol count over communication count.

Figure 3: Various test results, Intel Core i7-10750H CPU, Python 3.7 on Linux.

4. Conclusion

In this article, we study a ring topology-based quantum network with QKD and QDS for the quantum internet. We implemented the purposed network in the SQUANCH network simulator.

We will discuss about security on the suggested topology. While all nodes must run QKD protocol for any communication but we always thought QKD is penetrable protocol. The first aim is about how to always check for observers.

In the suggested topology first two ring of network should locally close to center nodes and always supervise. We prefer other rings expand exponential in distance since we wanted to maximize security for long distance. The nodes of first two ring should able to hold original signs from center node and their count in their ring also should lower than any ring in topology.

The other topic would be about generation of signs. Center node should never generate same sign from past. Sign holder nodes should throw old signs as new request made. Center node only approver of sign other then sign holder.

The other subject we need to discuss about is partition method of messages. We imagine dynamic and huge pool for partition methods. Center node should handle all syncing these pool over whole topology. In addition to classical complexity we prefer a dummy message to start a new communication between nodes.

The topology in our work may appear to much centralized but in the any communication other party nodes never gets the complete of message. Even in worst case scenario where center node decrypts all piece of message center node do not know partition method.

Attacker must penetrate QKD protocol and doing this for all channels of a target node. But being success with this method also drops if attacker tries to listen a high-level ring node since that node have much higher channel count. Attacker should perfectly listen $n+1$ quantum channel for a success penetration. Also success depends on attacker knows message partition method.

Acknowledgments

This work was supported by TUBITAK under the agreement no: 120E087.

References

[1] SQUACH *"A distributed simulation framework for quantum networks and channels,"*: B. Bartlett, arXiv: 1808.07047 [quant-ph], Aug. 2018.
[2] M. Peev, C. Pacher, R. Alléaume, C. Barreiro, J. Bouda, W. Boxleitner, T. Debuisschert, E. Diamanti, M. Dianati, J. Dynes et al. *"The secoqc quantum key distribution network in vienna,"*. New Journal of Physics, vol. 11, no. 7, p. 075001, 2009.
[3] M. Sasaki, M. Fujiwara, H. Ishizuka, W. Klaus, K. Wakui, M. Takeoka, S. Miki, T. Yamashita, Z. Wang, A. Tanaka et al. *"Field test of quantum key distribution in the tokyo qkd network,"*. Optics express, vol. 19, no. 11, pp. 10 387–10 409, 2011.
[4] D. Stucki, M. Legre, F. Buntschu, B. Clausen, N. Felber, N. Gisin, L. Henzen, P. Junod, G. Litzistorf,P. Monbaron et al. *"Long-term performance of the swiss quantum quantum key distribution network in a field environment,"*. New Journal of Physics, vol. 13, no. 12, p. 123001, 2011.
[5] S. Wang, W. Chen, Z.-Q. Yin, H.-W. Li, D.-Y. He, Y.-H. Li, Z. Zhou, X.-T. Song, F.-Y. Li, D. Wang et al. *"Field and long-term demonstration of a wide area quantum key distribution network,"*. Optics express, vol. 22, no. 18, pp. 21 739–21 756, 2014.
[6] J.-P. Chen, C. Zhang, Y. Liu, C. Jiang, W. Zhang, X.-L. Hu, J.-Y. Guan, Z.-W. Yu, H. Xu, J. Lin et al. *"Sending-or-not-sending with independent lasers: Secure twin-field quantum key distribution over 509 km,"*. Physical review letters, vol. 124, no. 7, p. 070501, 2020.
[7] D. Lago-Rivera, S. Grandi, J. V. Rakonjac , A.Seri, H. Riedmatten. *"Telecom-heralded entanglement between multimode solid-state quantum memories, "*. Nature, 594, p.37–40, 2021.
[8] R. Courtland. *"China's 2,000-km quantum link is almost complete,"* IEEE Spectrum, vol. 53, no. 11, pp. 11–12, 2016.
[9] J. Yin, Y. Cao, Y.-H. Li, S.-K. Liao, L. Zhang, J.-G. Ren, W.-Q. Cai, W.- Y. Liu, B. Li, H. Dai et al. *"Satellite-based entanglement distribution over 1200 kilometers,"*. Science, vol. 356, no. 6343, pp. 1140–1144, 2017.
[10] E. Gibney et al. *"Billion-eboost for quantum tech,"*. Nature, vol. 523, p. 426, 2016.

[11] E. Cartlidge. *"Europe's e1 billion quantum flagship announces grants,"*. 2018.
[12] *"Quantum internet alliance,"* https://quantum-internet.team, 2018.
[13] C. Monroe, M. G. Raymer, and J. Taylor. *"The us national quantum initiative: From act to action,"*. Science, vol. 364, no. 6439, pp. 440–442, 2019.
[14] S. K. Joshi, J. Pienaar, T. C. Ralph, L. Cacciapuoti, W. McCutcheon, J. Rarity, D. Giggenbach, J. G. Lim, V. Makarov, I. Fuentes et al. *"Space quest mission proposal: experimentally testing decoherence due to gravity,"*. New Journal of Physics, vol. 20, no. 6, p. 063016, 2018.
[15] S.-K. Liao, H.-L. Yong, C. Liu, G.-L. Shentu, et al. *"Long-distance free-space quantum key distribution in daylight towards inter-satellite communication,"*. Nature Photonics 11, 509 (2017).
[16] S.-K. Liao, W.-Q. Cai, W.-Y. Liu, L. Zhang, et al. *"Satellite-to-ground quantum key distribution,"*. Nature 549, 43 (2017).
[17] H. Takenaka, A. Carrasco-Casado, M. Fujiwara, M. Kitamura, M. Sasaki, and M. Toyoshima. *"Satellite to ground quantum-limited communication using a 50-kg class microsatellite,"*. Nature Photonics 11, 502 (2017).
[18] J.-G. Ren, P. Xu, H.-L. Yong, L. Zhang, S.-K. Liao, J. Yin, W.-Y. Liu, W.-Q. Cai, M. Yang, L. Li, et al. *"Ground-to-satellite quantum teleportation,"*. Nature 549, 70 (2017).
[19] S.-K. Liao, W.-Q. Cai, J. Handsteiner, B. Liu, et al. *"Satellite-relayed intercontinental quantum network,"*. Physical Review Letters 120, 030501 (2018).
[20] L. Calderaro, C. Agnesi, D. Dequal, F. Vedovato, et al. *"Towards quantum communication from global navigation satellite system,"*. Quantum Science and Technology 4, 015012 (2018).

Analysis of an algorithm used for finding Unipolar Orthogonal Codes

Vinutha J N[1], Hitesh Trivedi[2], Suraiya Tarannum[3] and Harshavardhana M N[1]
[1]Cambridge Institute of Technology, Bengaluru, India
[2]ARDC, HAL, Bengaluru, India
[3]HKBK, College of Engineering, Bengaluru, India

vinutha.19lde07@cambridge.edu.in

Abstract. An Auto-Correlation and Cross-Correlation property of an Orthogonal Code enables the transfer of data with less interference. With more Orthogonal Codes, an Optical Fiber Communication channel can support a large number of users efficiently and reliably with high bandwidth. This paper analyses an algorithm used for finding Unipolar Orthogonal Codes using four major parameters (n, w, λa, and λc) where n is the length of a binary string, w is the weight of the string n (number of one's in the string), λa is the Auto-Correlation, and λc is the Cross-Correlation. These four parameters jointly with the Johnson Bound (J_A) condition are used in the algorithm. The paper illustrates the analysis of a possible number of Orthogonal Code Sets calculated using Orthogonal Codes generated from a software code that is based on the algorithm in consideration. The data analysis is carried out and the findings are summarized.

Keywords. Data Analysis, Orthogonal Codes, DOP, Correlation Code Matrix, DOP Matrix, Auto-Correlation, Cross-Correlation, Constraint, and Johnson Bound (JA).

1. Introduction

Orthogonal Codes are generated using four parameters (n, w, λa, and λc) where n is the length of the binary string, w is the weight of the string (count of one's in the string), λa is the Auto-Correlation, and λc is the Cross-Correlation.

1.1. Auto-Correlation and Cross-Correlation

The λa and λc are calculated using DOP Matrix and used to form a Correlation Code Matrix. λa and λc are defined as for formula (1) and (2) presented in [1,3].
By considering, two binary strings P and Q,
P = (P0, P1. . . Pn-1)
Q = (Q0, Q1. . . Qn-1); where P_t, Q_t ∈(0,1),
Auto-Correlation is defined as,

$$\lambda a \geq \sum_{t=0}^{n-1} P_t \, P_{t \oplus \tau} \quad \text{and}$$

IOP Conf. Series: Materials Science and Engineering 1187 (2021) 012021 doi:10.1088/1757-899X/1187/1/012021

$$\lambda a = \sum_{t=0}^{n-1} Q_t \, Q_{t\oplus\tau} \quad \text{for } 0 < \tau \leq n-1$$

when $\tau = 0$, $\lambda a = w$, which is the peak of the Auto-Correlation. (1)

Cross-Correlation is defined as,

$$\lambda c \geq \sum_{t=0}^{n-1} P_t \, Q_{t\oplus\tau} \quad \text{and}$$

$$\lambda c \geq \sum_{t=0}^{n-1} Q_t \, P_{t\oplus\tau} \quad \text{for } 0 < \tau \leq n-1$$

where $t\oplus\tau$ is equal to the convolution of $(t + \tau) \bmod n$. (2)

There are two strings P and Q in which w is overlapped with the circular shifted version for the P string itself. The $t\oplus\tau$ is the reason for the circular shifts; it is represented in formula (1). Formula (2) is used to determine the w overlapped with a circular shifted version of the P and Q string [3].

The method of generation of maximum Orthogonal Codes uses four parameters (n, w, λa, and λc). By selecting Auto-Correlation (λa) and Cross-Correlation (λc) constraints from the Correlation Code Matrix, the maximum Unipolar Orthogonal Code generation is carried out. For the Orthogonal Code generation Johnson Bound (J_A) condition is used.

1.2. Johnson Bound

If $\lambda a=\lambda c=\lambda$, where $1\leq \lambda \leq w-1$, then J_A (n, w, λ), J_B (n, w, λ) and J_C (n, w, λ) (Johnson Bounds) [3,4], are used for computing Maximum Orthogonal Codes (Z) possible for different conditions of w, n, and λ.

If $w < n$ then J_A is given as,

$$Z \leq \left\lfloor \frac{1}{w} \left\lfloor \frac{n-1}{w-1} \cdots\cdots\cdots \left\lfloor \frac{n-\lambda}{w-\lambda} \right\rfloor \right\rfloor \right\rfloor = J_A\,(n, w, \lambda)$$

If $w > n\lambda$ then J_B is given as,

$$Z \leq \min\left(1, \left\lfloor \frac{w-\lambda}{w^2-n\lambda} \right\rfloor\right) = J_B\,(n, w, \lambda)$$

If $(w-k)^2 > (n-k)(\lambda-k)$, k an integer, $1\leq k \leq \lambda-1$ then J_C is given as,

$$Z \leq \left\lfloor \frac{1}{w} \left\lfloor \frac{n-1}{w-1} \cdots\cdots \left\lfloor \frac{n-(k-1)}{w-(k-1)}\, h \right\rfloor \right\rfloor \right\rfloor = J_C\,(n, w, \lambda)$$

$$h = \min\left(n-k, \left\lfloor \frac{(n-k)(w-\lambda)}{(w-k)^2-(n-k)(\lambda-k)} \right\rfloor\right)$$

$\lfloor x \rfloor$ = largest integer value less than 'x'.

There are three conditions in Johnson Bound J_A, J_B, and J_C as given above. The algorithm under analysis is limited to using only Johnson Bound J_A for which $w < n$.

IOP Conf. Series: Materials Science and Engineering 1187 (2021) 012021 doi:10.1088/1757-899X/1187/1/012021

2. Analyzed Algorithm

There is a method to generate a maximum number of Unipolar Orthogonal Codes. The number of users allowed to use the Optical Fiber Communication channel is equal to the total number of Orthogonal Codes generated.

2.1. Steps used to generate the maximum Orthogonal Codes are the following

- Generate DOP and DOP Matrix using n and w.
- Calculate λa and λc using DOP Matrix.
- Form a Correlation Code Matrix using λa and λc values. It will be of a size equivalent to the number of DOP.
- The λa and λc constraints are selected from the Correlation Code Matrix generated in step 2.
- Using (n, w, λa, and λc), the maximum number of Orthogonal Codes (Z) are generated with a Johnson Bound (J_A) condition.
- The maximum possible numbers of Orthogonal Sets are generated using Orthogonal Codes generated in step 4. While generating maximum Orthogonal Sets, the Johnson Bound (J_A) condition λa = λc = λ is used, given in introduction Para I.

Among the generated Orthogonal Sets, only one of these sets will be used for applications in incoherent Optical CDMA [1].

Example:

Table 1 shows some of the combinations of n and w with condition (**w < n**).

Table 1. Few examples for values n and w

n = 7 and w = 3	1110000
	1101000
	1100100
	0101010
n = 10 and w = 2	0011000000
	0101000000

2.2. DOP Generation

Firstly, compute all the **Differences of Positions**, DOP [2, 3] for given input values of n and w. The DOP ($x_1, x_2, x_3 ..., x_w$) presents combinations of n and w as shown in table 1. There are two standard formats followed for the generation of DOP for given n and w. Following conditions (i) and (ii) are used in DOP generation.

i. $x_w \geq (x_1, x_2, x_3 .... x_{w-1})$

ii. $\left\lceil \frac{n}{w} \right\rceil \leq x_w \leq (n\text{-}w+1)$

$\lceil x \rceil$ = ceiling integer > x

The following points are to be considered before generating DOP:

- Position of one's in a binary string, for example, 01101000, is (2,3,5)
- But the Differences of Positions (DOP) for the same string is (1,2,5)

Here DOP is defined as the numbers of positions that are there between two 1's from left to right and circular shift.

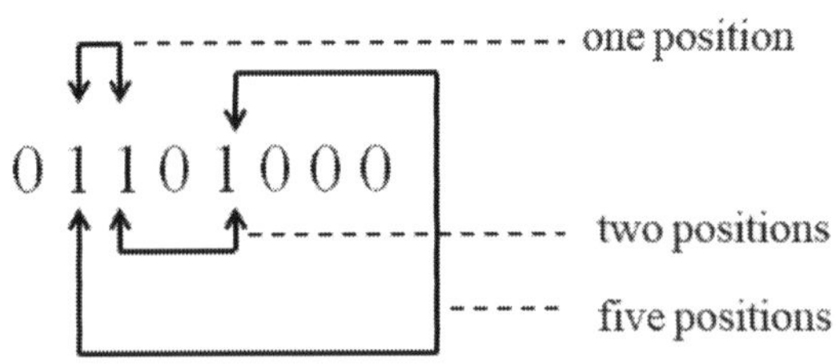

Figure 1. DOP for '01101000' is (1,2,5)

IOP Conf. Series: Materials Science and Engineering 1187 (2021) 012021 doi:10.1088/1757-899X/1187/1/012021

Figure 1 shows the DOP relation concerning for to position of 1's in the binary string. The table 2 shows the DOP generation for n = 8 and w = 3.

Table 2. DOP for inputs n and w

n = 8 and w = 3				Differences of Positions (DOP)			
01101000				(1,2,5)			
x_1	x_2	x_3	x_4	x_5	x_6	x_7	x_8
7th bit	6th bit	5th bit	4th bit	3rd bit	2nd bit	1st bit	0th bit
0	1	1	0	1	0	0	0

2.3. Correlation Code Matrix Generation

To generate a Correlation Code Matrix, initially, it is required to form a DOP Matrix. Using the DOP matrix, λa and λc are calculated, which are the elements of the Correlation Code Matrix. The DOP Matrix is formed using the conditions presented under DOP Generation and the following is the format of the DOP matrix.

$$\begin{bmatrix} x_1 & x_1+x_2 & \dots & x_1+x_2+\cdots.+x_{w-1} \\ x_2 & x_2+x_3 & \dots & x_2+x_3+\cdots.+x_w \\ x_3 & x_3+x_4 & \dots & x_3+x_4+\cdots.+x_1 \\ \dots & \dots & \dots & \dots \\ x_w & x_w+x_1 & \cdots & x_w+x_1+\cdots.+x_{w-2} \end{bmatrix}$$

Table 3 contains DOP Matrix which is computed using the above format.
The DOP Matrix for w = 3 are given as follows:

$$\begin{bmatrix} x_1 & x_1+x_2 \\ x_2 & x_2+x_3 \\ x_3 & x_3+x_1 \end{bmatrix}$$

For a binary string = 01101000, the DOP is (1,2,5) and the DOP Matrix is given as:

$$\begin{bmatrix} 1 & 3 \\ 2 & 6 \\ 5 & 7 \end{bmatrix}$$

2.3.1. Auto-Correlation (λa) computation

Table 3 shows the computation of Auto-Correlation using the DOP Matrix. Compare two different rows of the same DOP matrix. Note the common elements (N) between them.
Compute λa using condition: $i \neq j$ & $N_{ij} = N_{ji}$, where 'i' and 'j' are row indices of DOP Matrix.

λa = maximum number of common elements+1, [3]

For example, if string= 01101000, w=3, $\lambda_{11}=N_{ij}$, $i \neq j$

- N12 = compare 1st row (1 3) with 2nd row (2 7) of DOP Matrix = 0
- N13 = compare 1st row (1 3) with 3rd row (5 6) of DOP Matrix = 0
- N23 = compare 2nd row (2 7) with 3rd row (5 6) of DOP Matrix = 0

Therefore, $\lambda a = \lambda_{11} = 0+1 = 1$.

Table 3. Computation of λa and λc using DOP Matrix

Code (n=8,w=3)	DOP	DOP Matrix	λa	λc
11100000	(1,1,6)	$\begin{bmatrix}1&2\\1&7\\6&7\end{bmatrix}$	2	2
01101000	(1,2,5)	$\begin{bmatrix}1&3\\2&7\\5&6\end{bmatrix}$	1	2
11001000	(1,3,4)	$\begin{bmatrix}1&4\\3&7\\4&5\end{bmatrix}$	2	2

2.3.2. Cross-Correlation (λc) computation

Table 3 also shows the computation of Cross-Correlation using the DOP Matrix. Cross-Correlation is given by:

$$\lambda c = \text{Max}\ \{N_{ij}\} + 1\ \forall_{i,j}$$

It increases the complexity; hence formula can be used as:

$$\lambda c = \text{Max}\ \{N_{ij}\} + 1\ \forall_{i}$$

The Correlation Code Matrix is formed with λa and λc as its elements. λa are the diagonal elements and λc are the remaining elements of the Correlation Code Matrix. The format for the Correlation Code Matrix is as follows:

$$\begin{bmatrix} & code1 & code2 & code3 & codex & codeN \\ code1 & \lambda_{11} & \lambda_{12} & \lambda_{13} & \cdots & \lambda_{1N} \\ code2 & \lambda_{21} & \lambda_{22} & \lambda_{23} & \cdots & \lambda_{2N} \\ code3 & \lambda_{31} & \lambda_{32} & \lambda_{33} & \cdots & \lambda_{3N} \\ codex & \cdots & \cdots & \cdots & \cdots & \cdots \\ codeN & \lambda_{N1} & \lambda_{N2} & \lambda_{N3} & \cdots & \lambda_{NN} \end{bmatrix}$$

2.4. Maximum Orthogonal Codes (Z) Generation:

From the Correlation Code Matrix given above, Constraints λ = λa = λc are taken. Using parameters n, w, and λ, computation of a maximum number of orthogonal codes are carried out by following the Johnson Bound (J_A) condition.

2.5. Maximum Orthogonal Code Sets Generation:

The maximum possible number of Orthogonal Code Sets are calculated using the maximum number of Orthogonal Codes generated using n, w, λa, and λc [2].

3. Findings of the Analysis

To evaluate the performance of the considered algorithm, coding and data analysis are carried out using MATLAB R2018a. Also, a parallel analysis is carried out using Python 2.7 Language. **The maximum possible Orthogonal Sets generated are 375168 for 6 Orthogonal Codes for n = 39 and w =3.**

3.1. The following are the key points of the data analysis carried out:

A. The n < w is not a valid entry as per the Johnson Bound (J_A) condition which is used for the generation of maximum Orthogonal Codes (Z).

IOP Conf. Series: Materials Science and Engineering 1187 (2021) 012021 doi:10.1088/1757-899X/1187/1/012021

B. Case w = 1:

Based on condition $1 \leq \lambda \leq w-1$, giving w=1 as input is invalid as the above condition is not satisfied.

C. Case w = 2:

- For any value of n, constraint λ=1, the Z increases with the value of n.
- For any value of n, all the elements of the Correlation Code Matrix are 1. Therefore, the λa and λc constraint equal to 1 is used. It is observed that the resultant maximum Unipolar Orthogonal Code is equal to the Size of the Correlation Code Matrix.
- For any value of n, the constraint λ=2, cannot be used, if it is used Z becomes infinite, and the 'for loop' error arises in software code. The reason is that there is no element in Correlation Code Matrix which is equal to 2. So, it is important to observe the Correlation Code Matrix to select constraint λ for generating Z.
- For w=2, constraint λ=1, the code runs without any errors for any value of n. It is observed that Z is half of the value of n for most of the cases.
- The plot of Z is shown in figure 2 for various lengths of binary strings from n=3 to 101, w=2, and constraint λ=1. But the maximum number of possible Orthogonal Code Sets is computed to be only 1, for all the cases.

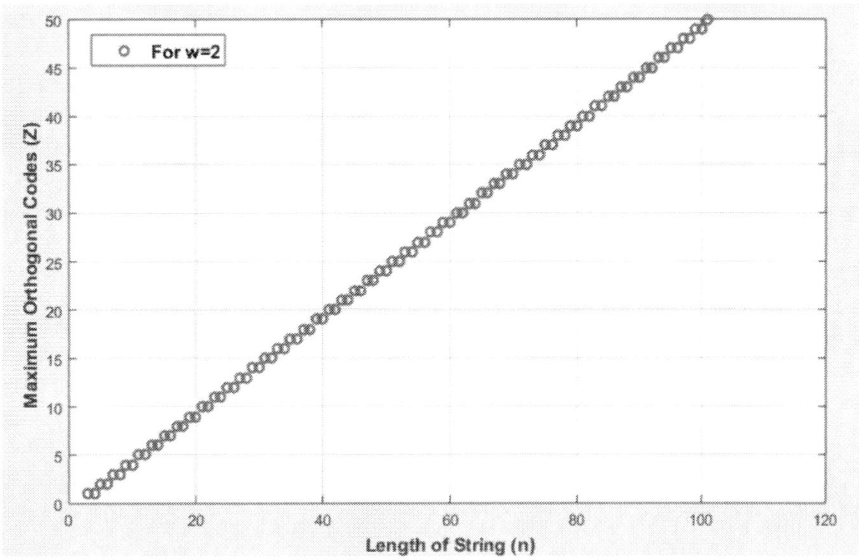

Figure 2. Maximum Orthogonal Codes for w=2

D. Case w = 3:

- The generation of Z for constraint λ=1 is carried out for n=4 to n=40.
- Whenever Z value exceeding 6 for constraint λ=2 the possible Orthogonal Code Sets are not generating as Code Sets Matrix is exceeding its index.
- The plot of Z for various lengths of binary strings from n=4 to 40, w=3, constraint λ=1 is shown in figure 3.

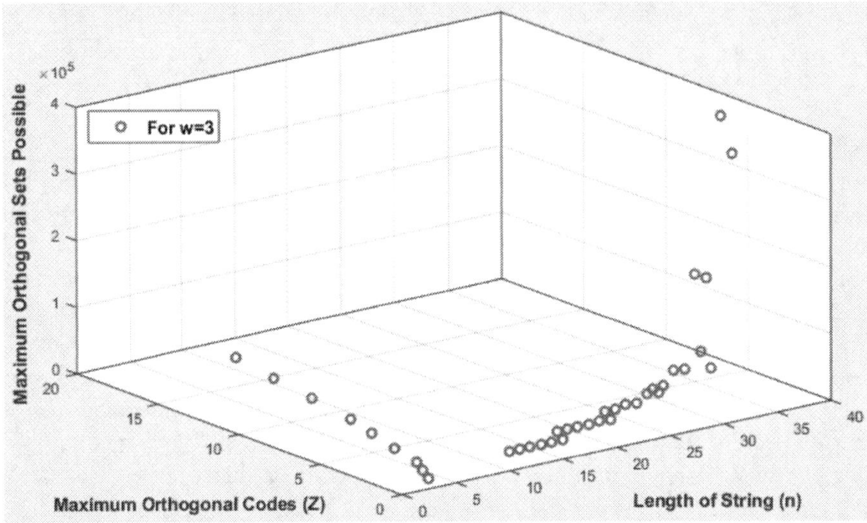

Figure 3. Maximum Orthogonal Sets Possible and Maximum Orthogonal Codes for w=3

E. Case w = 4:

- The generation of Z for constraint λ=3 is carried out for n=5 to n=30.
- The possible maximum Orthogonal Code Sets are generated only for n =5 and n=6.
- For other values of n from 7 to 30, the Orthogonal Code Sets are not generated as the Code Sets Matrix is exceeding its index.
- Figure 4 shows the plot of Maximum Orthogonal Codes for various lengths of binary strings from n=5 to 30, w=4 and constraint λ=3.

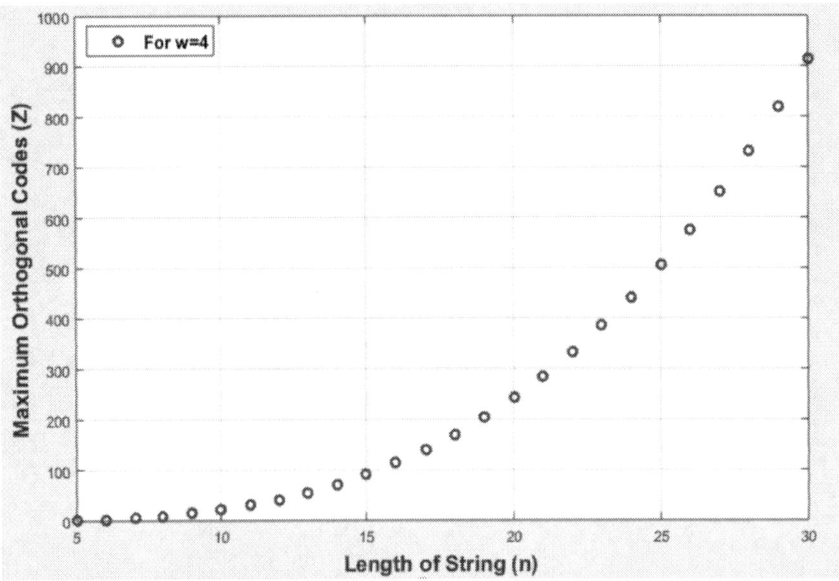

Figure 4. Maximum Orthogonal Codes for w=4

F. For larger values of w, eg. w=5,6, etc. the Z can be generated but Orthogonal Code Sets cannot be generated. Hence, the algorithm is analyzed for w=2 to 4.

G. It is also found that there is a need for an industrial-grade PC with high-speed computing hardware for this algorithm to be faster and efficient. This will help users to find out the maximum number of Orthogonal Code Sets possible rapidly.

IOP Conf. Series: Materials Science and Engineering 1187 (2021) 012021 doi:10.1088/1757-899X/1187/1/012021

H. The maximum Orthogonal Codes (Z) are equal to the highest possible number of users of the optical fiber communication channel. The maximum possible Orthogonal Code Sets are generated which can be used for applications in incoherent CDMA systems [3].

4. Conclusion

The considered algorithm presents a simplified way to obtain possible Orthogonal Code Sets from the maximum number of Unipolar Orthogonal Codes. The algorithm was coded in Matlab and Python separately. Python code is found to be sluggish compared to the Matlab code. The computer processor becomes idle for a long time while running the iterations in Python. It was assumed that Python code will give faster results, but it did not because the complexity of 'for loop' remains the same for an algorithm. Many iterations were carried out in both the software. Data Analysis of the output offered out many findings, which are mentioned in Para IV. These findings have also led to the unearthing of new problems which can be further explored and solved by engineers and researchers working in this technical field.

5. References

[1] Chung F R K, Salehi J A and Wei V K 1989 Optical orthogonal codes: design, analysis and applications in *IEEE Transactions on Information Theory* **35** pp 595-604

[2] Chauhan R C S, Asthana R and Singh Y N 2010 A general algorithm to design sets of all possible one dimensional unipolar orthogonal codes of same code length and weight *IEEE International Conference on Computational Intelligence and Computing Research* pp. 1-7

[3] Harshavardhana M N and Tarannum S 2019 Algorithm for Enhanced Design Set of all Possible Unipolar Orthogonal Codes *IEEE International Conference on Electrical, Computer and Communication Technologies (ICECCT)* pp. 1-3

[4] Omrani R, Kumar P V 2006 Codes for Optical CDMA. In: Gong G., Helleseth T., Song HY., Yang K. (eds) Sequences and Their Applications – SETA **4086**

IOP Conf. Series: Materials Science and Engineering 1187 (2021) 012022 doi:10.1088/1757-899X/1187/1/012022

Modelling of partial discharge pulse propagation in transformer windings using multi conductor transmission line modelling scheme

Annie Achankunju[1], Jineeth Joseph[1], Ancy Sara Varghese[1], K R M Nair[2]
[1]Department of EEE
Saintgits College of Engineering, Kottayam, Kerala
[2] Independent Consultant in Transformer Designing and Testing
annieachankunju97@gmail.com
jineeth.joseph@saintgits.org
ancy.sara@saintgits.org

Abstract. Power transformers are considered as an essential component in the electric power systems. They are placed in services in different circumstances like electrical, environmental and mechanical stresses, and through the progress of operation it can affect with various threats and hazards. Transformer failure should be diagnosed earlier in order to avoid its unsafe and unstable operation. The main reason for failure of insulation in power transformer is due to partial discharges. The partial discharge activity occurring in the transformer have to be identified earlier, afore it progresses into a complete discharge, otherwise it results in the maintenance of the device and causes loss of economic cost to the utilities. In this paper, multi conductor transmission line theory is used for the modelling of transformer windings and the partial discharge pulse propagation through the winding is simulated in the MATLAB Simulink. The effect of partial discharge on the windings are also discussed.

Keywords- Partial Discharge (PD), Transformer Winding Modelling, Pulse Propagation, Multiconductor Transmission Line (MTL), MATLAB Simulink

1. INTRODUCTION

Partial discharge can occur in the transformers due to degradation processes of the insulation systems and ageing. Several factors including thermal, mechanical and electrical overstressing are the reasons for the frequent occurrence of partial discharge activities in the transformer. This PD activity happening through primary stages of process can be due to the imperfections of design or processing during manufacture [1]. PD measurement is an authentic and efficient method and using this method the insulation condition of an electrical asset can be diagnosed at all times. Regular partial discharge measurements and continuous monitoring collects very sensitive information to efficiently identify the stresses and weak points occurring in the insulation structure. Systematic partial discharge measurement in the system can safeguard the reliable, efficient and long-term operation of electrical equipment without causing any economic loss to the utilities [2].

There are various methods available for partial discharge detection in transformer windings. The primary set contains studies about partial discharge measurement. It is divided into electric, audio and optical detection. The second type consist of data analysis measurement for identifying the kind of fault and its effect. The various technique employed for the analysis of data includes signal processing, statistical methods etc. Transformer winding models [3-4] are also very useful for partial discharge studies. For the analysis purpose, either the Lumped parameter or distributed parameter archetypal can be used. The MTL is very useful for the partial discharge studies. In MTL process, the winding is treated as parallel lines having the equivalent length. Each turns or winding of the transformer can be modelled in terms of a transmission line. The propagation of partial discharge pulses taking place in the transformer windings can be effectively analysed using this method of modelling. This method can also be suitably used for frequency up to several MHz [5].

In this paper, pulse propagation due to partial discharge occurring in the windings of transformer is investigated. A transformer winding is designed and the parameters required for the studies are calculated. In section 2, the structure of the transformer winding, the design steps and obtained values of design parameters are presented. The MTL concept is used for the propagation study of PD pulse in the windings. The third section gives an idea about the MTL theory and description about circuit used in MATLAB Simulink model. The obtained results are presented in section 3.

2. DESIGN AND ESTIMATION OF PARAMETERS OF TRANSFORMER WINDING

2.1. Structure of the transformer winding

The transformer winding is modelled as an equivalent RLC circuit. The winding includes a series connection of resistance and inductance in parallel with a capacitance. A shunt capacitance or parallel capacitance is also connected to the ground. From the first winding to the last, the structure remains the same. The equivalent circuit of a 42 - disc transformer winding model is depicted in Figure 1.

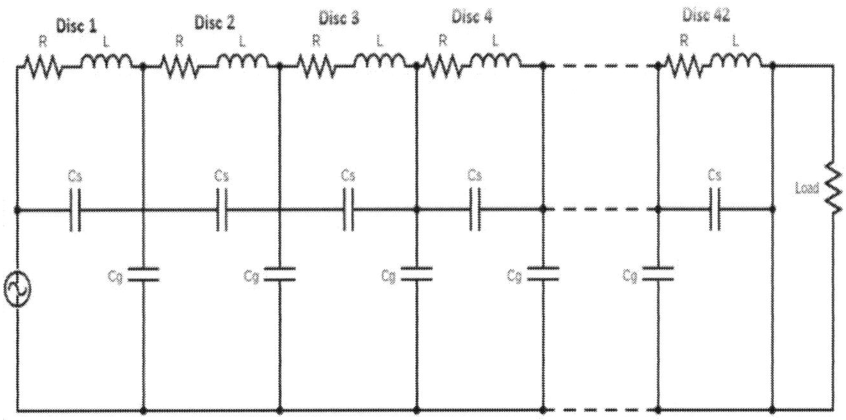

Figure 1. Equivalent circuit of a 42 disc transformer winding model

2.2. Design steps

The transformer winding modelling is a crucial part for the partial discharge propagation study. In the design part the specification of transformer is given. The specification of the transformer is 1 MVA,11KV/415V, delta-star connected dry type transformer. The core circle diameter is initially assumed. The number and width of core steps are selected and the gross area is calculated. The low voltage and high voltage currents and voltages are subsequently calculated. The maximum and rated tap turns are also estimated. An initial current density is selected to compute the low voltage net area and cross sectional area. Then the actual current density is determined. Similarly, the high voltage winding cross-sectional area and conductor size is calculated.

After finding the low voltage and high voltage parameters the calculation of number of discs are calculated. Then the inside and outside diameter of both high voltage winding and low voltage winding is calculated. Based on these values the resistance, inductance and capacitance values of the windings are obtained. The values obtained from the design are given in Table 1.

Table 1: Specifications and design parameters of 1MVA,11KV/415V transformer

Sl.No	Specifications	Values
1	KVA	1000
2	Phases	3
3	Frequency	50 Hz
4	Primary Volts	11,000 v
5	Secondary Volts	415 v
6	Percentage Impedance	5%
7	Tap Range	±5%
8	Core Circle Diameter	266 mm
9	Inside Diameter of LV	282 mm
10	Outside Diameter of LV	371 mm
11	Inside Diameter of HV	437 mm
12	Outside Diameter of HV	545 mm
13	HV rated tap turns	597 turns
14	HV highest tap turns	627 turns
15	Height of winding	800 mm
16	No of Discs	42
17	Flux Density	1.65 T

2.3. Parameters Calculation
For the parameters calculation a transformer winding having forty-two continuous disc type sections is designed. Each winding has a resistance, inductance, capacitance and conductance [9].

2.3.1. Resistance:
The resistance of the disc winding can be calculated using the following equations:

$$R=\rho\frac{l}{A} \quad (1)$$

Where 'R' is the resistance of the winding

'l' is the length

'A' is the area of the conductor

'ρ' is the resistivity

2.3.2. Inductance:
The inductance calculation is done by assuming that the winding comprises of loss-less multi-conductor transmission lines bounded via identical insulator. The inductance is given by:

$$L = 0.001n^2a\,\rho \quad (2)$$

Where 'n^2' is the number of turns in each winding

'ρ'= b/c

'b' = width of the conductor

'c' = radial thickness of the disc.

2.3.3. Capacitance

Based on permittivity of the insulation and geometry of the winding the capacitance is calculated. Inter-turn capacitance (CIT), Disc-Disc capacitance (C_d) and Ground capacitance(C_g) are the main three capacitance used in the calculation. The CIT and C_d comprises the total series capacitance of the winding (C_t) [6].

i) Capacitance between turns of winding(CIT)

$$CIT = 0.974\, D_m \frac{(h+e)}{e} \quad (3)$$

Where D_m- mean diameter of the conductor

h - Width of the bare conductor

e - insulation thickness between conductor

ii) Capacitance between Discs (C_d):

$$C_d = D_m\, 0.851 \frac{(B+t)}{(t+0.874\, e)} \quad (4)$$

Where D_m - mean diameter of discs

e - insulation thickness

B - radial thickness of Disc

t - distance between two discs

iii)Total series capacitance of Disc winding (C_t):

$$C_t = \frac{CIT}{N} + C_d \frac{4}{3} \quad (5)$$

Where CIT- Capacitance between turns of disc winding

C_d - Capacitance between Discs

iv) Ground Capacitance (C_g):

$$C_g = \frac{0.378\,(D_a+D_b)(L+\Delta)}{\Delta-0.43t} \quad (6)$$

Where D_a - outer diameter of low voltage winding

D_b - inner diameter of high voltage winding

L - winding height

Δ - radial gap between windings

t - total insulation thickness between windings

2.3.4. Conductance

As a result of capacitive loss in the insulation conductance(G_a) occurs. The frequency f, capacitance C and the dissipation factor tan δ are the factors the conductance depends on.

$$G_a = 2\pi f\,[C] \tan \delta' \quad (7)$$

Where G_a is the conductance

tan δ' denotes the dissipation factor

The parameters of the winding are calculated using the equations (1- 7) and the values are given in Table 2.

Table 2: Calculated values of transformer parameters

Sl.No	Parameters	Values
1	Resistance	1.03 Ω
2	Inductance	1.3703 mH
3	Capacitance between turns	1207.54 pF
4	Capacitance between Discs	711.031 pF
5	Total series capacitance	976.8 pF
6	Ground Capacitance	754.168 pF
7	Conductance	0.97087 S

3. MODELLING OF PARTIAL DISCHARGE PULSE PROPAGATION IN TRANSFORMER WINDINGS

3.1 Multi Conductor Transmission Line Model

The primary phase of partial discharge identification is to develop a model that can be used in high frequencies to analyse and monitored precisely the behaviour of the transformer windings [7]. One of the best model for evaluating the PD studies and analysis is the MTL model. In the MTL model each turn or winding of the transformer is modelled as a transmission line.

The boundary conditions to realise the multi conductor transmission line theory given in equations (8) and (9) is illustrated in Figure 2:

$$V_R(j) = V_S(j+1) \text{ for } j = 1 \text{ to } p-1 \tag{8}$$

$$I_R(j) = -I_S(j+1) \text{ for } j = 1 \text{ to } p-1 \tag{9}$$

where $V_S(j)$ and $I_S(j)$ are sending end voltage and current of jth transmission line. Similarly $V_R(j)$ and $I_R(j)$ are its receiving end voltage and current. The boundary conditions of equations (8) and (9) are applied to the wave equations, then the sending and receiving end current vectors can be defined in terms of voltage vectors [8].

$$\begin{pmatrix} I_S \\ I_R \end{pmatrix} = \begin{pmatrix} Y_0 \coth([D]l) & -Y_0 \operatorname{cosech}([D]l \\ -Y_0 \operatorname{cosech}([D]l) & Y_0 \coth([D]l) \end{pmatrix} \begin{pmatrix} V_S \\ V_R \end{pmatrix} \tag{10}$$

where $[D]^2 = [Z_a]\,[Y_a]$

$[Y_0] = [Z_a]^{-1} \cdot [D]$

$[Z_a] = R \cdot [In] + j \cdot 2\pi f \cdot [L]$

$[Y_a]=[G_a] + j \cdot 2\pi f \cdot [C]$

$[Z_a]$ and $[Y_a]$ are impedance and admittance matrices of the model respectively. The impedance and admittance matrix comprises of resistance, inductance, capacitance, conductance matrices, frequency (f) and the average length of the line (l) [8].

The matrix D have to be diagonalized using modal transform in order to find the value of coth([D)l) and cosech([D]l). After dual phases of simplification and an inversion of matrix, expanding the first and last rows and applying the terminal conditions all the internal voltages needed for the PD localization can be computed [8].

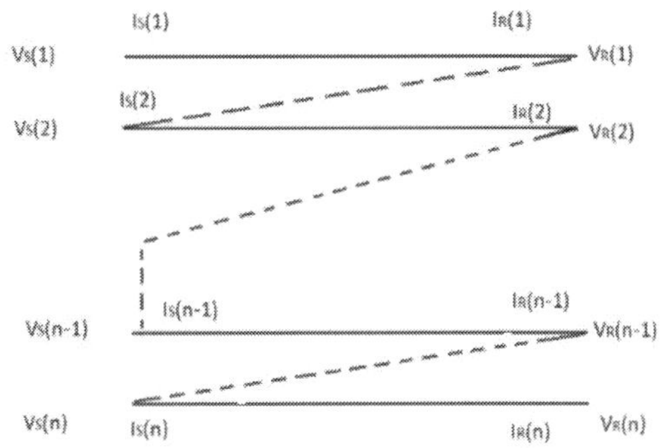

Figure 2. MTL model of Transmission Line

3.2. Simulation

A forty-two disc winding of a transformer is modelled using MTL theory in Simulink. The matlab software is used for carrying out the simulation process. The Simulink model of a disc winding is shown in Figure 3. A partial pulse is given as an input to the disc winding. The expression for the current pulse is given by:

$$\text{Pulse} = A_m \left(\exp^{(-\alpha t)} - \exp^{(-\beta t)}\right) \tag{11}$$

where 'A_m' is the amplitude of the wave and 't' denotes the time instant.

The PD pulse is modelled using a signal builder, and a controlled current source. The pi network is used for modelling the MTL model of the transmission line. Fig 4 depicts the model of a single disc of a transformer winding. Similarly, 42 disc winding model is developed in Simulink. The input PD signal is generated in the signal builder block and is fed to the different discs of the winding through a current controlled source.

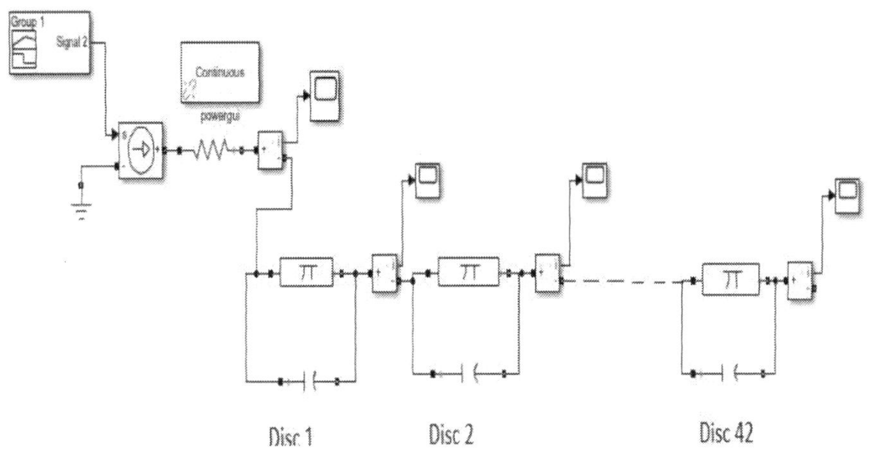

Figure 3. Simulink model of a 42 disc winding transformer

IOP Conf. Series: Materials Science and Engineering 1187 (2021) 012022 doi:10.1088/1757-899X/1187/1/012022

3.3. Results

In this paper, the PD pulse propagation through a 42 disc winding model is analysed. The input pulse is shown in Figure 4. The input partial discharge pulse generated from the signal builder is applied to the first disc winding of the transformer. As the PD pulse propagates along the transformer winding geometry, attenuation and distortion occurs and the magnitude of the pulse will be slightly small compared to the input. These pulses propagate through the whole winding of the transformer and on reaching the terminal the pulse magnitude will be very low.

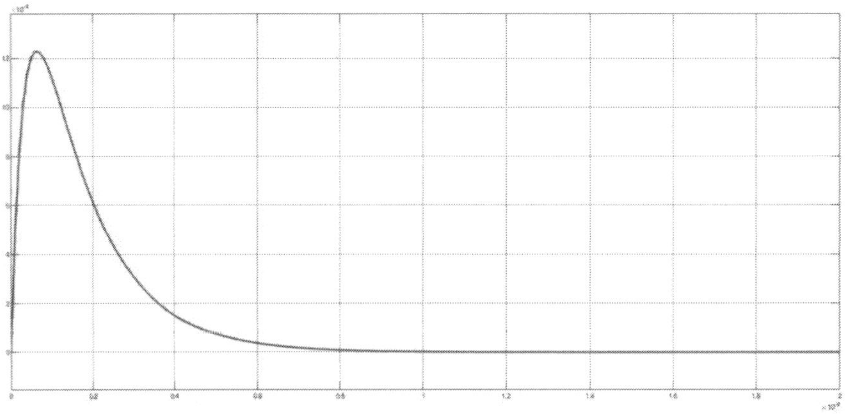

Figure 4. Input partial discharge pulse

This input PD signal in the form of a current pulse is given as a source to the first disc winding. The magnitude of the input PD signal is 0.012. The output obtained from the discs 1 and 2 are shown in Figure 5. The magnitude of the pulses starts decreasing as it travels through each of the disc of the winding.

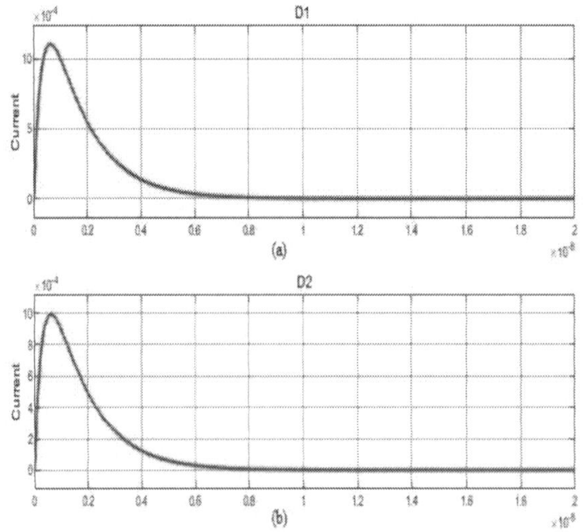

Figure 5: Output waveforms of propagated PD current pulses obtained from: (a) Disc 1 (b) Disc 2

Similarly, the outputs are obtained from discs 5 and 6 when the same input PD signal is given to the fifth disc winding are shown in Figure 6. Likewise, the same input can be applied to the different disc winding to find the characteristics of the PD pulse propagation.

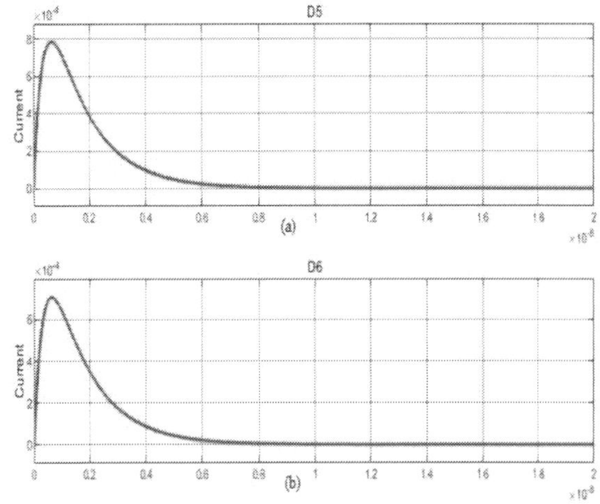

Figure 6: Output waveforms of propagated PD current pulses obtained from:(a) Disc 5 (b) Disc 6

Table 3: Peak value, rise time and settling time of pulse propagates from disc 1

Disc No	Peak Value (mA)	Rise Time (ns)	Tail Time (μs)
1	1.1	2.08	2.28
2	1	1.89	2.05
3	0.9	1.73	1.84
4	0.8	1.58	1.65
5	0.7	1.45	1.49
6	0.63	1.34	1.33

Table 4: Peak value, rise time and settling time of pulse propagates from disc 5

Disc No	Peak Value (mA)	Rise Time (ns)	Tail Time (μs)
5	0.8	1.47	1.46
6	0.7	1.36	1.31
7	0.62	1.27	1.18
8	0.58	1.20	1.06
9	0.51	1.14	9.51
10	0.45	1.09	8.54

Comparing the results of Figure 5 and Figure 6 with Figure 4, it shows that the peak amplitude of PD pulse goes on decreasing while travelling through the whole disc of the transformer. The results of time domain simulation of PD propagation when a PD signal is applied to disc number 1 is depicted in Table 3. The simulation of PD propagation when the PD signal is applied to disc number 5 is shown in table 4 for comparison. From Tables 3 and Table 4, it is clear that the peak value of pulse is decreasing while travelling through the whole disc. The rise time of the wave is also reduced when it propagates through the disc. The result shows that the peak time, rise time and tail time are reduced when the signal propagation from disc 5 is compared to the propagation from disc 1. The effect of partial discharge pulse occurring in the winding will be very high compared to that of other windings through which it propagates. This establishes that the proposed simulation method is a useful method for partial discharge analysis.

4. CONCLUSION

In this paper, the MTLM theory is used to find the partial discharge pulse occurring in the windings. A 1MVA,11KV/415V transformer is designed and the parameters required for the simulation are calculated. An input PD signal is generated and is fed to the various disc of the windings. As it propagates through the discs the magnitude of the pulses is reduced. The peak value, rise time and tail time of the pulses are also calculated for the analysis purpose. When the pulse propagates through the whole windings of the transformer variations were occurred in the rise time, peak time and tail time of the pulse. These variations can be further analysed to develop suitable methods for the partial discharge source detection and localisation.

REFERENCES

[1] M. S. Abd Rahman, P. L. Lewin' and P. Rapisarda, "*Comparative Study of Partial Discharge Localisation Techniques within Transformer Windings*" 2015 Electrical Insulation Conference (EIC), Seattle, Washington, USA, 7 -10 June 2015.

[2] Mithun Mondal & G.B. Kumbhar," *Detection, Measurement, and Classification of Partial Discharge in a Power Transformer: Methods, Trends, and Future Research*" IETE Technical Review 2017.741, 2008.

[3] S. M. H, Hosseini, M, Vakilian., G.B, Gharehpetian "*Comparison of Transformer Detailed Models for Fast and Very Fast Transient Studies*" IEEE Trans. Power Delivery., Vol. 23, No.2, pp.733.

[4] S. M. H. Hosseini; M, Vakilian, G.B, Gharehpetian "*Partial discharge location in transformers through application of MTL model*" Int'l. Conf. Power Systems Transients (IPST2009), pp.1-6, 2009 Study", J. Electr. Eng. Technology (JEET), Vol. 9, No.1, pp.154-161, 2014.

[5] S.M.H, Hosseini, P. R. Baravati, "*Transformer Winding Modelling Based On Multi-Conductor Transmission Line Model For Partial Discharge".*

[6] S. N. Hettiwatte, P. A. Crossley, MIEEE, Z. D. Wang, MIEEE, A. Darwin, SMIEEE and G. Edwards," *Simulation of a Transformer Winding for Partial Discharge Propagation Studies,*" *IEEE 2002.*

[7] S. M. Hassan Hosseini, Peyman Rezaei Baravati," *New High Frequency Multi-conductor Transmission Line Detailed Model of Transformer Winding for PD Study* ",IEEE Transactions on Dielectrics and Electrical Insulation Vol. 24, No. 1; February 2017.

[8] Ali Mazhab Jafari a,b, Asghar Akbari a," *Partial discharge localization in transformer windings using multi-conductor transmission line model* ", Electric Power Systems Research 78 (2008) 1028–1037.

[9] Zhiye Du, Ling Ruan, Chun Zhao, Jiangjun Ruan, Yunfei Wu, Tao Wang. *"Calculation of Electrical Parameters for Studies on Propagation Characteristic of PD along Transformer Winding"*, 2008 International Conference on Computer and Electrical Engineering.

Simulation Tests of Tokyo Quantum Network

Osman Ceylan[1] **& Ihsan Yılmaz**[2]
[1]Student of Department of Computer Engineering, Çanakkale 18 Mart University, Çanakkale, Turkey
[2]Department of Computer Engineering, Çanakkale 18 Mart University, Çanakkale, Turkey
E-mail: [1]osman.semi.ceylan@gmail.com
E-mail: [2]iyilmaz@comu.edu.tr

Abstract. In this modern era where secure communication is imperative, scientists have always been in search of a new solutions for this problem. One of the best solutions discovered in this century, uses the superiorities of the quantum world. This theorem, called no cloning, mainly uses the collapse of quantum state when observing attribute of qubits. In this study, we will examine the Tokyo Quantum Network [1], which is based on the protocols derived from the BB84 [2] and B92 protocols by implementing in three quantum network simulators.

Keywords: quantum internet, quantum key distribution, quantum network, network simulation

1. Introduction

New studies are added every day in the field of quantum networking. Researchers have shown tremendous work, although studies that have applied on a large scale quantum network applications have been less frequent. In this context, QKD networks are also currently possible for long-distance [3–5]. Nowadays QKD networks reach the distance of 500 km has on fibre based channels [6]. Some countries such as China [7, 8], European countries [9–11] and America [12] invests heavily in the quantum network field. It seems difficult to do such complex studies without simulation. So this reason push us to research network simulations of quantum networks. In this study, we will examine Tokyo Quantum Network more closely and run some tests with three different network simulation software named SQUANCH [13], QuNetSim [14] and SimulaQron [15]. The Tokyo Quantum Network is seen in the Figure 1.

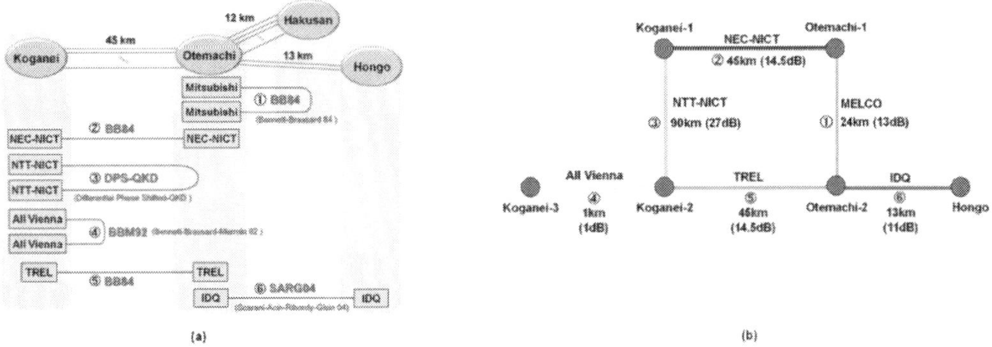

Figure 1: (a) Physical link configuration of the Tokyo QKD Network.
(b) Logical link configuration with 6 nodes [1, p. 4].

The paper is organized as follows: In Section 2, the structure and details of Tokyo Quantum Network is given. In Section 3, simulation software evaluation is given. In Section 4, we share our test results. In Section 5 concludes the article.

2. Tokyo Quantum Network

The Tokyo Quantum Network [1] is one of the principles of large-scale studies in this field. Quantum network was placed in 4 districts of Tokyo, security was always at the forefront, but specific studies were also made for other qualities such as speed communication. Different protocols were tested on different hardware by placing one or more stations in Koganei, Otemachi, Hakusan and Hongo districts in Tokyo. The network has a total of 6 stations, there is a classical network between the devices with the same hardware that connects each station to each other with quantum channels and the whole station can access each other. Along with different organizations such as the Austrian Institute of Technology and Mitsubishi, different qualities of secure communication with quantum technologies were presented at these stations. Some of these stations use the simpler BB84 protocol, while some others use more complex protocols like differential phase shifted quantum key distribution (DPS-QKD) in NTT-NICT machines.

3. Simulation Software Evaluation

Although there are more simulation software available than the original, we chose the best three due to their implementation limitations on this specific work. As we mention before these software or more like frameworks are SimulaQron [15], SQUANCH [13] ve QuNetSim [14]. They are all coded in Python 3.5+.

3.1. SimulaQron

SimulaQron is the only software among them that performs socket simulation using ports. But because of this feature make framework works slower. Although it leaves a lot of work that should be done in a network to the user, SimulaQron is the most professional coded work among them.

3.2. SQUANCH

SQUANCH probably is the simplest coded among them since it do not even support multiprocessing capabilities. But it works faster on networks with few nodes because it is simple. When node count on network goes beyond 10 SQUANCH performs slower then the others in same task. So SQUANCH depends heavily on host computer raw single core performance. But SQUACNH supports simple error patterns which make him more real simulation.

3.3. QuNetSim

We found QuNetSim for user appeal then others. Using QuNetSim framework much easier then the others. The main purpose of QuNetSim is to allow the user to focus on protocol generation as its developers mentioned. While it's not convenient for us that QuNetSim doesn't support noise patterns, some manual operations are not inoperable. unfortunately QuNetSim not assertive in terms of speed.

3.4. Software Implementation

This section is about how we code Tokyo Quantum Network in the mentioned simulation software. As we must mention this first, we did not include NTT-NICT and IDO stations due to the lack of common protocols on simulation software.

As for start from SimulaQron, we treat all stations as a device object and add all stations in the Tokyo Quantum Network. The methods included in the SimlaQron were not used for the classical network. For this job, a process that connects each device and runs in a separate process and is also used as a server for key distribution when needed. Again, this process handles routing when needed. In addition, we have created a monitor process to transfer instant and simulation data. As we mentioned before, SimulaQron expects to user to handle the must have actions in network.

Coding in SQUANCH more less complex than the others because of data locality. But it has the same shortcomings about the must have actions. With the using queues and more threads all stations coded in SQUANCH.

We coded the network on QuNetSim with already existing functions. But we coded the needed monitor process ourself.

IOP Conf. Series: Materials Science and Engineering 1187 (2021) 012023 doi:10.1088/1757-899X/1187/1/012023

4. Test of the Network

In this section, we share test results of network test among frameworks. Our tests involves around Koganei-1, Koganei-2 and Otemachi-1 stations as are fidelity test of Tokyo Quantum Network. We exchanged data between the stations mentioned in our test. Before we start we run all test in high performance AMD Threadripper 3950X CPU in Linux 5+ kernel environment with using Python version 3.7. All frameworks updated to most recent version as of May 2021 and these tests did not run at same time.

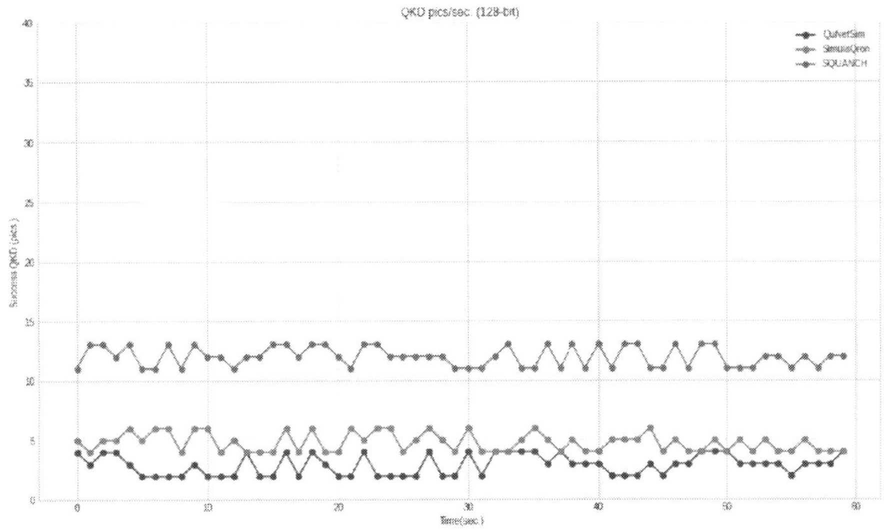

Figure 2: Successful 128-bit QKD counts among frameworks as time passes.

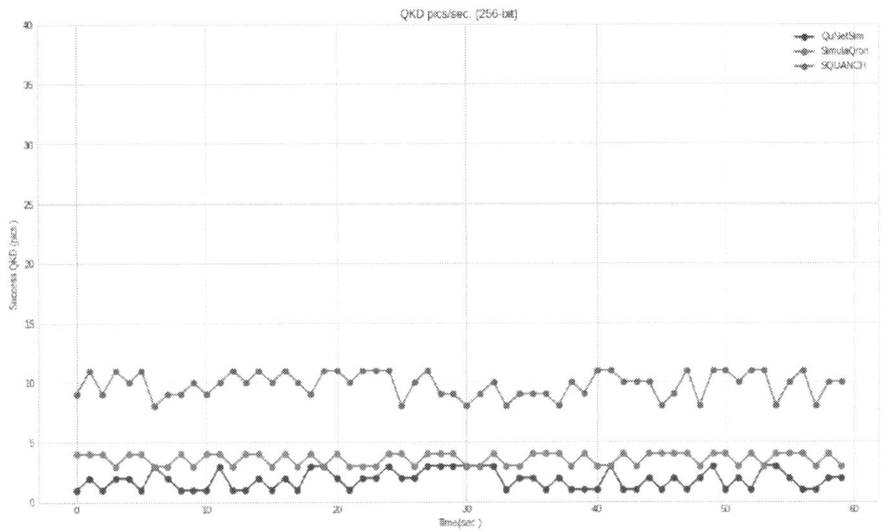

Figure 3: Successful 256-bit QKD counts among frameworks as time passes.

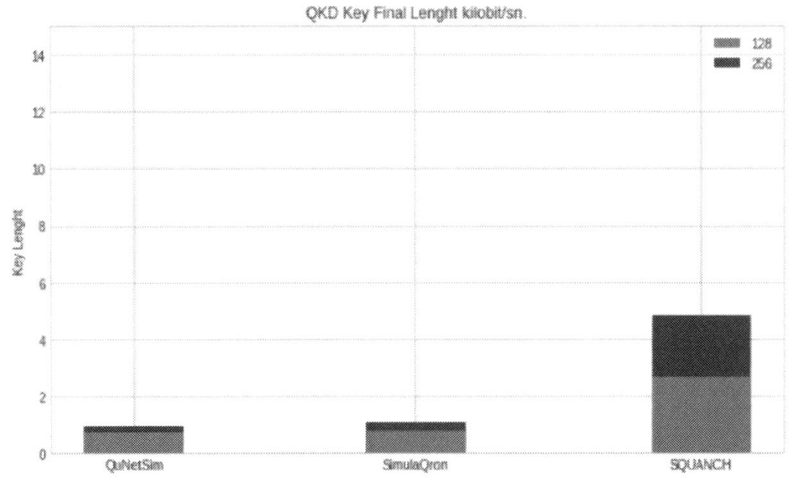

Figure 4: Final key length among frameworks.

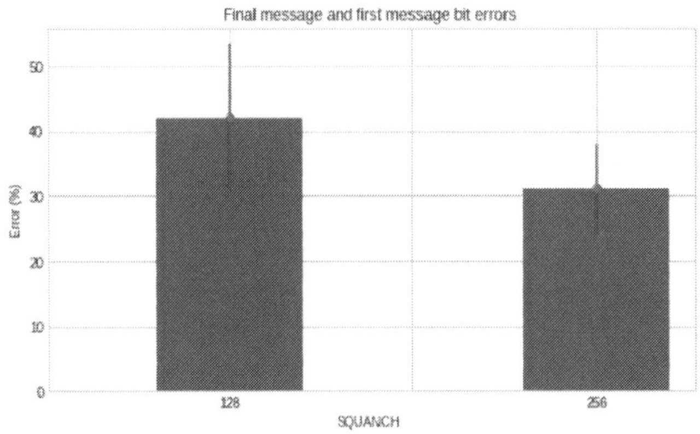

Figure 5: Message error rate in only SQUANCH.

5. Conclusion

The first we tested was total QKD counts on simulation process. If we look at the Figure 2, we see that SQUANCH is much faster. This happens most likely allocating quantum resources on a computer way faster then the others. Again this most likely because of the data locality coded in SQUANCH that lack of multiprocessing. The others performed identical. We believe the reason is they are using same backend called ProjectQ [16].

The next Figure 3 which presents the same test results as before is counts of successful QKD in whole simulation process. This 256-bit test hits hard to especially

IOP Conf. Series: Materials Science and Engineering 1187 (2021) 012023 doi:10.1088/1757-899X/1187/1/012023

QuNetSim. QuNetSim could not even hold 1 QKD/sec. sometimes. While we observed SimulaQron and QuNetSim performs identical, QuNetSim is stays behind most time. The reason for this maybe QuNetSim support of extra services like routing and acknowledge checks.

The next Figure 4 represents the final key length in simulation process. We observed the same act of QuNetSim and SimulaQron in this test too. As we examined further, SQUANCH yielded expected 256-bit results while the others did not. We found this quite interesting because the CPU we used had a high core count and we expected multiprocess supported frameworks should work better. But that was not in the case. Meanwhile the single process framework SQUANCH handled far better then the other. The reason for this most likely related with pre-allocation on quantum resources.

The final Figure 5 is about error rates between sended and received message. Only SQUANCH supports errors in simulation, so we continue with it. We The total quantum channel length for this test was $45 + 24 = 79km$. We used a special pattern to make the noise more computable. The relative result of 128-bit and 256-bit was expected.

As we investigate more Python multiprocessing and multithreading performance is far from ideal expectations, at least for simulation software that mimics quantum resources. But most of these kind of network simulation frameworks, maybe all, written by Python.

With this study, we tested the capacities and reality limits of simulation software. Such framework like the SimulaQron be able to distribute simulation targets to other computers to creates an opportunity to analysis communication. Like the QuNetSim framework, it can help us to understand more about quantum communication protocols. Such as framework the SQUANCH helps us to calculate errors in quantum memory and channels. While all these frameworks abstract most of the complexity of real quantum networks and run too much slow then a real one but we find these software are essential for who wants to built such a quantum network.

Acknowledgments

This work was supported by TUBITAK under the agreement no: 120E087.

References

[1] M. Sasaki, M. Fujiwara, H. Ishizuka, W. Klaus, K. Wakui, M. Takeoka, S. Miki, T. Yamashita, Z. Wang, A. Tanaka et al. *"Field test of quantum key distribution in the tokyo qkd network,"*. Optics express, vol. 19, no. 11, pp. 10 387–10 409, 2011.

[2] C.H.Bennett,G.Brassard,S.Breidbart,andS.Wiesner. *'Quantum Cryptography,or Unforgeable Subway Tokens', to appearin Advances in Cryptography: Proceedings of CRYPTOB2, PlenumPress.* [These CRYPTO82 Proceedings were published in 1983 and thispaper was on pages267–275.]

[3] M. Peev, C. Pacher, R. Alléaume, C. Barreiro, J. Bouda, W. Boxleitner, T. Debuisschert, E. Diamanti, M. Dianati, J. Dynes et al. *"The secoqc quantum key distribution network in vienna,"*. New Journal of Physics, vol. 11, no. 7, p. 075001, 2009.

[4] D. Stucki, M. Legre, F. Buntschu, B. Clausen, N. Felber, N. Gisin, L. Henzen, P. Junod, G. Litzistorf,P. Monbaron et al. *"Long-term performance of the swiss quantum quantum key*

distribution network in a field environment,". New Journal of Physics, vol. 13, no. 12, p. 123001, 2011.

[5] S. Wang, W. Chen, Z.-Q. Yin, H.-W. Li, D.-Y. He, Y.-H. Li, Z. Zhou, X.-T. Song, F.-Y. Li, D. Wang et al. *"Field and long-term demonstration of a wide area quantum key distribution network,"*. Optics express, vol. 22, no. 18, pp. 21 739–21 756, 2014.

[6] J.-P. Chen, C. Zhang, Y. Liu, C. Jiang, W. Zhang, X.-L. Hu, J.-Y. Guan, Z.-W. Yu, H. Xu, J. Lin et al. *"Sending-or-not-sending with independent lasers: Secure twin-field quantum key distribution over 509 km,"*. Physical review letters, vol. 124, no. 7, p. 070501, 2020.

[7] R. Courtland. *"China's 2,000-km quantum link is almost complete,"* IEEE Spectrum, vol. 53, no. 11, pp. 11–12, 2016.

[8] J. Yin, Y. Cao, Y.-H. Li, S.-K. Liao, L. Zhang, J.-G. Ren, W.-Q. Cai, W.- Y. Liu, B. Li, H. Dai et al. *"Satellite-based entanglement distribution over 1200 kilometers,"*. Science, vol. 356, no. 6343, pp. 1140–1144, 2017.

[9] E. Gibney et al. *"Billion-eboost for quantum tech,"*. Nature, vol. 523, p. 426, 2016.

[10] E. Cartlidge. *"Europe's e1 billion quantum flagship announces grants,"*. 2018.

[11] *"Quantum internet alliance,"* https://quantum-internet.team, 2018.

[12] C. Monroe, M. G. Raymer, and J. Taylor. *"The us national quantum initiative: From act to action,"*. Science, vol. 364, no. 6439, pp. 440–442, 2019.

[13] SQUANCH. *"A distributed simulation framework for quantum networks and channels, "*. B. Bartlett, arXiv: 1808.07047 [quant-ph], Aug. 2018.

[14] Stephen DiAdamo, Janis N ötzel, Benjamin Zanger, Mehmet Mert Beşe. *"QuNetSim: A Software Framework for QuantumNetworks, "*. http://arxiv.org/abs/2003.06397v4

[15] Axel Dahlberg1and Stephanie Wehner. *"SimulaQron - A simulator for developing quantum internet software, "*. arXiv:1712.08032v2

[16] S. Steiger, Thomas Häner, and Matthias Troyer. *"ProjectQ: An Open Source Software Frameworkfor Quantum ComputingDamian,"*. arXiv:1612.08091v2

Selection of an efficient wind turbine generator at Government science college premises Chitradurga, India

G H Smitha patil.[1,2], R Megha[3], Y T Ravikiran*[1,2], M Prashantkumar[4] K Udaykumar[5] and B G Kumaraswamy[6]

[1]Department of PG Studies and Research in Physics, Government Science College, Chitradurga 577 501, Karnataka, India
[2]Department of Physics, Visvesvaraya Technological University - Research Resource Centre, Belagavi 590 018, Karnataka, India
[3] Department of physics, P.E.S College of Enginerring Mandya 571401, Karnataka, India.
[4]Department of Physics, Government College (Autonomous), Kalaburagi 585105, Karnataka, India
[5]Department of of Physics, VSK University, Ballari 583102, Karnataka, India.
[6]Department of Electrical and Electroni , SJMIT Chitradurga 577501, Karnataka, India

Abstract. Wind speed modelling is very crucial in estimating the potential of wind energy for a particular site. In this paper, the wind energy potential at the premises of Government Science College area [Lat. 14°13′ N and Long. 76°27′ E] in Chitradurga town (Karnataka State, India) is statistically analyzed for the year 2019 – 2020. The speed of wind has been assessed at the height of 60 m from the ground level at the site. The seasonal and annual wind speed distributions (WSD) have been modelled using Weibull probability distribution functions (WPDF) i.e., shape parameter, k and scale parameters, c. With the use of them, the wind power density (WPD) has been determined. The WPD estimated from Weibull and normal methods has been found to be comparable. Further, at a height of 60 m above the ground, the capacity factors (CPs) of various marketablly useable wind turbine generators have been estimated to achieve the optimal power generation for the chosen site.

Keywords: Mean wind speed, Weibull distribution function, wind power density, capacity factor

1. Introduction

Across the globe, most of the countries have been encouraging the energy harvesting from wind energy resources as it is an affordable, inexhaustible and environmentally friendly alternative energy source [1]. The power output from wind energy sources has recently been observed to be on par with the power generated from conventional power generation methods. The enhancement in the power yield in wind energy can be attributed to the technological advancements that have taken place in recent times. Technology has now made the wind energy easily extractable and economically viable. The optimal installation of a wind energy system at a particular site crucially depends on several factors viz., variation of wind speed distribution, characteristic speeds of turbine and height of the hub. Therefore, it is very essential to employ the best available technologies to estimate these factors at a given site. In order to meet the required demands, the optimal matching of wind energy to the output energy can be achieved by incorporation of all the crucial factors mentioned above [2].

Many researchers [3-5] have employed simple WSD that are limited to the basis of arithmetic mean of the wind speed (WS). We, in this paper, have used the arithmetic mean of WS data computations for evaluation of the possibility of a given site for the wind power generation. Standard WPDF has been used for modelling WS distribution statistically. Further, the approximation method has been used to compute the Weibull shape parameter, *k* and scale parameter, *c* [6]. The annual CP has been estimated on the basis of actual WS measurements and typical speeds of installed wind turbines. In the present work, we have chosen the Government Science College, Chitradurga, a reputed higher education institution in Karnataka state (India) as a site for WS data measurements and also to validate analytical results obtained from the statistical model.

2. Wind speed measurements

The reliability of a wind energy system depends on long term measurement of WSs. The literature suggests that the annual mean WSs measured over a year is a good estimate to predict long term wind potentials up to an accuracy of 10 % and a confidence level of 90 % [7]. With this perspective, the WS measurements at a height of 60m above the ground level at the premises of Government Science College, Chitradurga has been carried out using an anemometer with three cone-shaped cups over the period 2019-2020. Using a 10W photovoltaic panel based data acquisition system, the average WSs have been recorded in the interval of 10 min.

3. Wind Data Processing

By processing the acquired data, the average and standard deviations of WSD have been determined [5]. The mean and deviation in WS have been estimated using the relations,

$$v_m = \frac{1}{N}\left[\sum_{i=1}^{N} v_i\right] \quad (1)$$

$$s = \sqrt{\frac{1}{N-1}\sum_{i=1}^{N}\left(v_i - v_m\right)^2} \quad (2)$$

Where v_m = mean WS in m/s, σ = deviation in WS, v_i = specific WS in m/s, , N = observed WS values.

4. Weibull Probability Distribution

The dimensionless quantity, shape parameter *k* and the scale parameter *c* (m/s)] satisfactorily explains the experimental WS data. These parameters have been estimated by the approximation method [6]. The probability of WS at any time interval is given by,

$$f(v) = \frac{k}{c}\left(\frac{v}{c}\right)^{k-1} e^{-\left(\frac{v}{c}\right)^{k}} \quad (k > 0,\ v > 0,\ c > 1) \quad (3)$$

Where, *k* is shape parameter and *c* is the scale parameter and these parameters given by [8],

$$k = \left(\frac{s}{\bar{v}}\right)^{-1.086} \quad (4)$$

$$c = \frac{\bar{v}}{\Gamma(1 + 1/k)} \quad (5)$$

Where Γ is gamma function

5. Normal Wind Power Density

The average specific power for a wind stream with velocity v (m/s) perpendicular to wind turbine is given by [9],

$$P_{wind} = \frac{1}{2} r \overline{v^3} \tag{6}$$

Where ρ is the standard air density in kg/m^3 (ρ = 1.225 kg/m^3 dry air at 1 atm and 15^{0}C), $\bar{v}$ is the mean WS in m/s.

6. Weibull Wind Power Density

As per WPD, the anticipated seasonal and annual wind power density is given by,

$$P_{Wei} = \frac{1}{2} r c^3 \Gamma\left[1 + \frac{3}{k}\right] \tag{7}$$

7. Capacity Factor

The ratio of the average power output to the rated power of the generator is called as the capacity factor which gives the percentile electric power that a turbine can produce at the given site with the available wind [2]. This factor can be estimated by the formula [10],

$$P_{e,avg} = P_{rated} \frac{e^{-\left(\frac{v_{in}}{c}\right)^k} - e^{-\left(\frac{v_r}{c}\right)^k}}{\left(\frac{v_r}{c}\right)^k - \left(\frac{v_{in}}{c}\right)^k} - e^{-\left(\frac{v_{out}}{c}\right)^k} \tag{8}$$

Where v_c, v_r and v_f are respectively cut in speeds, rated speeds and furling (cut out) speeds of a wind turbine generator.

8. Results and Discussion

Fig. 1 is a graphic representation of the monthly variation of WS data collected at a height of 60 m above the ground level at the premises of Government Science College, Chitradurga during 2109-2020. It can be noted from Fig.1 that the maximum WS was observed to be high during the monsoon i.e., between the months June and September. The variation in average WS observed from Fig.1 indicates that the WS in this area throughout the year is suitable for tapping wind energy at a height of 60 m above the ground level where the anemometer was installed.

Table 1. Comparison between seasonal and annual power densities calculated from normal and Weibull method.

Sl. No	Seasons	k	c (m/s)	Wind power density (Normal) W/m2	Wind power density (Weibull) W/m2
1	Summer	2.13	6.53	211.30	211.65
2	Monsoon	2.95	10.00	606.12	614.79
3	Winter	2.64	6.53	179.90	180.76
4	Annual	2.25	7.72	337.60	333.07

Fig. 2 and 3 depict seasonal and annual variation of Probability distribution functions with WS over the year 2019-2020. Using Weibull distribution function and Equation (4) & (5), the k and c were calculated and are tabulated in Table 1.

The Annual shape parameter k shows a value of 2.25 which indicates widely dispersed data with uniform distribution across wide range of WS. This implies that the selected area is suitable for wind power generation due to the higher average WSs. The smaller *k* in summer and monsoon seasons point out to the fact that the average WS at this location is sufficiently high enough for the efficient operation of a wind turbine generator. As the average WSs are observed to be higher, the *k* suggests that the chosen site is suitable for wind power generation [10]. The good match of the experimental annual power densities at this site with the corresponding values calculated using Weibull function shows lowest power density in winter and highest power density during monsoon. In this study, the wind turbine generator has been analyzed at the highest capacity factor. The annual capacity factor for various wind turbine generator have been listed in Table 2. The Table 2 shows that turbine 4 (Elecon T600-48) is giving highest capacity factor which is also depicted in Fig. 4. The Seasonal and annual power densities from normal and Weibull techniques are noticeably comparable (Table 1).

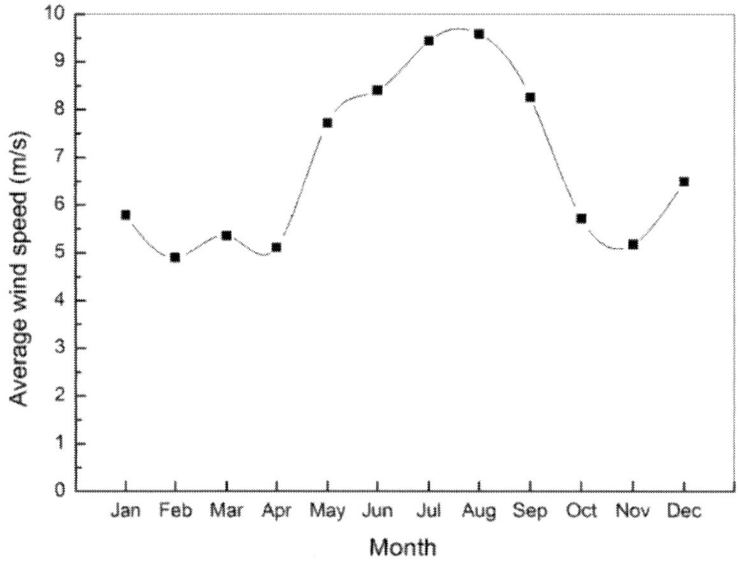

Fig. 1 Variation of average wind speed with month

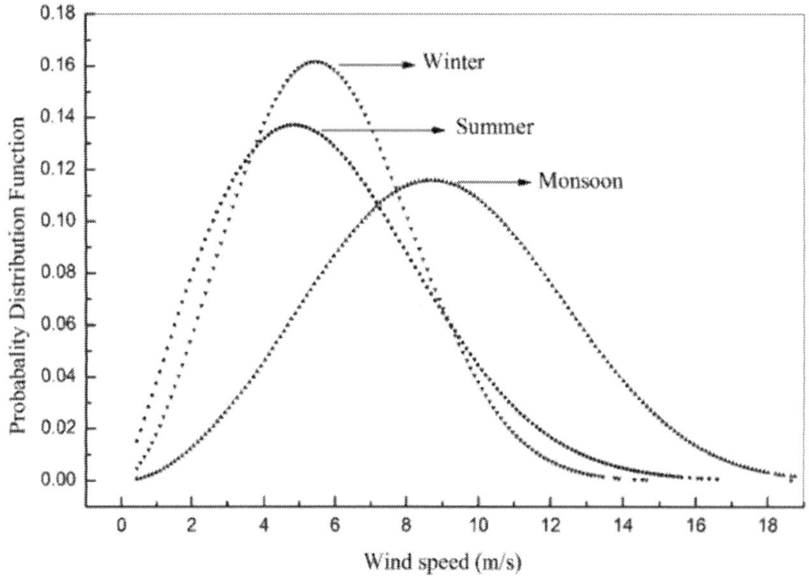

Fig. 2 Seasonal variations of Probability distribution functions with wind speed.

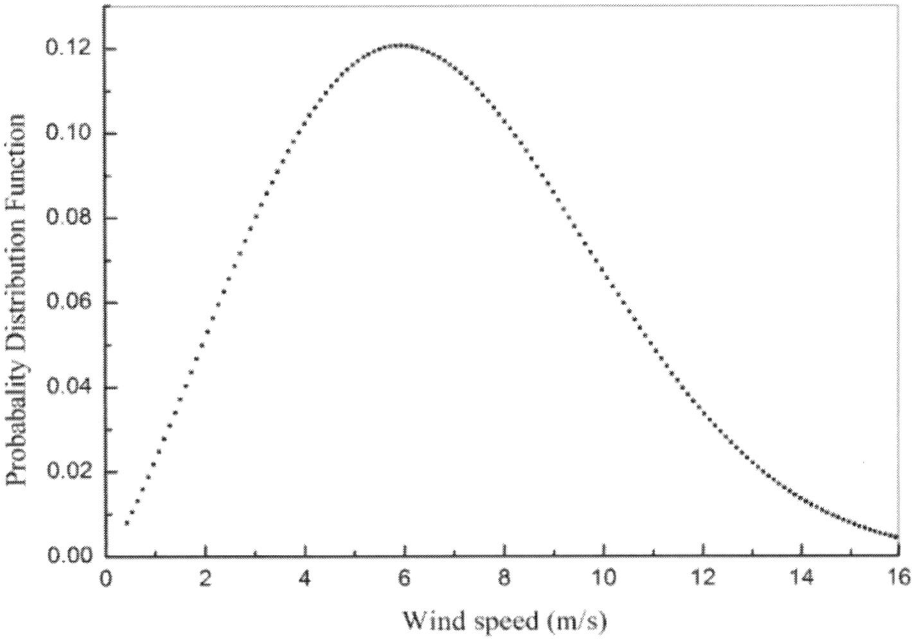

Fig. 3 Annual variation of Probability distribution functions with wind speed.

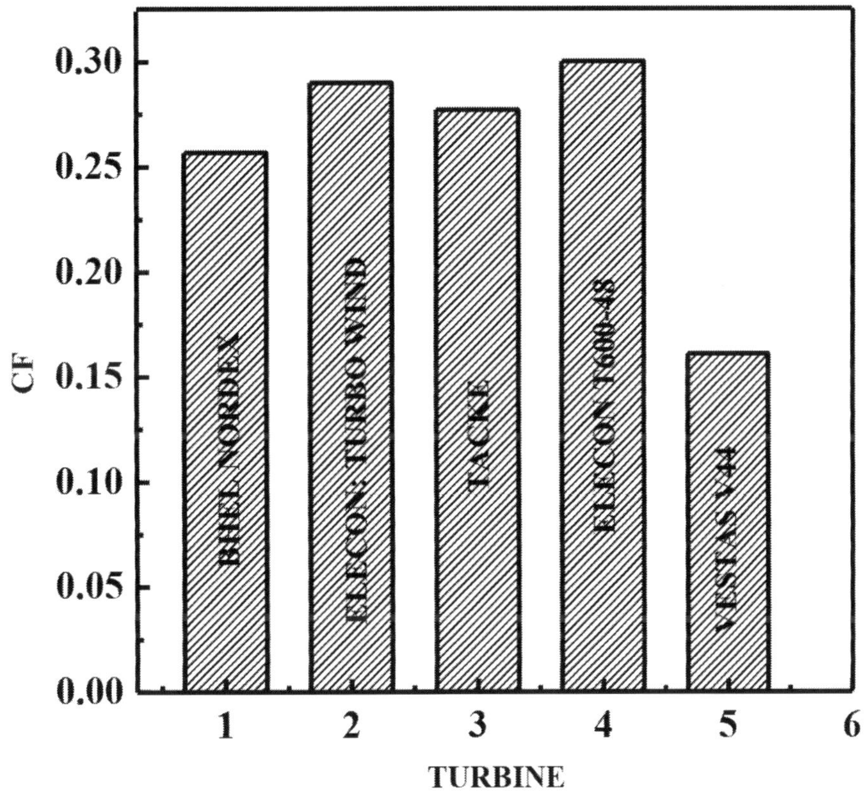

Fig.4. Annual variation of CP at elevation of 60 m for various wind turbines listed in Table 2

Table 2. Wind turbine data at 60 m height.

Sl. No	Turbine name and model	Rated power [KW]	Cut-in speed vc [m/s],	Rated speed vr [m/s]	Cut-out speed vf [m/s]	Capacity factor CF
1	BHEL NORDEX	600	4.0	13	25	0.25
2	ELECON: TURBO WIND	600	3.5	12.5	25	0.29
3	TACKE	600	3	13	20	0.27
4	ELECON T600-48	600	3	12.5	25	0.30
5	VESTAS V44	600	4	16	20	0.16

9. Conclusion

The WS and wind power density measurements have performed at the premises of Government Science College, Chitradurga for the year 2020-2021. Using the Weibull Distribution function, the measured data has been analyzed. The seasonal and annual WSD modeled using WDF. Using Weibull parameters, the WPD of the site has been determined. The WPD estimated from Weibull and normal methods has been found to be comparable. Also, the annual variations of CPs for optimum selection of wind turbine generators have been studied. Further, by approximating the capacity factors of different industrial wind turbine generators, the site matching has been done.

Acknowledgement

The authors acknowledge the support from Research Resource Centre, VTU, Belagavi.

References

[1] Yi Li, Xuan Huang, Kong Fah Tee, Qiusheng Li, Xiao-Peng Wu. 2020, *Sustainable Energy Technologies and Assessments* 39 100711.
[2] Tai-Her Yeh and Li Wang, 2008, *IEEE Trans Energy Convers*, 23, 2, 592-602
[3] Gholamreza Janbaz Ghobadi, Bahram Gholizadeh and Bagher Soltani. 2011, *International Journal of physical Sciences*, 19, 6, 4621-4628.
[4] Kumaraswamy B. G., Keshavan B. K. and Ravikiran Y. T., July 2011, 978-1-4577-1002-5/11, *IEEE PES General Meetin..*
[5] Seyit A. Akdag and Ali Ddinler, April. 2009, *Energy conversion and Management,* 50, 1761-1766,.
[6] Vladislovas Katinas, Giedrius Gecevicius, Mantas Marciukaitis.. 2018, *Applied Energy,* 218, 442-451.
[7] *Renewable Energy World. 'Wind Energy - A True Option for Developing Countries'*. November 1998, James and James (Science Publishers) Ltd, United Kingdom, 1, 3, 62.
[8] Gungor A., Gokcek M., Ucar H., Arabaci E., Akyuz A. 2020, *International Journal of Environmental Science and Technology*, 17 1011-1020.
[9] Fawzi A.L. Jowder, 2009, *Applied Energy*, 86, 538-545.
[10] Faleh H. Mahmood, Ali K. Resen, Ahmed B. Khamees. 2020, *Iraq. Energy Reports*, 6 79-87.

IOP Conf. Series: Materials Science and Engineering 1187 (2021) 012025 doi:10.1088/1757-899X/1187/1/012025

Low-compute facial expression recognition using fiducial feature-sets

A.V. Divekar and D.C. Gharpure

Dept. of Electronic Science, Savitribai Phule Pune University, Pune, India

E-mail: ankur.divekar@gmail.com, dcgunipune@gmail.com

Abstract. Facial Expression Recognition is an exciting area of affective computing. As mobile and embedded devices become increasingly ubiquitous, exploration of low-compute approaches for facial expression recognition is essential. Facial landmark points are fiducial features that are used to localize and represent salient regions of the face, such as eyes, nose and lips. Any facial expression can be expressed as an activation of facial muscles in specific parts of the face, thereby affecting the locations of the facial landmark points in those parts. This relationship can be captured concretely by deriving appropriate feature-sets from these points. In this paper, an approach for deriving three types of feature-sets from a set of facial landmark points detected on a face is discussed. Such feature-sets are derived for three standard facial expression recognition datasets with labelled expression classes. The derived features-sets are used as inputs for training computationally light machine-learning classifiers, yielding encouraging classification accuracy. These results are presented and discussed with both quantitative and qualitative observations.

1. Introduction

Facial expressions are an important modality of non-verbal communication during human interaction and can convey a wide array of emotions and intentions. For several years now, automated facial expression recognition (FER) has attracted significant research interest and activity in the field of computer vision. FER has been applied to numerous applications such as advanced driver assistance systems, gaming and virtual reality. Numerous promising approaches that utilize deep neural networks (DNNs) for FER have been reported. Training DNNs is a highly energy-intensive process with high computation requirements, and thereby less than optimal for large-scale deployments. On the other hand, an increasing number of mobile and embedded devices now run algorithms suitable for their relatively lower, but significant computational abilities. This makes the exploration of low-compute approaches essential, especially for applications like FER, which may be deployed on low-end devices at extremely large scales. Numerous commercial phone apps utilize facial landmark points (FLPs) for digital manipulation of facial images. Naturally, FLPs are a logical starting point for FER applications too.

In this paper, we present an approach that utilizes facial landmark points to derive fiducial feature-sets. We report the FER classification performance of multiple algorithms with low-compute requirements. The details of datasets used are given in section 3.1. In order to train

IOP Conf. Series: Materials Science and Engineering 1187 (2021) 012025 doi:10.1088/1757-899X/1187/1/012025

and test the facial expression classifier, the images in the dataset must be translated into a feature vector that can be applied as an input to the classifier. In sections 3.2 and 3.3, the process of deriving the fiducial feature-sets for each facial image has been described. Section 4.1 describes the process of training different classifiers using the derived feature-sets. Finally, section 4.2 provides details of the FER performance evaluation and observations thereof.

2. Previous work

A number of approaches towards FER have been demonstrated, using various hand-crafted features such as Local Binary Patterns, Singular Value Decomposition, Bag of Words (BoW), Histogram of Gradients (HoG), and Scale Invariant Feature Transform (SIFT). These methods have shown encouraging results on several standard databases.[1]

Numerous methods for estimating and localizing FLPs have been developed[2]. Various FER approaches that utilize facial geometric features have also been previously reported[5]. FLPs have been utilized as the intermediate step for numerous face-related tasks such as face alignment and facial image synthesis, apart from expression recognition.

3. Methodology

Ekman et. al. identified the six universal facial expressions[8] as anger, disgust, fear, happiness, sadness and surprise. Combined with a neutral expression, this creates seven fundamental, distinct classes of facial expression. A machine-learning system can be built to classify a given facial image into one of these pre-defined classes as shown in Figure 1

Figure 1: FER methodology

An appropriately labeled dataset with numerous instances of each expression class is required to train a machine-learning algorithm. End-to-end training approaches like convolutional neural networks generally utilize raw image data as input. However a classifier with significantly lower compute requirements can be built by performing feature extraction on the facial image. For a given facial image, the first step is to estimate the location of several facial landmark points which represent salient regions of the face such as eyebrows, eyes, nose and lips. A set of fiducial features is then derived by quantifying the positions of these facial landmark points relative to a common reference point. A feature vector representing various measurements of the face is thus extracted. The set of such extracted features from a labeled dataset is used to train a classifier. This trained classifier may then be utilized to make a prediction of expression class for an image that has not been seen by the classifier before.

3.1. Datasets

Curated datasets are one of the driving factors behind rapid progress in fields using machine learning and deep learning. Three standard FER datasets were utilized for the experiments. The details of these databases are as follows:

(i) ADFES: The Amsterdam Dynamic Facial Expression Set[9] has been created at the Amsterdam Interdisciplinary Centre for Emotion. The dataset consists of videos as well as still images of 12 male and 10 female subjects with 10 facial expressions. The still images were obtained by isolating a single video frame at the apex point of the model's expression.

(ii) TFEID: The Taiwanese Facial Expression Image Database[10] has been jointly by the Brain Mapping Laboratory at National Yang-Ming University and the Integrated Brain Research Unit at Tapei Veterans General Hospital in Taiwan. Only a subset of the entire dataset has been made available so far for research purposes. This subset consists of images of 20 male and 20 female subjects with eight posed facial expressions, captured in frontal view.

(iii) WSEFEP: The Warsaw Set of Emotional Facial Expression Pictures[11] dataset contains images from 14 male and 16 female models with 7 facial expressions. The images were carefully selected to fit criteria of basic emotions and then evaluated by independent judges.

In order to maintain consistency within the training data, only front-facing images were utilized, and images captured at other angles were ignored. Further, only images for the 7 expressions common to all three datasets were selected in order to minimize categorical imbalance. Data augmentation was done by executing a horizontal flip on every selected image, and adding the flipped image to the dataset, thereby doubling the number of samples in each expression class, as shown in Table 1.

Table 1: Expression-label counts by dataset

Expression	Dataset			Total
	ADFES	TFEID	WSEFEP	
Neutral	44	78	60	182
Anger	44	68	60	172
Disgust	44	80	60	184
Fear	44	80	60	184
Joy	44	80	60	184
Sadness	44	78	60	182
Surprise	42	72	60	174
Total	306	536	420	1262

While the three datasets were also utilized individually for the experiments, the samples of each expression class were merged across the three datasets to form a composite dataset, which was used to examine the feasibility of training a classifier capable of generalizing across racially diverse faces.

3.2. Feature Extraction

In order to ensure uniformity of extracted features, the facial images must be preprocessed before feature extraction. For each image in the database, facial detection was performed. The detected face was then aligned and resized to predefined dimensions. 68 Facial Landmark Points(FLPs) were localized on the aligned and resized facial image, as illustrated in Figure 2b. A set of such FLPs was generated for all the images in the database. Given this set of FLPs, feature extraction was performed through the following steps:

Step 1: Let F be a set of n chosen FLPs. Then for image i in the dataset, the individual FLPs can be denoted as $f_j^i = (x_j, y_j)$ where x_j is the x-coordinate, y_j is the y-coordinate and

j is the index number of the FLP, from 1 to n. The centroid $C^i_{x,y}$ of all the FLPs is defined as

$$C^i_{x,y} = (\frac{\sum_{j=1}^n x_i}{n}, \frac{\sum_{j=1}^n y_i}{n})$$

Step 2: Let the length of the line joining an FLP f_j to $C_{x,y}$ be denoted as l_j. Then,

$$l_j = \sqrt{(x_j - C^i_x)^2 + (y_j - C^i_y)^2}$$

Let L be the set of the respective lengths of all the lines joining each individual FLP to the centroid. Then,

$$L = [l_1, l_2, l_3, \ldots, l_n]$$

Let l_{max} denote the maxima of the set of lengths. Then the set of normalized lengths is represented as

$$L^i_{norm} = [\frac{l_1}{l_{max}}, \frac{l_2}{l_{max}}, \frac{l_3}{l_{max}}, \ldots \frac{l_n}{l_{max}}]$$

Step 3: Next, consider a coordinate system with origin at the centroid calculated earlier. Then the angle a_j created between x-axis and the line joining FLP f_i to the centroid is calculated as

$$a_j = \arctan\left(\frac{y_j - C^i_y}{x_j - C^i_x}\right)$$

Let A be the set of all angles thus calculated for the n FLPs. Then,

$$A = [a_1, a_2, a_3, \ldots, a_n]$$

Let a_{max} and a_{min} be the maxima and minima of A respectively. Then the set of angles normalized between 0 and 1 is obtained as:

$$A^i_{norm} = [\frac{a_1 - a_{min}}{a_{max} - a_{min}}, \frac{a_2 - a_{min}}{a_{max} - a_{min}}, \frac{a_3 - a_{min}}{a_{max} - a_{min}}, \ldots, \frac{a_n - a_{min}}{a_{max} - a_{min}}]$$

Step 4: The individual sets of length and angle features extracted for image i are L^i_{norm} and A^i_{norm} respectively.

Step 5: The composite feature vector AL^i for image i is created by concatenating the sets of normalized lengths and angles. Thus,

$$AL^i = [L^i_{norm}, A^i_{norm}]$$

Step 6: Let the set of length feature vectors for all the n images in the dataset be L_n. Similarly A_n and AL_n are the sets of angle and concatenated feature vectors respectively.

3.3. Feature-set Design
It is observed that face detection systems often generate a crop of the facial image that excludes the outer periphery of the facial area, such as the top of forehead and the chin. To check the feasibility of using these cropped images for FER, a subset of the 68 detected FLPs was also considered. The FLPs for the chin, the jaw-line, and the side of the face were excluded, to simulate a tightly cropped facial image. The remainder subset consisted of 46 FLPs for left eyebrow, left eye, right eyebrow, right eye, nose and mouth.

Considering the two sets of FLPs (68 and 48 FLPs) and the two types of features (length and angle) extracted, six feature-sets were created as follows:

(i) A_{68}: Angle features A_{norm} for the 68-FLP set.

(ii) L_{68}: Length features L_{norm} for the 68-FLP set (as shown in Fig. 2c).

(iii) AL_{68} Concatenated angle and length features for the 68-FLP set.

(iv) A_{46}: Angle features A_{norm} for the 46-FLP subset.

(v) L_{46}: Length features L_{norm} for the 46-FLP subset (as shown in Fig. 2d).

(vi) AL_{46} Concatenated angle and length features for the 46-FLP subset.

4. Experiments
The 6 feature-sets represent different types of fiducial measurements of the face. In order to examine the classification accuracy of each feature-set, they were tested on the 3 standard databases as well as the composite database described in 3.1.

4.1. Training and Testing
For sufficiently large datasets, it is often observed that multiple machine-learning classifiers are able to learn the internal patterns of the data, and give consistent and nearly equivalent results. However, for relatively small datasets such as the one utilized here, the appropriate choice of classifier algorithm can create a significant difference in the prediction. Hence, three classifiers that are known to be particularly efficient for small and medium-sized datasets were specifically selected for the experiments:

(i) Random Forest classifier[12]

(ii) Support Vector Machine(SVM)[13]

(iii) XGBoost Classifier[14]

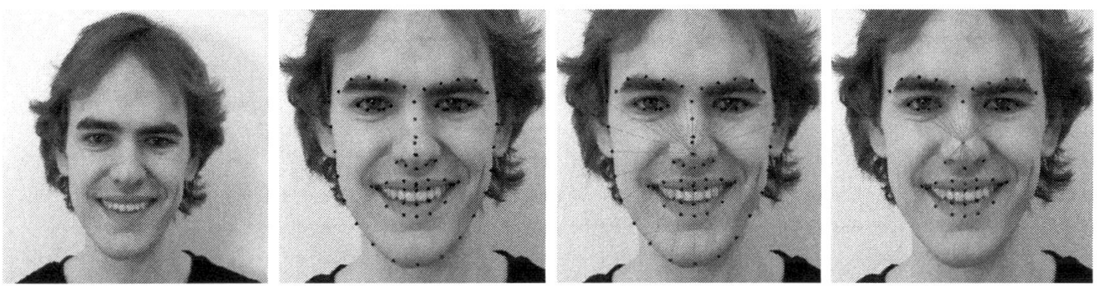

(a) Original image (b) FLPs on aligned face (c) L_{68} feature-set (d) L_{46} feature-set

Figure 2: Feature extraction

In order to get average accuracy figures for each classifier, k-fold cross validation was utilized, with k=5. The input feature-set for each experiment run was accordingly split into training(80%) and test(20%) sets. Each of the classifiers was fit on the training set, and then the predictions were generated for the test set and used to calculate accuracy for that experiment. As per k=5, the experiment was repeated 4 more times with different train-test splits of the feature-set, and the accuracies were averaged. This process was repeated for all six feature-sets, and the results obtained are tabulated in the next section.

4.2. Results

Table 2 lists the classification accuracies obtained for the three classifiers on all six feature-sets derived from the ADFES dataset. It is observed that the accuracy for for L_{68} is better than A_{68} for two of the three classifiers. Additionally the accuracy for AL_{68} is actually less than L_{68}. It may be concluded that the angle features do not contribute anything significant to the classification and perhaps only add noise to the feature-set with 68 FLPs. On the other hand, it can be observed that the average accuracy of L_{46} is less than that of A_{46}, and the angle features seem to have the greater contribution to the accuracy of AL_{46}

Table 2: Classification Accuracy for ADFES

Classifier	feature-set					
	A_{68}	L_{68}	AL_{68}	A_{46}	L_{46}	AL_{46}
RandomForest	0.85	0.81	0.81	0.85	0.81	0.85
SVM	0.81	0.85	0.81	0.85	0.78	0.81
XGBoost	0.85	0.85	0.85	0.81	0.78	0.89

Table 3 lists the classification accuracies obtained for the three classifiers on all six feature-sets derived from the TFEID dataset. Similar to the ADFES accuracy figures, it can be observed that length features have higher accuracies than angle features for the feature-set with 68 FLPs. In the case of the feature-set with 46 FLPs, the angle features have low accuracy by themselves, but exhibit a significant contribution to the accuracy of AL_{46}.

Table 3: Classification Accuracy for TFEID

Classifier	feature-set					
	A_{68}	L_{68}	AL_{68}	A_{46}	L_{46}	AL_{46}
RandomForest	0.76	0.76	0.8	0.76	0.72	0.76
SVM	0.80	0.83	0.87	0.76	0.76	0.83
XGBoost	0.76	0.83	0.83	0.74	0.76	0.85

Table 4 lists the classification accuracies obtained for the three classifiers on all six feature-sets derived from the WSEFEP dataset. A similar trend is observed for this dataset too. Angular

IOP Conf. Series: Materials Science and Engineering 1187 (2021) 012025 doi:10.1088/1757-899X/1187/1/012025

features have lower accuracy than length features, but clearly contribute to the accuracy of the composite feature vector, particularly in the feature-set with 46 FLPs.

Table 4: Classification Accuracy for WSEFEP

Classifier	feature-set					
	A_{68}	L_{68}	AL_{68}	A_{46}	L_{46}	AL_{46}
RandomForest	0.75	0.81	0.86	0.81	0.83	0.89
SVM	0.78	0.86	0.89	0.72	0.92	0.86
XGBoost	0.83	0.86	0.86	0.75	0.75	0.83

Table 5 lists the classification accuracies obtained for the three classifiers on all six feature-sets derived from the composite dataset created by merging the samples of each expression class in the three datasets above. Some very interesting observations may be made. The length features clearly outperform the angular features with all three classifiers in both the 68-FLP feature-sets as well as 46-FLP feature-sets. However, the more striking improvement in classification accuracy is seen in the AL_{68} as well as AL_{46}. This clearly indicates that training a classifier on multiple racially diverse databases actually helps the classifier to generalize better, and reduces overfitting to faces with a particular set of visual traits. The high accuracy of AL_{46} also indicates that FER may be successfully implemented on images with a close-cropped facial area.

Table 5: Classification Accuracy for composite dataset

Classifier	feature-set					
	A_{68}	L_{68}	AL_{68}	A_{46}	L_{46}	AL_{46}
RandomForest	0.77	0.86	0.91	0.78	0.86	0.94
SVM	0.84	0.93	0.94	0.85	0.91	0.95
XGBoost	0.80	0.87	0.92	0.81	0.83	0.91

5. Conclusion

As the use of systems executing face-related analysis becomes increasingly wide-spread, it is essential to explore methods that can be successfully deployed at scale without humungous power and computation budget requirements. Feature extraction based on fiducial points is a great candidate in this regard, as facial landmark point detection is well-tested and proven on fairly low end commercial-off-the-shelf hardware such as cell-phones and embedded boards like Raspberry Pi. In this paper, an analysis of FER using feature-sets generated through fiducial measurements based on facial landmark points has been presented. The classification accuracy has been evaluated on multiple datasets individually and collectively. It is observed that higher classification accuracy can be obtained by training a classifier on a diverse dataset. While the performance of the SVM classifier on the individual datasets is mixed, it scores the highest accuracy for the composite dataset. Notably, the fact that AL_{46} scores higher accuracy than AL_{68} seems to indicate that judicious exclusion of certain features can actually reduce the noise in the dataset, and help better classifier generalization. Further work on statistical feature selection in order to choose appropriate features is being studied. Additionally, methods of optimizing the classification accuracy as a function of specific expression classes are being explored, with respect to specific applications. While the current accuracy figures are lower than those obtained from deep-learning approaches, they are nevertheless encouraging from the perspective of large-scale deployment for consumer applications, considering their humble computational requirements.

References

[1] Roychowdhury S and Emmons M 2015 *International Journal of Computer Science & Engineering Survey* ISSN 09763252
[2] Johnston B and de Chazal P 2018 A review of image-based automatic facial landmark identification techniques
[3] Kotsia I and Pitas I 2007 *IEEE Transactions on Image Processing* **16** 172–187 ISSN 10577149
[4] Ghimire D and Lee J 2013 *Sensors (Switzerland)* **13** 7714–7734 ISSN 14248220
[5] Majumder A, Behera L and Subramanian V K 2014 *Pattern Recognition* vol 47 pp 1282–1293 ISSN 00313203
[6] Kowalski M, Naruniec J and Trzcinski T 2017 *IEEE Computer Society Conference on Computer Vision and Pattern Recognition Workshops* vol 2017-July pp 2034–2043 ISBN 9781538607336 ISSN 21607516 (*Preprint* 1706.01789)
[7] Yang X, Li Y, Qi H and Lyu S 2019 *IH and MMSec 2019 - Proceedings of the ACM Workshop on Information Hiding and Multimedia Security* pp 113–118 ISBN 9781450368216 (*Preprint* 1904.00167)
[8] Ekman P and Friesen W V 1971 *Journal of Personality and Social Psychology* ISSN 00223514
[9] van der Schalk J, Hawk S T, Fischer A H and Doosje B 2011 *Emotion* **11** 907–920 ISSN 1931-1516 URL http://doi.apa.org/getdoi.cfm?doi=10.1037/a0023853
[10] Chen L and Chen Y 2007 Taiwanese facial expression image database
[11] Olszanowski M, Pochwatko G, Kuklinski K, Scibor-Rylski M, Lewinski P and Ohme R K 2014 *Frontiers in Psychology* ISSN 16641078
[12] Prokhorenkova L, Gusev G, Vorobev A, Dorogush A V and Gulin A 2018 *Advances in Neural Information Processing Systems* ISSN 10495258 (*Preprint* 1706.09516)
[13] Cortes C and Vapnik V 1995 *Machine Learning* ISSN 15730565
[14] Chen T and Guestrin C 2016 *Proceedings of the ACM SIGKDD International Conference on Knowledge Discovery and Data Mining* ISBN 9781450342322 (*Preprint* 1603.02754)

IOP Conf. Series: Materials Science and Engineering 1187 (2021) 012026 doi:10.1088/1757-899X/1187/1/012026

Symmetric Key Cryptosystem based on Sequential State Machine

Basappa B Kodada
A J Institute of Engineering and Technology Mangalore, VTU Belagavi, India
E-mail: basappabk@gmail.com

Demian Antony D'Mello
Canara Engineering College Bantwal, VTU Belagavi, India
E-mail: demianantony@gmail.com

Abstract. Cryptography is a type of art that translate strings of plain text into strings of cipher text and is very important in sharing of secrete information in communication system. It is very essential to have cryptographic algorithms to provide high security by utilizing optimum computational resources during information sharing to achieve data confidentiality, privacy, integrity and other security services. The traditional conventional encryption algorithms (DES, AES) have high computational cost due to generating number of sub keys and bilinear concept. Hence this paper propose the sequential state machine (Mealy state Machine) based cryptosystem which is very light weight in nature and provide high security due to its weakly invertability concept with relatively smaller key size. The paper also presents the performance analysis on proposed algorithm and describes the security analysis on it.
Keywords: Data Security, Confidentiality, Privacy, Encryption, Decryption, Finite automata, Finite state Machine, Sequential state machine, Mealy Machine, Integrity, Cryptography, security.

1. Introduction

Cryptography is an art of language used for transmitting the secrete writing[1]. The main idea of cryptography is to translate meaningful message or sentence to uncommon message which is not understood by unauthorized person. This process is useful only when hard to retrieve the original message. The cryptography acts as middle man between sender and receiver who take care of secrecy of information being transmitted. In the cryptography technology, the original or human understandable message is called "Plain text", the translated message is called "cipher text or crypto text" and the technique is used for the translation process is called "Encryption and decryption" technique. The research scientists have proposed different hybrid encryption techniques of symmetric or one key encryption[2, 3] (DES, AES, Blowfish, Twofish etc.) and Asymmetric or public key encryption (RSA, ECC, Elgamal, ABE, IDE, FKE, QKE, HE, etc.). The basic objective of these techniques to provide confidentiality, privacy, integrity, authentication, authorization, auditing, non-repudiation services[4, 5] over the data transmission in insecure channel.

All these techniques are based on number theory which might cost heavily due to utilization of large number of computational resources, bi-linear theory and logarithmic concept. Due to this reason, many researchers have shown their interest towards the state machine based cryptosystem. The mathematical model of finite state machine based cryptography is widely studied and presented in [6]. Authors have presented the one-key and public key based encryption and decryption technique. The idea of state machine based cryptosystem is very simple. The encryption process will be depends on the state machine that acts as key space for encryption process and the inverse of state machine acts as key space at receiver end for decryption process. the security of the system relies on the difficulties in constructing the inverse machine to decrypt the information.

The traditional cryptography techniques are based on number theory[7] gives more computational cost interms of time complexity. The authors of [8, 9, 10, 11, 12, 13] have started proposing the cryptosystem based on cellular automaton and later finite state automaton based cryptosystem - a mathematical model has been presented by the author[6]. The author also described that, the applications of finite state machine can be used for one-key and two-key cryptosystem. In this paper we discusses the performance of one-key cryptosystem based on sequential state machine(Mealy state machine) for different size of plain text. This paper also gives an insight into security analysis on proposed methods.

1.1. Motivating Example

One key or shared key cryptosystem is a method of encryption and decryption process that is uded used for secrete communication between two parties[14]. This cryptosystem is faster than two-key cryptosystem - consumes the lot of network and CPU resources due to its bi-linear concept used in key generation algorithm and one-key cryptosystem is also efficient to provide confidentiality and privacy of information being transmitted in communication channel[14]. Due to the better efficiency and faster in time complexity than two-key cryptosystem, it is used in encrypting huge amount of data being exchanged between client and server.

The security of existing one-key cryptosystem[2, 3] (DES,AES,Blowfish,RC4) is depends on length of key, bi-linear theory and logarithmic concept of number theory. This will leads to significant delay in encryption and decryption process eventhough these cryptosystems are highly effective[15]. The researchers [16, 9, 10, 11, 12, 13, 17, 18] have concentrated only on two-key cryptosystem based on finite state machine. It is very essential to have one-key cryptosystem that consumes less resources and provide faster encryption and decryption mechanism during data in-transit. Consider the scenario of HTTP client server secure communication[19] as a motivating example where user as a client would like to make online money transaction through the online banking system. The customer enters url of netbanking in browser that will initiate HTTP connection between customer and bank server. The bank server takes the login credentials and verify. Once the customer credentials are accepted, the HTTP protocol(at Client side) shares the secrete key (shared key) to bank server as session key to have secure communication between customer and bank server. The all communication between customer and bank server will be encrypted and decrypted by using session key at both the ends[20]. It is very important that, the encryption and decryption algorithm must be faster and efficient one for large data set. Hence we propose a very light weight one-key cryptosystem for encryption and decryption process based on sequential state machine. The system has relatively small key size and higher security due to its weakly invertablity concept.

The rest of the paper is organized as follows: Section 1 briefs about the introduction on cryptography and its applications and about sequential state machine. The sequential state machine based one-key cryptosystem is explained in section 2 with proposed method and algorithm. The section 3 discusses the experimentation details and security analysis on proposed method. The section 4 gives you the related work on cryposystems based on state machines proposed by researchers and scientists. Finally we conclude the paper in section 5.

2. Cryptosystem based on Sequential state Machines

Sequential state machine changes their states and produces the output sequence for each input symbols[21]. The behavior of the sequential state machine are represented by set of components: Set of states, set of input symbols, set of output symbols, start state and final state. There are two types of sequential state machines - Moore Machine and Mealy machine and these machines are also called as finite state machine since the states in the machines are finite. These machine are used as cryptography techniques due to its generation of output sequence for each input sequence. Most of the proposed cryptography uses the applications of cellular automata, transducers (More and Mealy Machine) and state machines with accepting state and without output symbols. It means to say that, the system provides output sequence for each input sequence at different states are called sequential state machines. These Moore and Mealy machines are represented by graphical notation for digital circuits[22].

Definition 1 *The sequential state machines (Moore and Mealy) M consists of six tuples i.e. $M = (Q, I, O, \delta, \lambda, q_0)$ where the tuples Q, I, O, q_0 represents collection of states, input symbols, output symbols and an initial state respectively. The tuple δ, λ are one way mapping function where δ defines $\delta : Q \times I \to Q$ and λ defined as $\lambda : Q \to O$ for moore machine and $\lambda : Q \times I \to O$ for mealy machine. The mapping function are expanded as follows:*

(i) The Mapping function δ for any given state $q_0 \in Q$ and the input sequence $a_0a_1....a_{l-1} \in I$ with length $l > 0$, then we get next states $q_i \in Q$ i.e $\delta(q_0, a_0a_1a_2......a_{l-1}) = q_{i+1} \in Q$ where q_0 is the current state and q_{i+1} is the next states for each input sequence $a_0, a_1, a_2.....a_{l-1} \in I$. Consider the transition table of M shown in table 1 and string 10010, *then*

$$\delta(A, 1) = C \quad \delta(C, 0) = A \quad \delta(A, 0) = B \quad \delta(B, 1) = D \quad \delta(D, 0) = A$$

(ii) The mapping function λ for moore and Mealy machines are defined as

(a) The mapping function for mealy machine λ with any state $q_0 \in Q$ and the input sequence $a_0a_1a_2....al - 1 \in I$ with length $l > 0$, then we get sequence of output symbols $b_0, b_1, b_2.....b_{l-1} \in O$ i.e $\lambda(q_0, a_0a_1a_2.....a_{l-1}) = b_0b_1b_2......b_{l-1}$ where $b_0b_1b_2......b_{l-1} \in O$ are the set of output sequence for the input sequence $a_0a_1a_2.....a_{l-1} \in I$. This is also equivalent to $\lambda(\delta(q_0, a_0), a_1a_2.....a_{l-1}) = b_0b_1b_2......b_{l-1}$. In this length of output sequence must be same as length of input sequence. Consider the transition table of M shown in table 1 and string 10010, *then*

$$\lambda(A, 1) = 1 \quad \lambda(C, 0) = 0 \quad \lambda(A, 0) = 0 \quad \lambda(B, 1) = 0 \quad \delta(D, 0) = 1$$

$$i.e$$

$$\lambda(A, 10011) = 10001$$

(b) For moore machine, the mapping function λ is with any states $q_0, q_1.....q_n$, then we get sequence of output symbols on each state $b_0, b_1, b_2.....b_{l-1} \in O$ i.e $\lambda(q_i) = b_0b_1b_2......b_{l-1}$ where $b_0b_1b_2......b_{l-1} \in O$ are the set of output sequence for the input sequence $a_0a_1a_2.....a_{l-1} \in I$ on each state q_i.

We are using mealy state machine in our proposed method. The applications of Mealy state machine (MSM) is used for encoding (Encryption) and decoding (Decryption) of cryptosystem. It is very essential to obtain inverse of MSM $M = (Q, I, O, \delta, \lambda, q_0)$ can be used for decoding process. One of the most important issue in information coding and transmission is to reconstruct the input sequence (data transmitted) from output sequence at receiver end. We use

MSM based encoder and decoder to encrypt and decrypt the data at sender and receiver end respectively. That means, receiver must have inverse of MSM to reconstruct the input sequence from the received output sequence.

The inverse of MSM is a state machine that produces the input sequence with finite delay. If machine is information loss less then only inverse of MSM can be constructed. Therefore, first the MSM must be verified whether it is information lossless or not[23]. A MSM is said to be information loss less of $order(\mu)$, if lower testing table [23] does not consist of same repeating pair of states. Consider the states in the MSM are $Q = \{q_0, q_1, q_2, ...\}$ and pair of states $(q_0, q_0), (q_1, q_1)...etc$ said to be repeating same pair of states. The MSM with no repeating such states are called information loss less.

A MSM $M = (Q, I, O, \delta, \lambda, q_0)$ is invertible $[M' = (Q', I', O', \delta', \lambda', q_0')]$ with delay τ if input sequence $a_0a_1a_2.......a_n$ is reconstructed at the receiver end for the output sequence $b_0b_1b_2.......b_n$ i.e $\lambda'(q, b_0b_1b_2......b_n) = a_0a_1a_2.......a_n$, the output mapping function can be written as $\lambda'(q, b_i) = a_i$. Consider the MSM with four states (A, B, C, D) where A is initial state and B is final state and is also a loss less, the inverse of M is shown in the table 1 for understanding purpose. The table 2 shows the encoding(encryption) and decoding(Decrption) process by using MSM for the sample input sequence 11100101 and the same sequence is reconstructed at receiver end from the output sequence 11000011. The transition movement along with output bit generation for each input bit is also presented in table 2.

The plain text (input sequence) $a_0a_1a_2....a_i$ will be represented over the $GF(2)$ or sequence of binary bits. These binary bits will be input symbols to MSM $M = (Q, I, O, \delta, \lambda, q_0)$ that encrypts it and generate the cipher text (output sequence). Before generating output sequence, the permutation function is applied to plain text to construct the randomness of cipher text as shown in figure $1(a)$. The encoder in figure $1(a)$ takes plain text a_i and apply the permutation function $g(a_i)$ on plain text a_i and give it to MSM M. The MSM M generates the cipher text(output sequence) b_i with help of output mapping function λ and the same is shown in figure $1(a)$. The plain text will be reconstructed at the receiver end by using inverse of MSM $M' = (Q', I', O', \delta', \lambda', q_0')$ as shown in figure $1(b)$. The decoder in figure $1(b)$ takes cipher text b_i and apply the inverse permutation function $g^{-1}(b_i)$ on b_i and give it to inverse of MSM M'. The M' generate the corresponding output sequence as plain text for input sequence(cipher text)b_i received from encoder and the same is shown in figure $1(b)$.

The algorithm 1 presents the step wise step for encryption and decryption process by using MSM and the size of key space is used in the algorithm is length of states in MSM. If size of each state is one byte then the size of key space will be $K = |Q| \times 8$. Consider the same scenario discussed in motivating example, the plain text over $GF(2)$ is communicated between bank customer and bank server for transmission of 111011011100101....... The plain text 111011011100101...... is permuted by using function[6] g, get the intermediate plain text and use the mealy state machine for encryption. The state machine shown in table 1 for example purpose will encrypt intermediate plain text 101101101010001...... with 32 bit key size and produce the cipher text as follows

$$C_i = E_M(101101101010001....)$$

$$\lambda(A, 101101101010001....) = 101111111001000.....$$

The cipher text for the intermediate plain text is 101111111001000.... and client at the sender end will transmit it to the bank server. The bank server use the inverse of MSM M' as shown in table 1 just for reader understanding and decrypt the cipher text to retrieve the plain text as follows and The pictorial representation of the same is shown in figure 2

$$MSM = E_{M'}^{-1}(C_i) = D_{M'}(C_i)$$

$$\lambda'(A, 101111111001000....) = 101101101010001....$$

The inverse permutation function g^{-1} is applied on message 101101101010001.... to get the plain text i.e

$$g^{-1}(101101101010001.....) = 111011011100101....$$

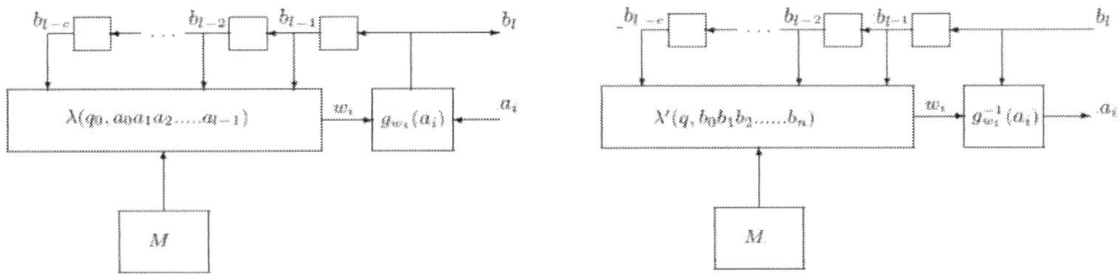

(a) Encoder M (b) Decoder M'

Figure 1. Encryption and Decryption process based on Mealy State Machine

---

Algorithm 1: Encipher and Decipher Algorithm

1. Let P be plain text over $GF(2)$;
2. Select $M = (Q, I, O, \delta, \lambda, q_0)$ is MSM and is weakly invertible;
3. Define the Key space K with length $l = |Q \times 8|$ and $K = A$, $|K| = l$;
4. Encryption:
   i. Apply the permutation function g [6] on P, $MSG = g(P)$
   ii. Apply MSM M to generate Cipher text: $C_i = E_M(MSG)$
   $E_M = \lambda(q, a_i) = \lambda_{a_i}(q) \quad a_i \in MSM$
   $C_i = \lambda(q, a_i) = \lambda_{MSM}(q)$
5. Decryption:
   i. Get C_i cipher text
   ii. Compute inverse of M i.e $M' = (Q', I', O', \delta', \lambda', q_0')$
   iii. Apply M' to get Message
   $MSG = E_M^{-1}(C) = \lambda'(q, C_i) = \lambda'_{C_i}(q)$
   iv. Apply the inverse Permutation function g^{-1} on MSG to get Plain Text P
   $P = g^{-1}(MSG)$

---

3. Experimental setup and results discussion

The performance analysis of proposed method based on sequential state machine (mealy state Machine) have been implemented and tested in DELL Inspiron 14 3000 series having Intel(R) Core(TM) i3-4005U CPU@1.70GHz with installed primary memory(RAM) of 4GB(3.41GB usable) under 32 bit operating system running in Windows 7 Home basic of service pack1. The designed encryption algorithm is implemented in python 3.6 version. The mealy state machine used in proposed method that consist of set of states $Q = \{q_0, q_1, q_2, q_3, q_4, q_5, q_6, q_7\}$, input and

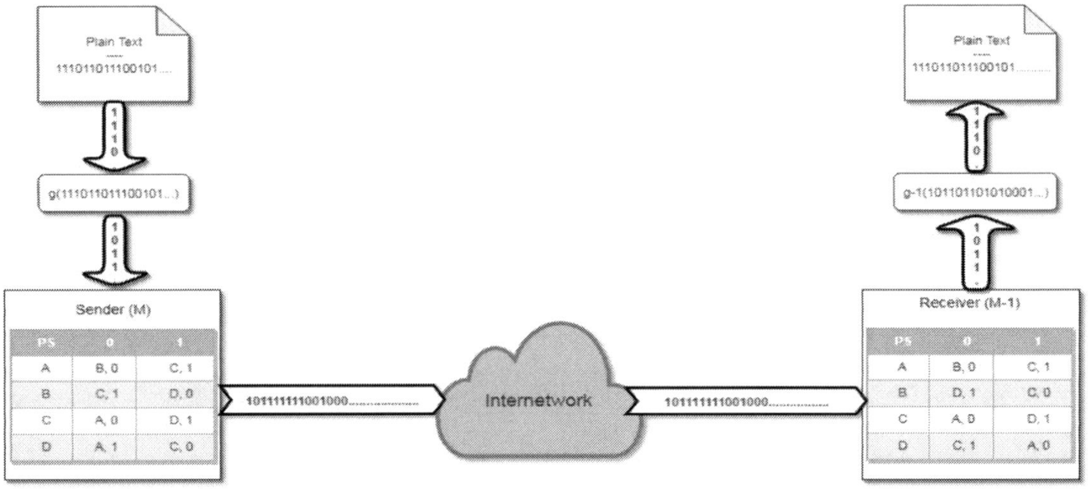

Figure 2. Pictorical Representation of motivating example

output symbols $I = O = \{0, 1\}$ and the start state q_0. That means to say that, we considered $|Q| \times 8 = 64$ bit key space for the encryption and decryption process and the number of rounds will be depending upon number of plain text block size.

The plain text(Input sequence) $a_0a_1a_2......a_i$ will be represented over $GF(2)[2, 3]$ or sequence of binary bits. These binary bits are inputs to the MSM that encrypt plain text(msg in figure 3) and produce cipher text(cipher in figure 3) and original plain text(in figure 3) can be retrieved back from cipher text, the same is as shown in figure 3. It also apply the permutation function $g(a_i)$ to binary bits of plain text. The plain text is divided into number of blocks of size 64 bits. The performance has been conducted for both DES and designed algorithm in python 3 interpreter with above mentioned system configuration and speed of proposed method has been observed for different size of plain text and same is presented in the table 1 and figure 4. The computer system with above configuration was running with 97 process with CPU usage 51% and physical memory 73%. The table 3 and figure 4 shows that, the proposed method has better time complexity compare to DES algorithm in encryption and decryption process.

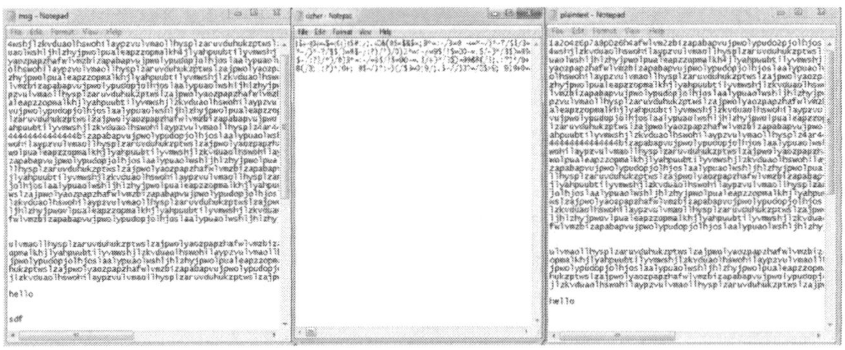

Figure 3. Snapshot of results:Input message, cipher text and Plain text

3.1. Security Analysis

The proposed cryptography based on mealy state machine has variable length of key space is depends on the number of states in the machine. If number of states in the MSM is 8, then size of key will be 64 bit. The key space is of variable length depends on the number of states present in the state machine and same is used in the proposed method. The security of algorithm will be increased if MSM has more number of states. Since entire MSM is a key space in the encryption and decryption process, it is very difficult to find the decryption key for eavesdropper due to its weakly invertable concept. In our implementation, we define 8 states, that means to say that size of key is 64 bit in the designed algorithm. The eavesdropper need $2^{|Q|\times 8}$ time to get encryption key. Even if eavesdropper is able to get the encryption key, but it is very hard to get inverse of MSM. Hence the security of proposed method is completely depends on the Mealy state machine and number of states used in it.

4. Related Work

The sequential state machines are the models that takes the sequence of inputs and produces the sequence of outputs. These models are used as cryptographic techniques due to its generation of outputs sequence for the input sequences. The primary concept of sequential state machine based cryptosystem is very simple: The encryption key will be a state machine itself and inverse of it is the key for decryption. The security of whole system is depends on the difficulties in identifying state machine and inverse of given state machin[24]. Most of the proposed cryptography uses the applications of cellular automata, transducers (More and Mealy Machine) and state machines with accepter and without output symbols. The author of [8] has proposed the cellular automata based cryptosystem in 1985 to generate pseudorandom stream of bits P_i, then later the sequence of bits will be added to plain text a_i for cipher text generation i.e $C_i = P_i \oplus a_i$. The decryption process is found in [16] and in this method, if plain text sequence a_i is known to adversory, then known-plain text attack can be possible. To overcome this problem, the author of [25] presented cryptosystem based on reversible cellular finite automata. The cipher text will be generated by using cellular finite automata, but the key size used in this method is big problem[24]. The other cryptography techniques based on cellular automata is found in [9, 10, 11, 12, 13]

The author of [26] has presented novel encryption technique for sharing of secrete information by using finite state machine. The author has used the recurrence relation matrix to get the cipher text, but there is no performance testing of proposed automata based cryptography. These systems depends on constructing the inverse of it. The author of [27] has developed and presented one-key cryptosystem by using extended mealy state machine. However, it is observed that, the brute force attack is possible on the proposed method and also it is slower than the DES encryption alogirthm. Then later the author [6] presented a mathematical model of cryptography (one-key and two key) based on extended finite state automata. The authors presented different mathematical models of public key cryptosystem like FAPKC, FAPKC2 and FAPKC3 based on linear and non-linear finite state automata.

The stream cipher technique based on Rabbin-scott model of finite state machine is presented in [17] i.e state machine without output symbols. The proposed method suffers from the practical difficulties that effects the performance of algorithm. The author of [7] has presented DES encryption technique based on finite automata for plain text as well as images. The author have used finite state machine for generating 16 sub-keys for encryption and decryption process and also presented the performance testing for different texts size and images. The author of [18] have proposed finite state machine based block cipher to exploit the idea of parallel computation and introduce the new block cipher that perform well with high security. This block cipher depending upon the moore machine that is used for encryption and decryption process. The Moore machine based cryptography is also presented in [28]. The paper presents the secret sharing technique by using fibonacci recurrence matrix to provide higher security so that, no adversary can exploit the cipher text. This paper lacks of proof and experiment results on the proposed method.

Based on the observations made in related work, the most of the authors have concentrated only on public key based cryptosystem based on finite state machine which are relatively slow compare to DES encryption algorithm and exploited to cryptanalytic attacks. Hence we propose symmetric key (one key) encryption technique based on sequential state machine (Mealy Machine) which is faster than DES and provide higher security due to weakly invertible state machine concept.

5. Conclusion

The one-key cryptosystem is faster than two-key cryptosystem consumes the lot of network and CPU computational resources. One-key cryptosystem is also efficient to provide confidentiality and privacy of information being transmitted in communication channel. The traditional conventional one-key cryptosystems are in high computational cost due to security dependency on size of key, key generation algorithm and bilinear concept. It is very essential to reduce the time complexity of cryptosystem during data communication system. The proposed sequential state machine - Mealy State Machine(MSM) based one-key cryptosystem that reduces the time complexity 30% w.r.t DES algorithm that utilizes the optimal computational resources and provide the higher security due to difficulties in identifying MSM M as well as inverse of MSM M'. The proposed algorithm is explained with help of motivating example and then implementation results are compared with DES block cipher algorithm with 64 bit key size, the proposed method is faster than the DES algorithm.

This work can be further expanded by using extended mealy machine with additional customize excitation function to provide higher randomness and higher security.

[1] Valtteri Niemi. Cryptology: language-theoretic aspects. In *Handbook of Formal Languages*, pages 507–524. Springer, 1997.

[2] William Stallings. *Cryptography and network security, 4/E*. Pearson Education India, 2006.

[3] Behrouz A Forouzan. *Cryptography & network security*. McGraw-Hill, Inc., 2007.

[4] Dorothy Elizabeth Robling Denning. *Cryptography and data security*. Addison-Wesley Longman Publishing Co., Inc., 1982.

[5] Douglas R Stinson. *Cryptography: theory and practice*. Chapman and Hall/CRC, 2005.

[6] T Renji. Finite automata and application to cryptography, 2009.

[7] Sarshad Abubaker and Kui Wu. Dafa-a lightweight des augmented finite automaton cryptosystem. In *International Conference on Security and Privacy in Communication Systems*, pages 1–18. Springer, 2012.

[8] Stephen Wolfram. Cryptography with cellular automata. In *Conference on the Theory and Application of Cryptographic Techniques*, pages 429–432. Springer, 1985.

[9] Joan Daemen, René Govaerts, and Joos Vandewalle. A framework for the design of one-way hash functions including cryptanalysis of damgård's one-way function based on a cellular automaton. In *International Conference on the Theory and Application of Cryptology*, pages 82–96. Springer, 1991.

[10] Ivan Bjerre Damgård. A design principle for hash functions. In *Conference on the Theory and Application of Cryptology*, pages 416–427. Springer, 1989.

[11] Howard Gutowitz. Cryptography with dynamical systems. In *Cellular Automata and Cooperative Systems*, pages 237–274. Springer, 1993.

[12] S Nandi, BK Kar, and P Pal Chaudhuri. Theory and applications of cellular automata in cryptography. *IEEE Transactions on computers*, 43(12):1346–1357, 1994.

[13] P Guan. Cellular automaton public-key cryptosystem. *Complex Systems*, 1(1):(51–56, 1987.

[14] Dawn M. Turner Peter Smirnoff. Symmetric key encryption - why, where and how it's used in banking, January 2019.

[15] Symmetric cryptography, 2019. [Online; accessed 12-June-2019].

[16] Willi Meier and Othmar Staffelbach. Analysis of pseudo random sequences generated by cellular automata. In *Workshop on the Theory and Application of of Cryptographic Techniques*, pages 186–199. Springer, 1991.

[17] P Dömösi. A novel cryptosystem based on finite automata without outputs. In *Automata, Formal Languages And Algebraic Systems*, pages 23–32. World Scientific, 2010.

[18] Gh Khaleel, SHERZOD TURAEV, MI Mohd Tamrin, et al. A new block cipher based on finite automata systems. *International Journal on Perceptive and Cognitive Computing*, 2(1), 2016.
[19] National Academies of Sciences Engineering and Medicine. *Decrypting the Encryption Debate: A Framework for Decision Makers*. The National Academies Press, Washington DC, 2018.
[20] Basics of cryptography: the practical application and use of cryptography, April 7,2018. [Online; accessed 12-June-2019].
[21] J.S. Dhillon D.P. Kothari. *Digital Circuits and Design*. Pearson, 2016.
[22] John M Yarbrough and John M Yarbrough. *Digital logic: Applications and design*. West Publishing Company St. Paul, MN, 1997.
[23] Zvi Kohavi and Niraj K Jha. *Switching and finite automata theory*. Cambridge University Press, 2009.
[24] G Khaleel, S Turaev, I Al-Shaikhli, and MI Mohd Tamrin. An overview of cryptosystems based on finite automata. *Journal of Advanced Review on Scientific Research*, 27(1):1–7, 2016.
[25] Jarkko Kari. Cryptosystems based on reversible cellular automata. *Manuscript, August*, 1992.
[26] PA Jyotirmie, S Srilakshmi, A Chandra Sekhar, and S Uma Devi. A novel encryption for secret sharing using finite state machine. *International Journal of Mathematical Archive EISSN 2229-5046*, 4(6), 2013.
[27] Marc Gysin. A one-key cryptosystem based on a finite nonlinear automaton. In *International Conference on Cryptography: Policy and Algorithms*, pages 165–173. Springer, 1995.
[28] B Krishna Gandhi, A Chandra Sekhar, and S Srilakshmi. Cryptographic scheme for digital signals using finite state machine. *International journal of computer applications (September 2011)*, 2011.

Table 1. Mealy state machine with its inverse

	M - Transition Table		M' - Transition Table		
δ	$a=0$	$a=1$	δ	$b=0$	$b=1$
A	$B,0$	$C,1$	A	$B,0$	$C,1$
B	$C,1$	$D,0$	B	$D,1$	$C,0$
C	$A,0$	$D,1$	C	$A,0$	$D,1$
D	$A,1$	$C,0$	D	$C,1$	$A,0$

Table 2. Process of Enciphering and Deciphering

Simple Enciphering and Deciphering Process								
Input Sequence a_i to M	1	1	1	0	0	1	0	1
Transitions of M on a_i	$A->C$	$C->D$	$D->C$	$C->A$	$A->B$	$B->D$	$D->A$	$A->C$
Output Sequence b_i from M to M'	1	1	0	0	0	0	1	1
Transitions of M' on b_i	$A->C$	$C->D$	$D->C$	$C->A$	$A->B$	$B->D$	$D->A$	$A->C$
Input sequence a_i from M':	1	1	1	0	0	1	0	1

Table 3. Speed of DES and MSM algorithms for different size of plain texts

	ENC SPEED IN SEC		DEC SPEED IN SEC	
PT in Bytes	*DES*	*PROPOSED*	*DES*	*PROPOSED*
64	0.046	0.0024	0.0623	0.0019
128	0.093	0.0078	0.1	0.0071
256	0.18	0.0153	0.171	0.0156
384	0.249	0.0156	0.265	0.0156
512	0.343	0.0468	0.327	0.0348
1024	0.655	0.0557	0.684	0.0498
4024	0.243	2.683	0.187	2.605
12288	7.21	4.913	7.132	4.780

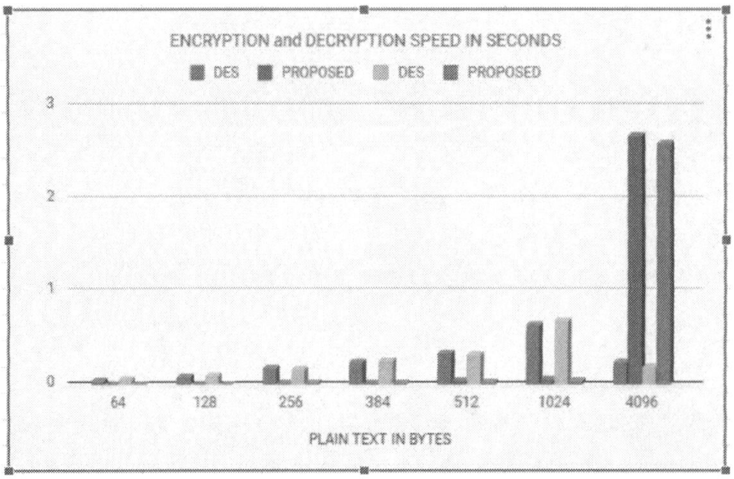

Figure 4. Performance comparison of DES and MSM encryption and decryption

UWB dual notch implementation using folded bi-section stepped impedance resonator

Thomas George[1], B Lethakumary[2] and M A Abdul Hakeem[2]

[1] Mar Athanasius College of Engineering, Kothamangalam, Kerala, India
[2] University College of Engineering, Thodupuzha, Kerala, India

Abstract. This paper proposes a method for dual notch implementation in ultra wideband (UWB) using a simple stepped impedance resonator (SIR). The impedance ratio and electrical length ratio of a bi-section stepped impedance resonator (BSIR) are optimized to have the transmission zeros in the UWB. The high impedance section is folded producing multipath leading to the formation of notches. Transmission zero frequencies are at which the differential phase becomes equal to 180°. Simulations are carried out using ANSOFT HFSS and are fabricated using conventional FR4 PCB board. Prototype developed is tested and the obtained results are used to validate the simulations. Designed filter can be transformed into bandpass filter (BPF) in UWB by enhancing the widths of lower and upper rejection bands.

Keywords: Ultra wideband; Bandpass filter; Dual notch; Multimode resonator; Bi-section stepped impedance resonator.

1. Introduction

Federal Communications Commission (FCC) legitimized the use of license free UWB signals in 2002 [1]. UWB has emerged as an alternate technology for the IEEE 802.15.3a (TG3a) standard and is defined as any signal that has a bandwidth of 500 MHz in the 3.1 To 10.6 GHz band [2]. The purpose of this standard is to provide a specification for a less complex, low cost, low power consuming wireless connectivity with high data rate among devices within or entering the personal operating space. Significant advantages of UWB include large throughput, low power requirement, along with wide bandwidth in the range of GHz. As per FCC standard, the indoor and hand held UWB system operation must be strictly contained in the UWB band ranging from 3.1 to 10.6 GHz.

A variety of methods are available in the literature for the design of UWB filters. A simple UWB BPF using co-planar waveguides at the broadside is proposed in [3]. In the ground plane an open ended coplanar waveguide (CPW) is formed. Zhu and Wang [4] presented a BPF based on an aperture in the ground plane. It functions as a non-uniform resonator and its first three resonances constitute the UWB band. In [5] a UWB filter is designed using a transition structure between CPW and microstrip. The multimode resonator (MMR) includes a wide central strip and two identical narrow side strips. A UWB filter with extended stopband is reported in [6] combining two lowpass filters on both sides of a BPF. UWB BPF using SIR is developed in [7], which uses the first four resonant modes of SIR for wide bandwidth. The resonator has five high-low impedance quarter wavelength sections. In [8] a stub loaded MMR is used for designing a UWB BPF. Three open ended stubs are loaded on an SIR to form the MMR.

This paper proposes a very simple structure using a single Bi-section SIR. The high impedance sections on both sides are folded to form the filter. Multimode resonant characteristics of the SIR is applied to get a wideband response. Folding of SIR introduces multiple travelling paths for the signal from input to output and transmission zeros are generated in the frequency response. The filter designed has two transmission zeros and can be used as a basic building block for the design of UWB BPFs.

2. Stepped impedance resonator

SIR is a logical extension of uniform impedance resonator (UIR) which was originally proposed by Makimoto and Yamashita in 1980 [9]. Simple BSIR has a single step discontinuity with two impedance sections Z_1 and Z_2 having electrical lengths θ_1 and θ_2 connected together as shown in Figure1.

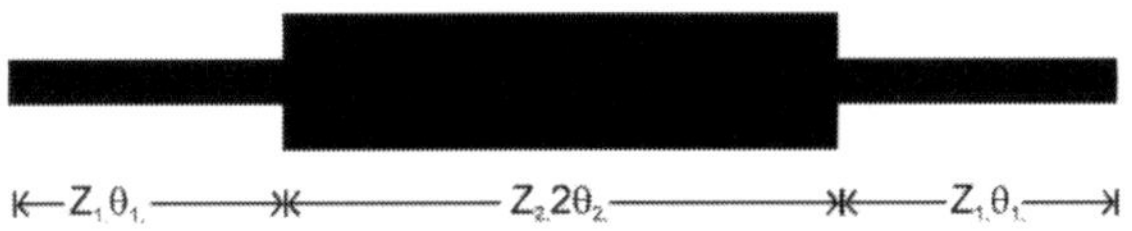

Figure 1. Structure of BSIR.

The two important attributes of an SIR are k the impedance ratio and α the electrical length ratio expressed as,

$$k = \frac{Z_1}{Z_2} \tag{1}$$

$$\alpha = \frac{\theta_1}{\theta_1 + \theta_2} \tag{2}$$

Electrical length ratio α is different from physical length ratio. It is due to the fact that, the two impedance sections are of different widths resulting in different effective dielectric constants. Input admittance of SIR is expressed as,

$$Y_{in} = j\frac{Z_2 \tan\theta_1 \tan\theta_2 - Z_1}{Z_1(Z_1 \tan\theta_1 + Z_2 \tan\theta_2)} \tag{3}$$

Resonance condition is obtained by setting Y_{in} to zero. From (1), (2) and (3), it is clear that the SIR resonances are greatly dependent on k and α, the two key SIR parameters.

A folded BSIR is shown in Figure 2. Both high impedance sections of the SIR are folded and are kept anti-parallel to the body. When input is applied to one side, the signal can travel through multiple paths towards the output. Multipath signal leads to phase difference between received signals, altering the insertion loss (IL) response. SIR folding leads to an added advantage of structure compactness.

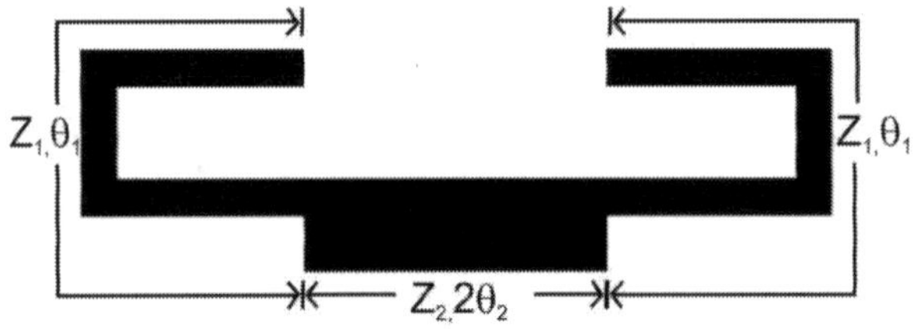

Figure 2. Folded BSIR.

3. Folded SIR Filter

Proposed structure is obtained by folding the SIR and connecting it to 50 Ω feedlines as shown in Figure 3. When the SIR is folded, one transmission zero each is formed at either sides of the frequency response, which is due to multiple signal path from input to output.

IOP Conf. Series: Materials Science and Engineering 1187 (2021) 012027 doi:10.1088/1757-899X/1187/1/012027

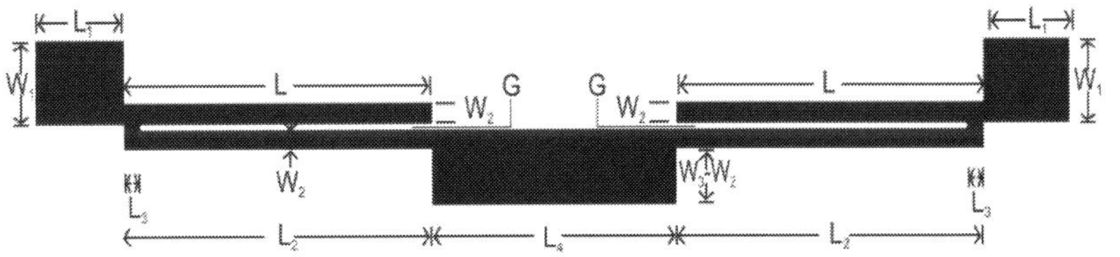

Figure 3. Filter using folded BSIR.

Filter dimensions are optimized using HFSS as follows, L_1 = 4 mm, L_2 = 14 mm, L_3 = 0.75 mm, L_4 = 11 mm, L = 14 mm, W_1 = 3 mm, W_2 = 0.75 mm, W_3 = 2.75 mm and G = 0.2 mm. Various SIR parameters are: $Z_1 = 98.04\ \Omega$, $Z_2 = 53.48\ \Omega$, $\theta_1 = 403.47^0$, $\theta_2 = 82.15^0$, k = 1.83 and α = 0.83. The filter passband is realized based on the 2nd to 4th resonances of the SIR. Figure 4 shows the frequency responses of the above folded BSIR filter.

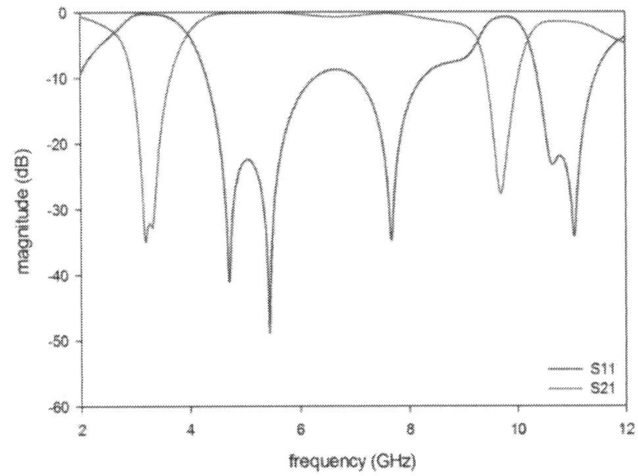

Figure 4. Frequency responses of filter using folded BSIR.

When the length of SIR folded arm is increased above 10 mm keeping all parameters constant, a pair of transmission zeroes are obtained. Under this condition, the multipath electrical length of the resonator will be equal to 180^0. When L = 14 mm, two transmission zeroes are obtained at 3.19 and 9.71 GHz. The multipath electrical lengths at these frequencies are found to be 182^0 and 181^0 respectively. Differential phase becoming equal to 180^0 causes cancellation of the signals leading to transmission zeroes. The lower and upper roll-off ratios of the filter are 41 and 70.5 dB/GHz respectively.

4. Results and discussions

Filter simulations are done using ANSOFT HFSS 3D full-wave electromagnetic field simulation software. For experimental demonstration, the proposed filter is fabricated using conventional FR4 substrate having a dielectric constant of 4.4, loss tangent 0.02 and thickness 1.6 mm. SMA connectors are connected to input and output ports for measurement. Photograph of the fabricated filter is shown in Figure 5. Area of the filter ground plane is 517 mm^2. Testing of fabricated filter is done using KEYSIGHT E5063A Series Network Analyzer (100 KHz – 18 GHz).

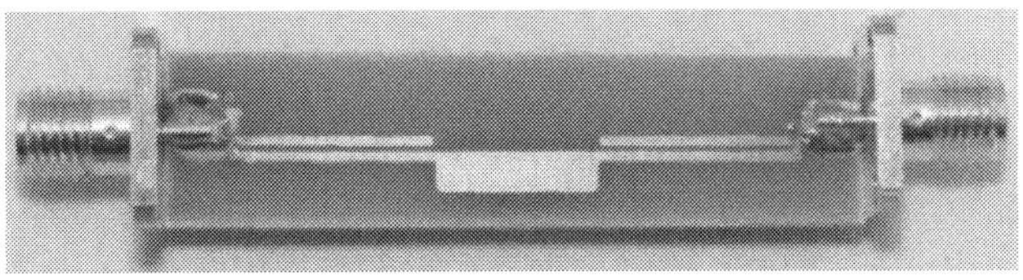

Figure 5. Photograph of the fabricated filter.

Frequency responses of simulated and validated filters are shown in Figure 6. Simulated filter passband is from 3.94 to 9.40 GHz (5.46 GHz) and that of validated filter is from 4.36 to 8.61 GHz (4.25 GHz).

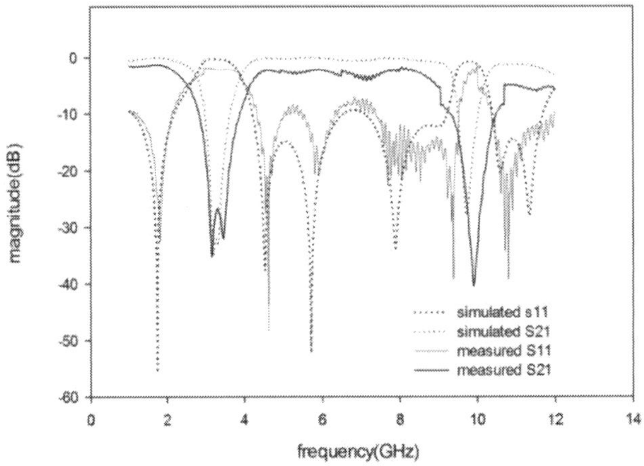

Figure 6. Responses of simulated and validated filters.

Table 1 summarizes important parameters of both simulated and validated filters. Some of the reasons for validated filter performance deviation are tolerance in the filter fabrication method, radiation losses and the use SMA connectors. Another important reason is increased insertion loss at high frequencies due to substrate characteristic variation.

Table 1. Comparison of simulated and validated filters.

Parameter	Simulated	Measured
Lower 3 dB frequency	3.94 GHz	4.36 GHz
Upper 3 dB frequency	9.40 GHz	8.61 GHz
Center frequency	6.67 GHz	6.49 GHz
Passband	5.46 GHz	4.25 GHz
First transmission zero	3.19 GHz	3.17 GHz
IL at first transmission zero	34.8 dB	34 dB
Second transmission zero	9.71 GHz	9.89 GHz
IL at second transmission zero	27.8 dB	40 dB
Maximum passband IL	0.62 dB	3.99 dB
Minimum passband return loss	-9.4 dB	-7.8 dB

5. Conclusion

This paper proposes a simple method for the design of dual notch UWB filter using a folded BSIR. Ultra wide passband is obtained by tuning of SIR parameters. SIR folding leads to multiple signal path and results in transmission zeros. Differential phase difference at transmission zeros are found to be 180^0. For experimental verification, the filter is fabricated and tested. Measured results justify the proposed design. Designed filter has the advantages of being simple, low cost, small size due to the use of folded SIR, steep passband selectivity, low insertion loss and wide passband. It is seen that, the filter performance is degraded by narrow lower and upper stopbands. This filter can be converted into UWB BPF by extending the widths of both lower and upper rejection bands.

IOP Conf. Series: Materials Science and Engineering 1187 (2021) 012027 doi:10.1088/1757-899X/1187/1/012027

6. References

[1] Federal Communications Commission 2002 *Revision of Part 15 of the Commission's Rules Regarding Ultra-Wideband Transmission Systems: ET Docket 98-153* (Washington: Federal Communications Commission)

[2] Aiello G R and Rogerson G D 2003 Ultra-wideband wireless systems *IEEE Microwave Mag.* **4** 36-47

[3] Li K, Kurita D and Matsui T 2005 An ultra-wideband bandpass filter using broadside-coupled microstrip-coplanar waveguide structure *IEEE MTT-S Int. Microwave Symp. Digest* (Long Beach) pp 678-81

[4] Zhu L and Wang H 2005 Ultra wideband bandpass filter on aperture backed microstrip line *Electronic Lett.* **41** 1015-16

[5] Wang H and Zhu L 2005 Ultra-wideband bandpass filter using back-to-back microstrip-to-CPW transition structure *Electronics Lett.* **41** 1337-38

[6] Wong S W, Sun S, Zhu L and Chen Z N 2006 Ultra-wideband (UWB) bandpass filters with improved upper-stopband performance *Proc. of the 2006 Asia-Pacific Microwave Conf.* vol 1 (Yokohama) pp 29-32

[7] Cai P, Ma Z, Guan X, Kobayashi Y, Anada T and Hagiwara G 2007 A novel compact ultra-wideband bandpass filter using a microstrip stepped-impedance four-modes resonator *IEEE MTT-S Int. Microwave Symp. Digest* (Honolulu) pp 751-54

[8] Li R and Zhu L 2007 Compact UWB bandpass filter using stub-loaded multiple-mode resonator *IEEE Microwave and Wireless Components Lett.* **17** 40-42

[9] Makimoto M and Yamashita S 2000 *Microwave Resonators and Filters for Wireless Communication Theory: Design and Application* (Kawasaki: Springer)

A Novel approach in Hybrid Median Filtering for Denoising Medical images

R. Seetharaman[1*], **M. Tharun**[1], **K. Anandan**[2]

[1]Department of ECE, CEG Campus, Anna University, Chennai-600025, India
[2] PS Division of E & E, TUV SUD South Asia Pvt Ltd, Chennai-600041, India

*Corresponding author email: rseetharamanece@gmail.com, rseetharamanece@annauniv.edu

Abstract. Image denoising is a key pre-processing step in medical image analysis. At current, deep learning-based models have shown a great promise, which outperformed many conventional methods over the past three decades. Speckle noise removal is a major issue in preserving all the delicate details and the edges in ultrasound image processing as it degrades the visual evaluation of ultrasound images. The multiplicative behavior of speckle-noise is converted into additive by using log transform as the additive noise removal is easy as compared to multiplicative noise. An innovative approach for denoising highly distorted images affected by speckle noise is proposed. This paper presents a result of significant work in image denoising and exploring several thresholding methods of denoising images such as SureShrink, VisuShrink and BayesShrink. The results of different approaches of wavelet-based image denoising methods are tabulated to find the best method. The main aim is to show the result of wavelet coefficients on a new basis, so that the noise can be minimized or removed from the data.

1. Introduction

Wavelet changes assist us to address signals with a serious level of sparsity. The rule behind a non-straight wavelet-based sign assessment technique is called wavelet denoising. In Image denoising, the most difficult is to secure the information-bearing designs, for example, edges and surfaces to get great visual quality [1]. During transmission, the picture is tainted by clamor which isn't handily dispensed within image handling. The space field is an information activity technique carried on the first picture and cycles the picture dim worth, for instance, area normal strategy, wiener channel, etc. Clamor is the most irritating issue in picture preparing. One approach to dispose of this issue is the improvement of such a hearty calculation that can play out the handling errands in presence of clamor. The alternate way is to plan a filtration interaction to kill the commotion from pictures. while protecting its components, edges and subtleties. Clamor brings arbitrary varieties into picture that vacillate the first qualities to some various qualities. Causes which might acquaint commotion with pictures remember defects for information transmission, flawed optics, sensor failing, handling methods and electronic impedance. The primary point of denoising is to eliminate clamor from the uproarious picture is accomplished by opposite wavelet change [2]. There are two fundamental kinds of wavelet change ceaseless wavelet change and discrete wave-let change. The PCs are discrete in nature, so PC programs utilize the discrete wavelet change. The discrete change is generally effective according to the computational perspective [3]. The cycle generally manages the demonstrating of the wavelet change coefficients of unique pictures. The denoising of unique pictures undermined by Gaussian clamor is a broad issue in picture handling. The wavelet change has turned into a significant apparatus to conquer the issue and the wavelets are favorable to vide a structure for signal disintegration as a grouping of signs. The strategies dependent on wavelet portrayals yields very basic calculations that are regularly more remarkable and simpler to work with than customary techniques for work assessment. A full reference picture quality appraisal measurement is utilized for assessing denoising quality across different clamor levels in various manufactured and genuine images [4]. The computerized picture components are known as picture components or picture components or pixels. Comparable pixels are looked across a non-nearby locale, by the reality discovering intently comparative pixels is more plausible than comparative patches [5]. Pixel is the most broadly utilized term to address the components of a computerized picture. A computerized picture is a discrete two-dimensional capacity (2D capacity), work f(x, y), which has been quantized over its area and reach. Without loss of consensus, it will be viewed as that the picture is rectangular fit, comprising of Y rows & X segments.

SAR Speckle commotion evacuation is a superb issue in ultrasound picture preparation utilized for acquiring necessary analytic information for the human body. Numerous picture denoising has generally been drawn closer through the viewpoint of old-style applied science dynamic ideas [6]. The ultrasonic pictures are visually debased by Speckle commotion, to dispose of dot clamor in the ultrasonic pictures, the edges and the fine subtleties of the ultrasonic pictures should be protected. Protecting every one of the fine subtleties of the picture makes testing one. From bounteous approaches to dispose of dot commotion, there is a special strategy that changes over the multiplicative conduct of dot clamor into added substance commotion by utilizing log change because the added substance commotion expulsion is a lot simpler than multiplicative commotion. The channel is changed in accordance with the control picture denoising impact, so the first picture includes data is more safeguarded while the im-age is denoised [7]. The wavelet parcel decay is utilized to spot and eliminate the commotion from influenced pixels. The Wiener channel is utilized for a pre-preparing reason. The profound neural organization has been created to lessen commotion in pictures [8]. In this paper, an adjusted mean middle channel is proposed to eliminate commotion and antiquities from the ultrasound liver picture. Test results show that the proposed channel is predominant as far as eliminating spot clamor and curios and safeguarding the edges. Picture preparing has a wide scope of uses in medication, biomedicine and different areas of science and designing [9].

Ultrasound imaging of the liver is a non-intrusive instrument utilized in medication for the determination of liver infections. The nature presence of spot clamor in ultrasound pictures shames the picture quality and influences edges and fine subtleties of pictures which make demonstrative more troublesome. Hence, the picture pre-handling is important for the expulsion of spot commotion which supports the improvement of processed helped finding for programmed grouping of liver infections. In this paper, an adjusted mean middle channel is proposed to eliminate commotion and curios from the ultrasound liver picture. The denoising attributes are assessed as far as picture quality measurements alongside visual quality and the exhibition of the proposed strategy beat a few existing channels. Test results show that the proposed channel is unrivaled as far as eliminating spot commotion and curios and furthermore safeguarding the edges.

2. Novel approach to noise reduction in ultrasound images based on geodesic paths

In this paper, a substitution technique for multiplicative commotion decrease in ultrasound pictures is proposed. The motivation behind our undertaking is to discover exact outcomes for the biomedical pictures. We have done it for different clinical pictures. We have taken out the spot clamor from the clinical pictures. We recognized the commotion in the edges which were extremely simple to eliminate. In the event that we eliminate the clamor in the edges, we can decide the precise outcomes which were exceptionally valuable to the patients. One of the basic difficulties in the field of picture handling and PC vision is picture denoising, where the hidden objective is to assess the first picture by smothering clamor from a commotion-polluted rendition of the picture. Picture denoising assumes a significant part in a wide scope of utilizations, for example, picture rebuilding, visual following, picture enlistment, picture division, and picture characterization, where getting the first picture content is urgent for solid execution. The issue of picture clamor concealment stays an open test, particularly in circumstances where the pictures are procured under helpless conditions where the commotion level is extremely high. The expanding number of patient information in clinical pictures forces an exploration challenge for the logical treatment for diagnosing, identifying, and foreseeing the infections. With the assistance of this venture, we can be ready to discover precise outcomes. The expanding number of patient information in clinical pictures forces an examination challenge for the logical treatment for diagnosing, recognizing, and anticipating the illnesses. The channel procedure denoises the pictures in the most proficient way. here we have utilized a half and half middle channel to demonstrate it. The sets of boisterous information and clean objective pictures are prepared by discriminative profound learning techniques is right now ruled in the field of picture handling [10]. The clever method is a change of the reciprocal denoising plan, which considers the similitude of pixels and their spatial distance [11]. The channel yield is determined as a weighted normal of the pixels which are in the local connection with the focal point of the sifting window, and the loads are elements of the negligible association costs between encompassing pixels. Customarily, the models are prepared in an extremely administered design with debased pictures as sources of info and clean pictures as targets, so the organization adjusts to eliminate the defilement [12]. Exploratory outcomes show that the proposed strategy yields essentially preferable outcomes over the contrary strategies simply in the event of ultrasound pictures sullied by medium and incredible multiplicative commotion unsettling influences [13]. Profound learning is an arising AI method from gained information for better denoising [14].

The term wavelet was initially utilized in the area of seismology to depict the aggravations that radiate and continue outward from a sharp seismic motivation [15]. A wave is an oscillating capacity of time or space and is occasional. Interestingly, wavelets are limited waves. In wavelet investigation, the sign to be broke down is duplicated with a wavelet capacity and afterward the change is figured for each fragment created. The Wavelet Transform, at high frequencies, gives a happy time goal and helpless recurrence goal, while at low frequencies the Wavelet Transform gives great frequencies the helpless time goal. A discretionary sign can be examined as far as scaling and interpretation of a single parent wavelet work (premise). Wavelets permit both time and recurrence analysis of signs at the same time, thus, wavelet portrayal gives an adaptable mathematical instrument to break down transient, time variation (non-fixed) signals that are not genuinely predictable particularly districts of discontinuities include that is common in pictures having discontinuities at the edges. Picture denoising is a significant examination region in the picture handling field that has its wide reach, for example, clinical picture denoising, satellite picture denoising and pressure commotion denoising [16].

3. Mathematical representation of wavelet

The idea of wavelet was stowed away underway of mathematicians significantly over a century prior. In 1873, Karl Weirstrass numerically depicted how a group of capacities can be built by superimposing scaled forms of a given premise work. The term wavelet was initially utilized in the area of seismology to depict the unsettling influences that radiate and continue outward from a sharp seismic motivation. Wavelet implies a "little wave". The diminutiveness alludes to the condition that the window work is of limited length minimal upheld. A wave is a wavering capacity of time or space and is occasional. Interestingly, wavelets are restricted waves. They have their energy amassed on schedule and are fit for the investigation of transient signs. In wavelet investigation, the sign to be dissected is duplicated with a wavelet capacity and afterward the change is processed for each section created. The Wavelet Transform, at high frequencies, gives fun time goal and helpless recurrence goal, while at low frequencies; the Wavelet change gives great frequencies the helpless time goal. Wavelets are capacities produced from one single capacity (premise work) called the Proto-type or mother wavelet by enlargements (scaling) and interpretations (changes) on schedule (Frequency) space. The mother wave condition $f_{a,0}(t)$ is time scaled by an and abundancy scaled by $\sqrt{|a|}$. The boundary „a" called enlargement or scaling boundary causes withdrawal of (t) in the time pivot. When $a < 1$ and extension or extending $a > 1$. For $a < 0$, the capacity $f_{a,b}(t)$ brings about time inversion with the widening. Numerically substitute t as (t-b) for an interpretation or change in the time hub as in eqn 1. Wavelet's ability to give spatial recurrence data is the primary justification for this examination. This property guarantees opportunities for better separation between the commotion and the information. There are two sorts of wavelets that change in particular Continuous Wavelet Transform (CWT) and Discrete Wavelet Transform (DWT). DWT of picture signals creates a nonredundant picture portrayal, which gives better spatial and otherworldly limitation of picture development contrasted and another multi-scale portrayal, for example, Gaussian and Laplacian pyramid [17]. This is called disintegration or examination. The parts can be related once more to the first sign without loss of information. This interaction is called reproduction or amalgamation. The numerical control, which im-employs investigation and amalgamation, is called Discrete Wavelet Transform and Inverse DWT.

HAAR. Utilized for picture pressure. HAAR has been utilized for picture preparation and design recognition.

Daubechies. The Daubechies wavelets are a group of symmetrical wavelets described by a maximal number of evaporating minutes for some given support. There is a scaling capacity that creates an orthogonal multi-rearrangement examination. The principle contrast with the traditional development of Daubechies is that the multiresolution spaces are gotten from scale-subordinate creating capacities [18].

Symlets. The Symlet are almost symmetrical, even, biorthogonal wavelets proposed by Daubechies as adjustments to db family. The properties of the two wavelet families are comparable. sym 6 was utilized in the hard thresholding strategy. The balanced wavelet (symlet) is proposed as a premise work in a multi-goal investigation (MRA) utilizing the discrete wavelet change (DWT) to break down power load utilization signals, notwithstanding, are commonly defiled by non-fixed and non-Gaussian clamor cycles of which no models exist [19].

Coiflets. Coiflets wavelets are discrete wavelets planned by Ingrid Daubechies in line with Donald coifman to have scaling capacities. Properties are balanced, symmetrical, biorthogonal. Sorts of coiflets are 1,2,3,4,5.

CWT is an execution of the wavelet change utilizing self-assertive scales and practically discretionary wavelets. Nonorthogonal wavelets are utilized for its advancement in the information got by this change for profoundly corresponded. CWT works by registering a convolution of the sign with the scaled wavelet and it is carried out in the CWT module that can be gotten to with Data Process> Integral Trans-forms>CWT.

4. Hybrid Median Filter

Drive commotion decrease or expulsion is a profoundly dynamic examination space of picture preparing. A median filter having great denoising power has been the most famous channel. Its varieties have been proposed in the multistate median channel, middle channel dependent on homogeneity data choice based managed middle channels to work on its presentation. As of late the expansions of numerical morphology to shading pictures have been proposed. The proposed crossover channel consolidates the solid preparing capacities of numerical morphology and a choice-based managed middle channel for commotion concealment or undoing. By utilizing the idea of ASF, an arrangement of disintegration and widening joined with the managed middle channel is made out of line organizing components. It accomplishes extraordinary outcomes as well as produces outwardly palatable remade pictures even in exceptionally high commotion densities. A nonlinear mixture channel for eliminating fixed drive commotion from shading pictures depends on numerical morphology and the middle channel has been proposed. The mixture middle channel, the windowed channel are utilized to eliminate im-beat clamor and to safeguards the edges of the ultrasonic pictures. The fundamental thought behind the channel is for any component of the sign (picture) to apply the middle procedure a few times changing window

IOP Conf. Series: Materials Science and Engineering 1187 (2021) 012028 doi:10.1088/1757-899X/1187/1/012028

shape and afterward take the middle of the got middle qualities. The underlying advance is the decaying of 2D wavelets of single-level processes with Dwt2. The deterioration is finished as for either a specific wavelet (see w-channels for more data) or specific wavelet disintegration channels. The subsequent stage figures the single level 2D discrete wavelet change (DWT) of the information X utilizing the wname wavelet where dwt2 returns the estimate coefficients network and detail coefficients frameworks. Then, at that point, idwt2 order plays out a solitary level two-dimensional wavelet remaking concerning either a standard particular wavelet or specific wavelet recreation channels either low pass channel and high pass channel. The idwt2 utilizes the wavelet to process the single level recreated estimate coefficients mama trix X, in light of guess grid and subtleties frameworks. At long last, BMF performs half breed middle separating of the lattice utilizing aNxN box. Stage 6: The mistake execution of these filters is determined utilizing Mean Square Error (MSE) and Peak Signal to Noise Ratio (PSNR) of the out-put picture as for the information uproarious picture. The proportion is utilized as a quality estimation of the original and packed picture. The distances between the focuses and the relapse line are estimated and afterward determined by figuring out them. For this situation, just DWT is applied as a denoising technique to acquire subgroups (Approximation band, A, and detail groups H, V, and D). This is utilized to estimate the clamor in the sub-band D, condition 6 is utilized to decide standard deviation for the detail subgroups, then, at that point used to discover the limit and incentive for each sub-band. Pixel upsides of each sub-band are recorded and afterward contrasted with the band limit esteem, on the off chance that the pixel esteem is more noteworthy than the edge esteem, it is set to nothing. Something else, the next pixel is prepared (This interaction is rehashed until every one of the pixels is taken care of. Presently, IDWT is applied to denoised subgroups. MSE and PSNR esteem are registered to utilize on the denoised and the first pictures. For this situation, just the median filter is utilized to handle denoising method on the noisy picture, thus getting the denoised picture then, calculating MSE and PSNR values.

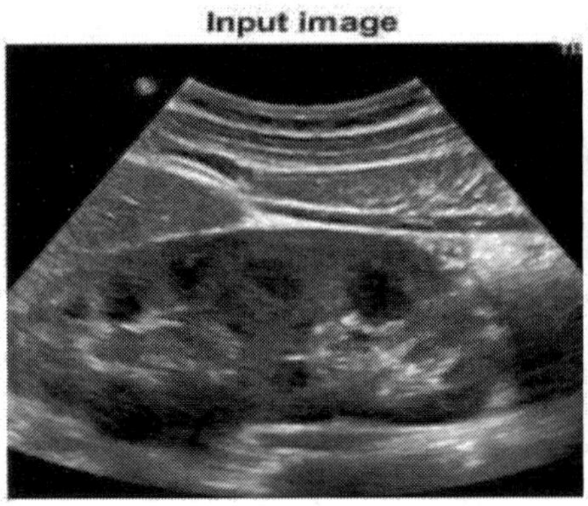

Figure 1. Input image

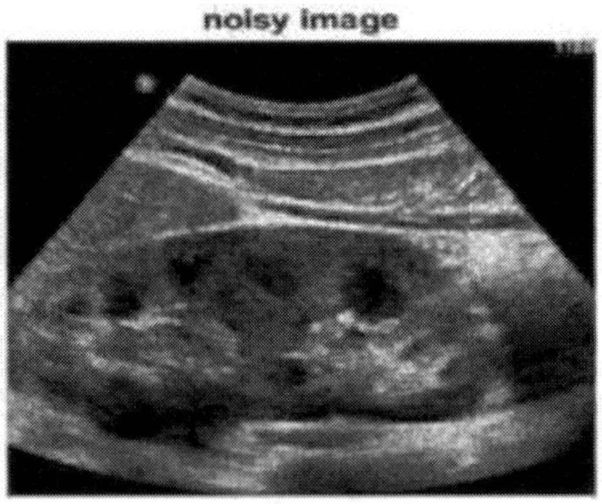

Figure 2. Noisy image

IOP Conf. Series: Materials Science and Engineering 1187 (2021) 012028 doi:10.1088/1757-899X/1187/1/012028

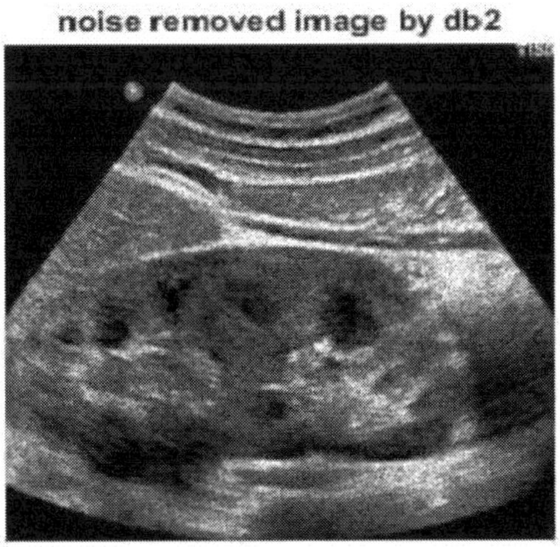

Figure 3. Noise removal by db2

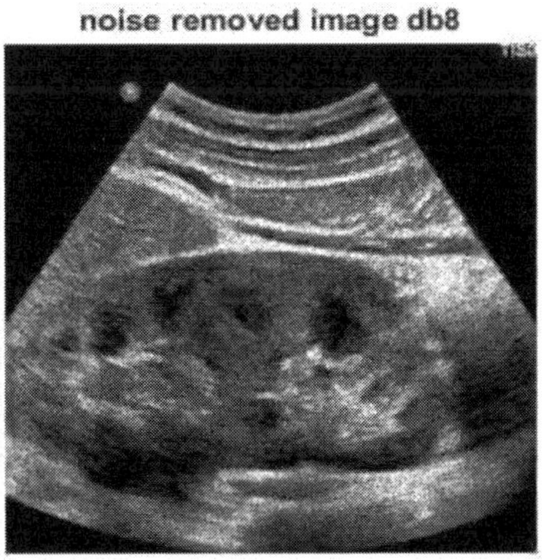

Figure 4. Noise removal by db8

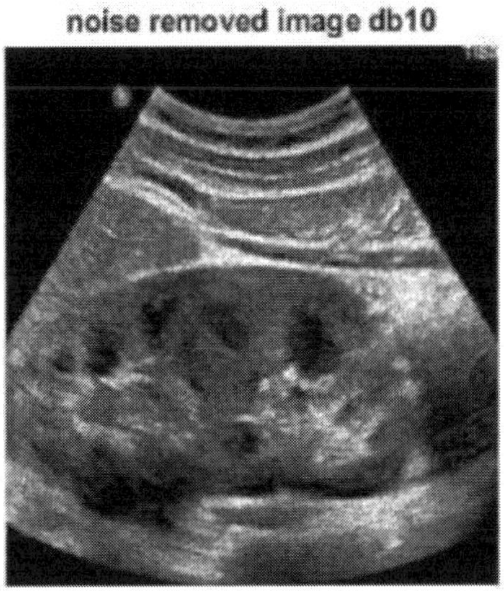

Figure 5. Noise removal by db10

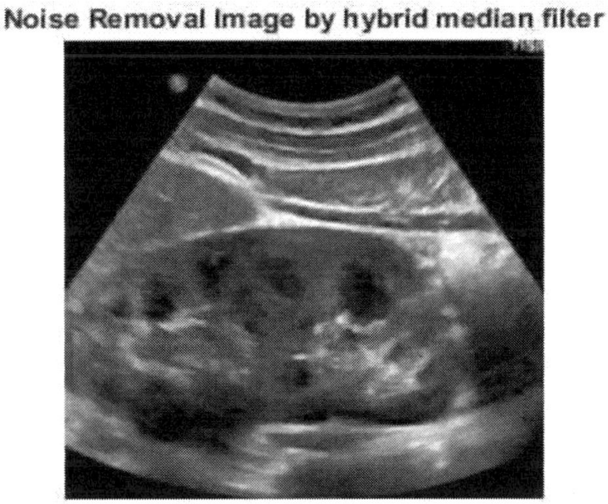

Figure 6. Noise removal by HMF

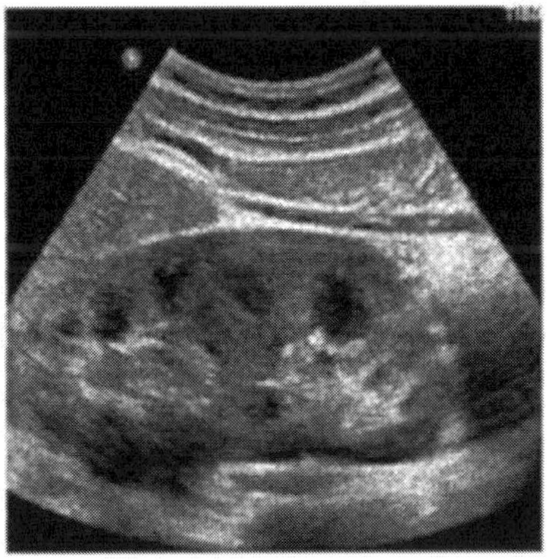

Figure 7. Noise removal by symlets

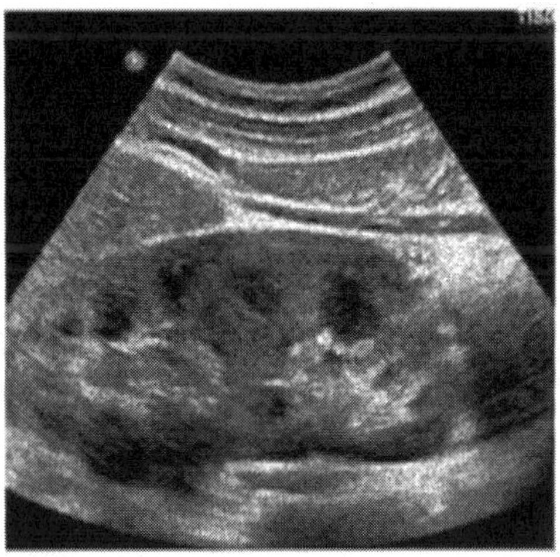

Figure 8. Noise removal by coiflets

IOP Conf. Series: Materials Science and Engineering 1187 (2021) 012028 doi:10.1088/1757-899X/1187/1/012028

We have applied six different methodologies such as db2, db8, db10, Hybrid Median filter, Symlets, Coiflets to remove noise in the noisy images. Figure 1 and Figure 2 are the input image and noisy image, we will perform various methodologies to noisy image and comparing the results of six methods to conclude the best method. The Daubechies wavelets are a family of orthogonal wavelets characterized by a maximal number of vanishing moments where we applied Daubechies 2, Daubechies 8, and Daubechies 10 in Figure 3, Figure 4, and Figure 5. The db2, db8, and db10 results are similar to each other where these methodologies have removed noise slighter and are less effective. The symlets are nearly orthogonal, symmetrical quite similar to bi-orthogonal properties proposed by Daubechies and this method is applied in Figure 7. The coiflets are discrete wavelets and share the same properties of symlets which is applied in Figure 8. The results of symlets and coiflets are similar to each other and slight variations in the images. The results are better than Daubechies but not effective. In the Figure 6, the noise removed by Hybrid Median Filter and compared with other methodologies. We can able to see the clear image and the result by the Hybrid Median Filter are the effective one.

Table 1. PSNR & MSE values of denoised image

	MSE	PSNR	FOM	MSSIM	NK
DB2	$5.5107e^{-22}$	26.0753	0.5345	0.5654	0.8023
DB8	$1.0978e^{-20}$	24.7760	0.5675	0.5543	0.8123
DB10	$1.7007e^{-20}$	24.5858	0.5467	0.5232	0.8223
Symlets	$1.8964e^{-20}$	24.7897	0.5176	0.5012	0.7412
Coiflets	$1.9873e^{-20}$	24.9878	0.5433	0.5113	0.7623
Hybrid Median Filter	5.2959	40.9254	0.6243	0.5213	0.8501

The mean squared error tells how close a relapse line is to a bunch of focuses. It does this by taking the good ways from the focuses to the relapse line (these distances are the "blunders") and figuring out them. The PSNR block registers the peak signal-to-noise ratio, in decibels, between two pictures. This proportion is utilized as a quality estimation between the first and a compacted picture. The Structural Similarity Index measure is a technique for foreseeing the apparent nature of realistic pictures, just as different sorts of advanced pictures and recordings. MSSIM is utilized for estimating the comparability between two pictures. The Normalized cross-correlation (NCC), typically its 2D adaptation, is regularly experienced in format coordinating with calculations, for example, in facial acknowledgment, movement following, enrollment in clinical imaging, and so forth Its quick calculation becomes basic in time-delicate applications. The MSE and PSNR values are calculated for all the six methodologies and tabulated in Table 1. From the tabulation, we can clearly assert that the values of five parameters results are better for Hybrid Median Filter.

5. Conclusion

In this paper, denoising has been done with various ultrasound images. The image has been resized and denoising has been performed using various wavelet transform techniques such as Haar, Daubechies, symlets, coiflets and hybrid median filter. Its efficiency has been calculated using MSE and PSNR values and it is found that hybrid median filter has been proved to be efficient for eliminating speckle noise in the medical images and it also enhances the visual worth of the medical images. It helps to select the finest wavelet transform for denoising the medical image and it will help in diagnosis.

5.1 Discussion

We have also done with the different wavelet transformation techniques such as db 2, db 8, db 10, haar wavelets. We have added another method called the hybrid median filter. One of the filters is the Hybrid median filter which is an improved version of the median filter, which removes the noise better than the median filter. Image filters produce a new image by finding edges and locate features from an original by operating on the pixel values. There are various filtering techniques in image processing to enhance the quality of the image by removing noise and preserving the image details. To find edges and locate features filters are used, even they help to increase the contrast in the image. On analyzing the various parameters, we have come to know that the hybrid median filter was best compared to the wavelet transformation techniques. The hybrid median filter shows the best result for PSNR and MSE values. In the base paper, they have compared only two parameters. Here we have done for five parameters. Hence compared with all the five parameters we can able to see hybrid median filter was best for the image denoising technique.

5.2 Future work

Our future work is to work with other denoising techniques such as spatial filtering, wavelet threshold- ing method. On working with this we can find any better way for the fulfilled image denoising. We have also planned to work with many other denoising technique for the betterment of the patients and doctors.

References

[1] Elhoseny M, Shankar K. *Optimal bilateral filter and convolutional neural network based denoising method of medical image measurements.* Measurement. 2019 Sep 1;143:125-35.

[2] Abdelhamed A, Afifi M, Timofte R, Brown MS. Ntire 2020 challenge on real image denoising: Dataset, methods and results. InProceedings of the IEEE/CVF *Conference on Computer Vision and Pattern Recognition* Workshops 2020 (pp. 496-497).

[3] Honzátko D, Kruliš M. Accelerating block-matching and 3D filtering method for image denoising on GPUs. *Journal of Real-Time Image Processing.* 2019 Dec;**16(6)**:2273-87.

[4] Thanh DN, Hien NN, Prasath S. *Adaptive total variation L1 regularization for salt and pepper im- age denoising.* Optik. 2020 Apr 1;208:163677.

[5] Hou Y, Xu J, Liu M, Liu G, Liu L, Zhu F, Shao L. NLH: *A blind pixel-level non-local method for real-world image denoising.* IEEE Transactions on Image Processing. 2020 Mar 19;29:5121-35.

[6] Mildenhall B, Barron JT, Chen J, Sharlet D, Ng R, Carroll R. *Burst denoising with kernel predic- tion networks.* InProceedings of the IEEE Conference on Computer Vision and Pattern Recogni- tion 2018 (pp. 2502-2510).

[7] Tournier JD, Smith R, Raffelt D, Tabbara R, Dhollander T, Pietsch M, Christiaens D, Jeurissen B, Yeh CH, Connelly A. MRtrix3: *A fast, flexible and open software framework for medical image processing and visualisation.* Neuroimage. 2019 Nov 15;202:116137.

[8] Yuan N, Zhou J, Qi J. *Low-dose CT image denoising without high-dose reference images.* In15th International Meeting on Fully Three-Dimensional Image Reconstruction in Radiology and Nucle- ar Medicine 2019 May 28 (Vol. 11072, p. 110721C). International Society for Optics and Photon- ics.

[9] Thanh DN, Hien NN, Prasath VS, Hai NH. *Automatic initial boundary generation methods based on edge detectors for the level set function of the chan-vese segmentation model and applica- tions in biomedical image processing.* InFrontiers in Intelligent Computing: Theory and Applica- tions 2020 (pp. 171-181). Springer, Singapore.

[10] Krull A, Buchholz TO, Jug F. *Noise2void-learning denoising from single noisy images.* InProceed- ings of the IEEE/CVF Conference on Computer Vision and Pattern Recognition 2019 (pp. 2129- 2137).

[11] Malik K, Smolka B. *Modified bilateral filter for the restoration of noisy color images.* InInternational Conference on Advanced Concepts for Intelligent Vision Systems 2012 Sep 4 (pp. 72-83). Springer, Berlin, Heidelberg.

[12] Laine S, Karras T, Lehtinen J, Aila T. *High-quality self-supervised deep image denoising.* Ad- vances in Neural Information Processing Systems. 2019;32:6970-80.

[13] Rosa R, Monteiro FC. *Performance analysis of speckle ultrasound image filtering.* Computer Methods in Biomechanics and Biomedical Engineering: Imaging & Visualization. 2016 Jul 3;4(3- 4):193-201.

[14] Xie D, Li Y, Yang H, Bai L, Wang T, Zhou F, Zhang L, Wang Z. *Denoising arterial spin labeling perfusion MRI with deep machine learning.* Magnetic resonance imaging. 2020 May 1;68:95-105.

[15] Kaur J, Kaur M, Kaur P, Kaur M. *Comparative analysis of image denoising techniques.* Interna- tional journal of Emerging Technology and Advanced engineering. 2012 Jun;2(6):296-8.

[16] Park B, Yu S, Jeong J. *Densely connected hierarchical network for image denoising.* InProceed- ings of the IEEE/CVF Conference on Computer Vision and Pattern Recognition Workshops 2019 (pp. 0-0).

[17] Bnou K, Raghay S, Hakim A. *A wavelet denoising approach based on unsupervised learning model.* EURASIP Journal on Advances in Signal Processing. 2020 Dec;2020(1):1-26.

[18] Vonesch C, Blu T, Unser M. Generalized Daubechies wavelet families. IEEE Transactions on Signal Processing. 2007 Aug 20;55(9):4415-29.

[19] Swee EG, Elangovan S. Applications of symlets for denoising and load forecasting. InProceedings of the IEEE Signal Processing Workshop on Higher-Order Statistics. SPW-HOS'99 1999 Jun 16 (pp. 165-169). IEEE.

Comparison of word embeddings in text classification based on RNN and CNN

*Merlin Susan David[1], and Shini Renjith[1]

[1]Department of Computer Science and Engineering, Mar Baselios College of Engineering and Technology, Thiruvananthapuram, Kerala, 695015, India
merlindavid92@gmail.com, shini.renjith@mbcet.ac.in

Abstract. This paper presents a comparison of word embeddings in text classification using RNN and CNN. In the field of image classification, deep learning methods like as RNN and CNN have shown to be popular. CNN is most popular model among deep learning techniques in the field of NLP because of its simplicity and parallelism, even if the dataset is huge. Word embedding techniques employed are GloVe and fastText. Use of different word embeddings showed a major difference in the accuracy of the models. When it comes to embedding of rare words, GloVe can sometime perform poorly. Inorder to tackle this issue, fastText method is used. Deep neural networks with fastText showed a remarkable improvement in the accuracy than GloVe. But fastText took some time to train when compared to GloVe. Further, the accuracy was improved by minimizing the batch size. Finally we concluded that the word embeddings have a huge impact on the performance of text classification models

Keywords: Convolutional neural network, recurrent neural network, fastText, text classification, word embeddings.

1. Introduction

We have both structured and unstructured data at our fingertips.Chats, web pages,email, and other soci al media platforms are examples of unstructured data. Since the unstructured data is a rich source of information, extracting information or knowledge from these messy data is challenging and time consuming.

Humans can easily comprehend different languages but when it comes to computers, it is difficult for the machines to understand natural languages. But nowadays we have Siri, Alexa, Cortana., which will get activated at our command and perform actions accordingly. This is only possible due to Natural Language Processing (NLP) along with elements like deep learning and machine learning. NLP allows the machine to comprehend human language, measure its sentiments, and identify the important prominent features from it.

Basic NLP tasks include tokenization, stemming, lemmatization, and so on. In natural language processing, text classification is crucial. It is the process of applying labels to the input document or categorising it (Fig. 1). It can also be used to organize, store, and manage information effectively. Different applications of text classification include tag recommendation, spam detection, sentiment analysis, and intent analysis.

Fig. 1. Text classification

Manual categorization of the corpus is challenging and time-consuming. Automated text classification include Rule-bas\d system, Machine learning-based system, and Hybrid approaches. Manual categorization of the corpus is challenging and time-consuming. Automated text classification include Rule-based system, Machine learning-based system, and Hybrid approaches.

2. Related Work

2.1 Text classification using string kernels [4].

The author of this study introduced a different kernel for comparing two text documents. The kernel is an inner product of sub-sequences of length k in feature space. In the text, a sub-sequence is an ordered sequence of k letters. The advantages are that once a proper kernel function is chosen, we can work in any dimensional space without incurring any additional costs. The most notable feature is that they use sequence alignment techniques to map the corpus to vectors without explicitly representing them. The disadvantages are, The results reported in the paper are very precursive (Fig. 2) and yet, there are lot of questions that still remains to be answered. The calculation of the new kernel takes longer, and more investigation is required to figure out how to proceed with this step of the calculation [4]. Other works in this area include [10, 11].

	F1	Precision	Recall	#SV
3 S-K	0.925	0.981	0.878	138
5 S-K	0.936	0.992	0.888	237
6 S-K	0.936	0.992	0.888	268
W-K	0.925	0.989	0.867	250

Fig. 2. Performance measures of different string kernels [4]

2.2 Comparison of SVM and some older classification algorithms in text classification tasks [5].

KNN and Naive Bayes are two well-known classifiers that are compared to the SVM. The performance of the classifier was assessed using 10-fold cross validation. Initial tests on 20Newsgroups to determine the optimal parameter settings. The value of C was also altered for a number of kernel functions. As a result, the default value C=100 is employed, along with a linear kernel. The benefits include the fact that the SVM does not require high precision to train. As the size of the feature space changes, so does the training time. The benefits include the fact that the SVM does not require high precision to train. As length of feature space alters, so does the training time. KNN and Naive Bayes have outstanding performance and are much quicker than SVM. The disadvantages are computational cost for training SVM for a huge training dataset [5]. Similar works in this area include [12, 13, 14].

2.3 Text classification based on multi-word with support vector machine [6].

This paper's major objective is to study the impact of employing multi-word textual in-cooperation on text classification. First, extract the multi-word from the document and it can be done using a linguistic method or by using a statistical method.If there are fewer terms in the corpus, the terms will have no discriminative capability in classifying the documents.After extracting the multi-word from the document, documents are represented as multi-word. So we have two strategies

1) Decomposition strategy
2) Combination strategy

Information retrieval and text mining in multi-word is a recently investigated feature, and it has a number of advantages. There are three advantages to employing, Multi-word: It has a little size when compared to individual words, but it performs well. MCL, for example, has a precision of up to 0.8673. It is simple to extract many words from a corpus using corpus learning without using a dictionary. Multi-word phrases have more semantics and are more powerful than single words. As a result, if multi-words are used for information finding, patterns will be more interpretable and comprehensible.

The disadvantages are that due to a lack of theoretical proof, a broad judgement was not reached during the assessment. The mathematical examination of the statistical aspects of multi-words on text categorization is a significant problem that has to be addressed further. For more trustworthy work, additional inspection and research are required [6]. Other attempts in this are ainclude [15, 16, 18].

2.4 Comparative effectiveness of CNN and RNN architectures for radiology text report classification [7].

Other published NLP experiments show that the accuracy measures are either superior or comparable. The results indicated the practicability of CNNs and RNNs in the classification tasks of imaging test reports and also supported the use of these techniques at a large scale in classifying text reports for other applications, including radiology patient prioritisation and clinical trial eligibility screening and imaging utilization.

They are also utilized in other areas of research like large scaled data production for automatic medical image interpretation. And perfect scores are yet to be achieved by neural nets.

RNN model accurately predicted the classes and situated the relevant sentences in the reports, but it is hard to agree upon how the model made that inference. Many of these errors require subtle reasoning to reach the correct conclusion, which may be a limitation of the models [7].

3. Convolutional Neural Network

Computers visualize things in a different way than we humans do. They only see things in terms of numbers and are called pixels. Because of their architecture, CNN differ from other types of neural networks. Other Neural Nets process their input by passing it through a sequence of layers. Each of the layer consists of a group of nodes, each of which is linked to all nodes in the previous layer. The predictions are represented by a fully linked layer and an output layer.

3.1 Architecture

Convolutional Neural Networks are a unique type of neural network. To begin, CNN comprises of 3 dimensions and they are depth, height and width. Furthermore, the nodes in one of the layer don't link to all of the nodes in the following layer, merely a little portion (Fig. 3). Finally, the result is condensed into probabilistic scores grouped with depth axis. There are two halves to CNN. The portion about feature extraction. To detect the features, the network will perform a multiple convolutions operation and pooling operations in this stage. On top of these retrieved features, fully linked layers would function as a classifier in the classification section. They will give the object on the screen a likelihood of being what the algorithm believes it is.

The most significant characteristics of a CNN is convolution. It is a mixture of two sets of data. The combination of two other functions to form a third function was referred to as convolution. Convolution is then applied to the input data with the help of a filter or kernel in the case of CNN to build a feature map. Then, sliding the filter function on the input, we perform a convolution. Matrix multiplication is also calculated at each point, and then the resulting is added to the feature map. The receptive field, is another term for this part of the filter. This filter is 3x3 in size. [19, 20].

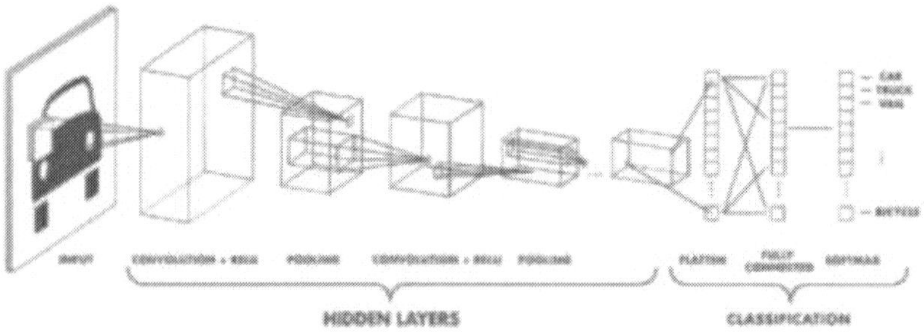

Fig. 3. Employed CNN architecture

We performed multiple convolution operation on the input, using different filter each time. This will result in various feature maps. We aggregate all feature maps at the conclusion of the process and output them as the convolution layer's final result. To introduce non-linearity, we employ an activation function. Convolution's final output is processed through an activation function, the most often utilized of which is Rectified Linear Unit (ReLu). The stride of a convolution filter is the number of steps it moves each time. A stride in most case is 1, that means the filter will slide pixel by pixel. As the size of the stride increases, eventually it also the filter will cover larger interval of the input and there will be less overlap between cells.

Fig. 4. Max pooling

And Since the size of the feature map is less when compared to the input, we should prevent further shrinkage of the feature map. This is where padding comes into play [19, 20].To prevent the feature map from being lowered, a layer of zeros is applied to result. Padding also increases performance by ensuring that the kernel and stride sizes are correct for the input. After convolutional layer there will be pooling layer between them. Convolution layer and pooling layer can be repeated many times. Dimensionality reduction is one of the advantage of pooling operation and hence the number of parameters, lowering computing costs and preventing over-fitting. Max pooling (Fig. 4) is the most often used pooling layer, that projects the highest value from every window. However, this window size must be selected ahead of time. This reduces the size of the feature map while preserving the important information [7,19].

After feature extraction part, the classification part which consist of fully connected layer that only accommodate 1-D information. For converting 3-D data to 1-D, the flatten function is used which is

IOP Conf. Series: Materials Science and Engineering 1187 (2021) 012029 doi:10.1088/1757-899X/1187/1/012029

available python. This will convert the 3-D data into a 1-D data. Last of the layers in CNN are fully connected. There will be nodes that are fully connected to the preceding layers. Training of CNN is very much similar to other neural networks, using back propagation or gradient descent. Our ultimate objective is to input photos into it, so that CNN may assign a probability to an target it believes it sees or to interpret an image with the words.

4. Recurrent Neural Network

The output from the previous step is provided as an input to the current step in recurrent neural networks. The inputs and outputs of other neural nets are independent of one other, but when it comes to word prediction, they are not. We need prior knowledge of the word that comes before it. RNN (Fig. 5) addresses this with the help of a hidden layer [8,19].

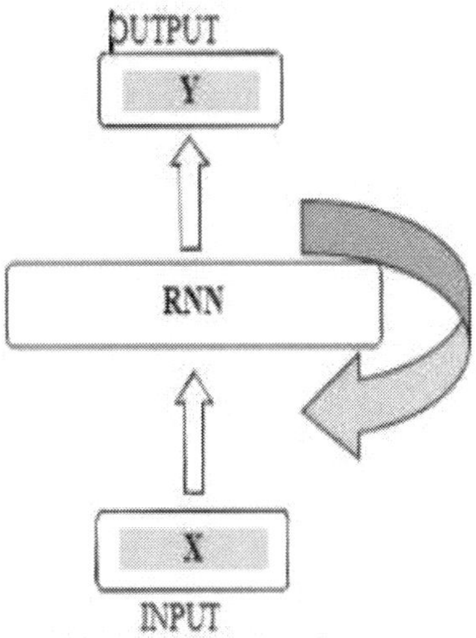

Fig. 5. Recurrent Neural Network

The prominent characteristic of Recurrent neural net is its hidden state, that help it to remember the information of the sequence [8]. Recurrent neural networks are of two common variants which include Gated Recurrent Unit (GRU) and Long Short Term Memory (LSTM). RNN stores all the information that was previously calculated. It uses the same parameter for all the input and hidden layers so that it reduces the complexity of parameter.

Because each layer has individual weights and biases, all of the layers in classic neural networks are independent of one another. RNN, on the other hand, applies the same weights and biases to all layers, reducing the complexity of rising parameters and memorising past outputs by feeding each output into the next hidden layer.

4.1 Bidirectional LSTM

The idea behind Bidirectional LSTM is from bidirectional recurrent neural network. They process sequential data coming in both forward pass and backward pass and also have two hidden layers. It is already established that bidirectional networks are best when compared to unidirectional network. Unfolded bidirectional LSTM it consists of both forward and backward LSTM layers. The BiLSTM layer generates an output vector, y, in which each element is calculated using the equation 1 [9]

$$y_t = \sigma(\overrightarrow{h_t}, \overrightarrow{\mathrm{h_t}}) \quad (1)$$

where sigma function is used for combining two output sequences.

5. Feature Engineering

Feature engineering is the process by which the text information will be changed into feature vectors and new features will be made utilizing the dataset. The way towards representing the words and documents utilizing a dense vector is called word embedding [1, 2, 3]. Word embeddings can be prepared with the given input document/corpus or with the pre-trained word embeddings like GloVe, fastText, and Word2Vec. After making a vocabulary that is made out of words. We create a matrix of low-dimensional word vectors. Furthermore, the matrix is provided as input to Neural Net [21].

GloVe is a method for collecting vector representations of words using an unsupervised learning algorithm.The final representation exhibits an intriguing linear substructure of the word vector space, which is based on collective global word-word co-occurrence insights from corpus [22].

FastText is a package that allows you to learn word representations. It is used for both supervised and unsupervised representations of sequence of words and the sentences to be trained. These embeddings are then incorporated for a variety of purposes, including data compression and transfer learning initializers. It is largely written in C++ and allows for multiprocessing during training. fastText also supports training models that use negative sampling, and softmax, such as the continuous bag of words or Skip-gram models [17].

Embeddings from Language Models (ELMo) is for learning words and their context, it uses a deep biLSTM language model. The deep BiLSTM architecture also allows ELMo to capture more of context-dependent aspects of words in the higher layers along with their syntax aspects in lower layers. This results in better word embeddings, and different representation of words depending on the context in which they are used.

Bidirectional Encoder Representations from Transformers (BERT) is built on the bidirectional concept from ELMo, but uses new transformer architecture to calculate word embeddings. It has been shown to produce excellent embeddings of word, achieving good performance on different NLP problems [24].

6. Dataset

TREC dataset collection has a question categorization consisting of domain, fact-based questions separated into several groups or labels. TREC dataset consist of 6 labels and has 5,452 training set data and 500 test set data. Models are assessed based on accuracy. The labels associated with TREC-6 are Abbreviation (ABBR), Entities (ENTY), Locations (LOC), Numeric values (NUM), Human beings (HUM), and Description and abstract concepts (DESC).

7. Classifier Comparison and Result

fastText: When it comes to embedding of rare words, it can sometime perform poorly. Inorder to tackle this issue, fastText method is used. It uses sub-word information so that rare words can be represented well. It still works based on skip-gram model. This model enhance the performance on syntactical problems fundamentally, yet very little in semantic questions. fastText consider individual words as made out of character n-grams. Since training is based on character n-gram, it takes more time to generate fastText embeddings compared to word2Vec [17].

GloVe: Glove's goal is rather straightforward - it executes word vectors in order to obtain sub-linear correlations in vector space. Thus, it proved that GloVe pull off better than Word2Vec. Besides, GloVe incorporate more reasonable significance into word vectors by considering the relation between the word pair rather than the word to word. GloVe takes up a lot of memory since it is trained on the co-occurrence matrix of words [22].

Word2Vec: Word2Vec regard each word in the document as single entity and generates a vector for each word. Word2Vec is similar to GloVe. When it comes to rare words Word2Vec perform poorly because, in Word2Vec a rare word has fewer neighbor, compared to a word that commonly occurs.The model is hard to train if we employ the softmax function, since the number of categories is enormous [23].

Results show that fastText has comparatively higher accuracy in both RNN and CNN when it is applied to the TREC dataset (Tables 1 and 2). Result may vary for other dataset based on attributes like dataset size, and so on.

Table 1. Performance analysis of CNN using fastText and GloVe

Word Embeddings	Accuracy	Loss
GloVe	89.09%	10.91%
FastText	90.93%	9.07%

Table 2. Performance analysis of RNN using fastText and GloVe

Word Embeddings	Accuracy	Loss
GloVe	24.59%	74.51%
FastText	83.33%	16.67%

8. Conclusion

In natural language processing, text classification is crucial. It can be used for many applications like sentiment analysis, spam detection, review analysis, and more. For all these applications a faster, efficient, and logical system is required. A text classification model must be capable of classifying text data even if it is small or large, must be able to identify meaningful words from the text. For this work, the comparison of the text classification using deep neural networks like CNN and RNN is performed. Both fastText and GloVe word embedding techniques are applied to both. Deep neural networks with fastText showed a remarkable improvement in the accuracy of than glove. But fastText took more time to train than a GloVe. Further, the accuracy was improved by minimizing the batch size.Finally we concluded that the word embeddings have a huge impact on the performance of text classification models.

References

[1]. Guo, Bao, Chunxia Zhang, Junmin Liu, and Xiaoyi Ma. "Improving text classification with weighted word embeddings via a multi-channel TextCNN model." Neurocomputing 363 (2019): 366-374. https://doi.org/10.1016/j.neucom.2019.07.052
[2]. Helaskar, Mukund N., and Sheetal S. Sonawane. "Text Classification Using Word Embeddings." In 2019 5th International Conference On Computing, Communication, Control And Automation (ICCUBEA), pp. 1-4. IEEE, 2019. https://doi.org/10.1109/ICCUBEA47591.2019.9129565
[3]. Rabut, Benedict A., Arnel C. Fajardo, and Ruji P. Medina. "Multi-class document classification using improved word embeddings." In Proceedings of the 2nd International Conference on Computing and Big Data, pp. 42-46. 2019. https://doi.org/10.1145/3366650.3366661

[4]. Lodhi, Huma, Craig Saunders, John Shawe-Taylor, Nello Cristianini, and Chris Watkins. "Text classification using string kernels." Journal of Machine Learning Research 2, no. Feb (2002): 419-444. https://doi.org/10.1162/153244302760200687
[5]. Colas, Fabrice, and Pavel Brazdil. "Comparison of SVM and some older classification algorithms in text classification tasks." In IFIP International Conference on Artificial Intelligence in Theory and Practice, pp. 169-178. Springer, Boston, MA, 2006. https://doi.org/10.1007/978-0-387-34747-9_18
[6]. Zhang, Wen, Taketoshi Yoshida, and Xijin Tang. "Text classification based on multi-word with support vector machine." Knowledge-Based Systems 21, no. 8 (2008): 879-886. https://doi.org/10.1016/j.knosys.2008.03.044
[7]. Banerjee, Imon, Yuan Ling, Matthew C. Chen, Sadid A. Hasan, Curtis P. Langlotz, Nathaniel Moradzadeh, Brian Chapman et al. "Comparative effectiveness of convolutional neural network (CNN) and recurrent neural network (RNN) architectures for radiology text report classification." Artificial intelligence in medicine 97 (2019): 79-88. https://doi.org/10.1016/j.artmed.2018.11.004
[8]. Kim, Jihyun, Jaehyun Kim, Huong Le Thi Thu, and Howon Kim. "Long short term memory recurrent neural network classifier for intrusion detection." In 2016 International Conference on Platform Technology and Service (PlatCon), pp. 1-5. IEEE, 2016. https://doi.org/10.1109/PlatCon.2016.7456805
[9]. Cui, Zhiyong, Ruimin Ke, Ziyuan Pu, and Yinhai Wang. "Deep bidirectional and unidirectional LSTM recurrent neural network for network-wide traffic speed prediction." arXiv preprint arXiv:1801.02143 (2018).
[10].Gonçalves, Teresa, and Paulo Quaresma. "Text classification using tree kernels and linguistic information." In 2008 Seventh International Conference on Machine Learning and Applications, pp. 763-768. IEEE, 2008. https://doi.org/10.1109/ICMLA.2008.78
[11].Zheng, Shiqiang, Yujiu Yang, Haiping Wu, and Wenhuang Liu. "Chinese Text Classification Using Key Characters String Kernel." In 2009 Fifth International Conference on Semantics, Knowledge and Grid, pp. 113-119. IEEE, 2009. https://doi.org/10.1109/SKG.2009.59
[12].Kalcheva, Neli, Milena Karova, and Ivaylo Penev. "Comparison of the accuracy of SVM kemel functions in text classification." In 2020 International Conference on Biomedical Innovations and Applications (BIA), pp. 141-145. IEEE, 2020. https://doi.org/10.1109/BIA50171.2020.9244278
[13].Basu, Atreya, Christine Walters, and M. Shepherd. "Support vector machines for text categorization." In 36th Annual Hawaii International Conference on System Sciences, 2003. Proceedings of the, pp. 7-pp. IEEE, 2003. https://doi.org/10.1109/HICSS.2003.1174243
[14].Dilrukshi, Inoshika, and Kasun De Zoysa. "Twitter news classification: Theoretical and practical comparison of SVM against Naive Bayes algorithms." In 2013 International Conference on Advances in ICT for Emerging Regions (ICTer), pp. 278-278. IEEE, 2013. https://doi.org/10.1109/ICTer.2013.6761192
[15].Silva, Catarina, and Bemardete Ribeiro. "Text classification from partially labeled distributed data." In Adaptive and Natural Computing Algorithms, pp. 445-448. Springer, Vienna, 2005. https://doi.org/10.1007/3-211-27389-1_107
[16].Aggarwal, Charu C., and ChengXiang Zhai. "A survey of text classification algorithms." In Mining text data, pp. 163-222. Springer, Boston, MA, 2012. https://doi.org/10.1007/978-1-4614-3223-4_6
[17].Yao, Tengjun, Zhengang Zhai, and Bingtao Gao. "Text Classification Model Based on fastText." In 2020 IEEE International Conference on Artificial Intelligence and Information Systems (ICAIIS), pp. 154-157. IEEE, 2020. https://doi.org/10.1109/ICAIIS49377.2020.9194939
[18].He, Ming, Jianjun Sun, and Ying Cheng. "Text Classification Based on Naive Bayes: A Review." Information Science 34 (2016): 147-154.
[19].Liu, Tengfei, Shuangyuan Yu, Hongtao Zhang, and Hongfeng Yin. "Recurrent Neural Networks and Convolutional Neural Networks for Text Classification." Computer Engineering &

Software 39, no. 01 (2018): 64-69.
[20].Cai, Jingjing, Jianping Li, Wei Li, and Ji Wang. "Deeplearning model used in text classification." In 2018 15th International Computer Conference on Wavelet Active Media Technology and Information Processing (ICCWAMTIP), pp. 123-126. IEEE, 2018. https://doi.org/10.1109/ICCWAMTIP.2018.8632592
[21].Wang, Bin, Angela Wang, Fenxiao Chen, Yuncheng Wang, and C-C. Jay Kuo. "Evaluating word embedding models: Methods and experimental results." APSIPA transactions on signal and information processing 8 (2019). https://doi.org/10.1017/ATSIP.2019.12
[22].Pennington, Jeffrey, Richard Socher, and Christopher D. Manning. "Glove: Global vectors for word representation." In Proceedings of the 2014 conference on empirical methods in natural language processing (EMNLP), pp. 1532-1543. 2014. http://dx.doi.org/10.3115/v1/D14-1162
[23].Goldberg, Yoav, and Omer Levy. "word2vec Explained: deriving Mikolov et al.'s negative-sampling word-embedding method." arXiv preprint arXiv:1402.3722 (2014).
[24].Gao, Zhengjie, Ao Feng, Xinyu Song, and Xi Wu. "Target-dependent sentiment classification with BERT." IEEE Access 7 (2019): 154290-154299. https://doi.org/10.1109/ACCESS.2019.2946594

Automatic Detection of New Vessels in Retinal Photographs using Machine Learning

[1] M Moorthi., [2] Dr M.Gopikrishnan., [3]V Pooja., [4]M R Ragavi Priya., [5]D K Vasanthakumar

[1,3,4,5]Department of Medical Electronics, Saveetha Engineering College, Chennai, India.
[2]Department of Computer Science Engineering, Prathyusha Engineering College, Chennai, India.
moorthidmp@gmail.com

Abstract. Diabetic Retinopathy is a rare micro-vascular complication that leads to severe ocular disorder or visual impairment. This complication is unidentified due to the mild progression of its symptoms that is associated with Diabetic Mellitus. The disease is categorized by the development of abnormal new vessels developing the optical disc of the retina of the eye. In this paper, an automatic detection system for detecting new vessels in the optical disc region is detected by using image processing techniques in advance. Segmenting the vessels from the optic disc by extracting the desired region of interest. Based on this, the features are extracted and detected as the presence of new vessels in the optic disc by using a Support Vector Machine (SVM) classifier. This automatic system is trained and tested by using 65 normal images that are categorized as 33 normal images and 32 abnormal images. The discrimination performance of the system shows an accuracy of 95.38% in detecting the new vessels. It obtained sensitivity and specificity of 96.8% and 93.9% respectively.

Keywords: Diabetic Retinopathy, Optic disc, Image processing, Fundus image, SVM

1. Introduction

The extensive accessibility of the image processing approach in the medical field plays a vital role in screening human abnormalities by a frequent visual perception with transient and less workload. Thus, image processing in the medical field plays a significant role in planning procedures of diagnosis and surgical treatments performed within patients suffering from different health issues. Images have the greatest density of morphological information that helps to communicate with humans. The recent development in automatic medical imaging techniques helps to analyze quickly to understand about the life and diseases for accurate diagnosis and quality treatment. The process of the human and the machine's understanding of images involves the selection of images for extracting the desired features from the images by the pre-trained model. The automatic processing technique plays a major part in medical image processing. Recently, for better analysis, many innovations are taken over in computer-based techniques to screen the abnormalities in humans. Specifically, the huge complexity in medical imaging is based on a host of factors such as the knowledge bases, noise, incompleteness, redundancy, inconsistency in data and knowledge, imprecision, different paths of reasoning such as qualitative subjective reasoning and heuristic reasoning [1].

Diabetes Mellitus (DM) is a serious medical complication that occurs due to metabolic disorder of improper secretion of insulin results in increased blood glucose level or hyperglycemia. When diabetes reaches the uncontrolled level it results in coma or even death if not treated immediately [2]. This disease is significantly related to microvascular and macro-vascular difficulties causing concurrent diseases like Diabetic Retinopathy, Diabetic Nephropathy, and Diabetic Neuropathy are conferred in 50% of type 1 and 30% of the type 2 diabetes mellitus patients [3]. The people affected by diabetes mellites show an 80% chance of Diabetic Retinopathy after 10 years [4,13,14].

Diabetic retinopathy is characterized as microvascular complications that result in eye blurring or vision loss. An increase in the glucose level leads to damages in the blood vessels of the light-sensitive tissue at the back of the eye. The symptoms of this disease are silent and cause mild eye problems in the early stages eventually it results in blindness. Therefore, identifying patients at risk for subsequent vision blurring due to diabetes mellitus is considered to be a crucial part of the treatment of diabetic retinopathy. Regarding Diabetic retinopathy diagnosis, the eye fundus image is captured by using the dilated fundoscopy, it will reflect the present status of the eye. The eye of the DR patients is observed with improper growth of new blood vessels that niish the retina and bulges, cutting the blood supply. This difficulty is easily recognized by using the image processing technique. The growth of new vessels is easily identified in the fundus images by extracting them by using the required image processing technique. Consequently, this will help to identify diabetic retinopathy during the early stages of the disease.

In this paper, an automatic method for detecting retinal photographs which show the development of new vessels on the optic disc region is analyzed and evaluated. The features of proliferative retinopathy, the development of new vessels at the optic disc helps in the identification and prediction of DR at the earliest and most likely helps in adding value to an automated diagnostic system.

2.Background Studies:

Ramya [5], in this paper, explained a method to classify DR. The system recognized the normal eye image by the rate of 86% and the DR eye image by the rate of 82%. Goatman et.al [6] in this paper, designed a system to classify the growth of new blood vessels in the optic disc region by using fifteen features that are oriented with the blood vessel, and the features are classified as normal and abnormal by using SVM classifier. Aswale and Shaik [7] in this paper, described a system to classify DR by using an SVM classifier. Priorly the fundus image is pre-processed, segmented and the feature is extracted and classified as DR or normal eye image. This system produced an accuracy of 93.33% respectively.

Chandrakala and, Sindhu Madhuri [8], this is another method performed by pre-processing the retinal image by using geometric transformation technique, used point-based technique for segmentation and by using performance matrix based on Principal Component Analysis (PCA) and SVM the DR is determined. The accuracy of prediction by PCA and SVM are 89% and 95% respectively. Gandhimathi et al. [9] in this paper used two different classifiers - Multi-Layer Perceptron (MLP) and Support Vector Machine (SVM). The retinal image is pre-processed using a filter, segmented using the watershed algorithm, and required features are detected and required features are extracted based on shape, contrast, position, orientation, brightness, and line density. The proposed methodology produced an accuracy of 93.7%. Hosanna et. al [10] had presented a new automatic vessel segmentation technique by using an SVM classifier. In this, the optic disc from the retinal image is segmented by using the discrete anisotropic filter with a bee colony algorithm. The features of the blood vessels are extracted by using Gray Level Co-occurrence Matrix (GLCM) and trained by using an SVM classifier to predict true or false blood vessels.

Revathy et al. [11] proposed another machine learning approach to predict features of exudates, hemorrhages, and microaneurysms by designing Hybrid classification techniques by combining various machine learning methods that include a combination of support vector machine, k nearest neighbor, random forest, logistic regression, multilayer perceptron network. By this hybrid approach, the technique obtained the highest accuracy of 82%.

Enrique et. al [12] utilized 400 retinal images of non-proliferative diabetic retinopathy classified based on different grades. The process is carried out by detecting the features of blood vessels, microaneurysms, and hard exudates to train the SVM classifier based on the different grades of retinal images. The automatic system designed using an SVM classifier obtained an accuracy of 94%, maximum sensitivity, and predictive capacity of 94.6% and 93.8% respectively. The paper is outlined as followed, Section 3 discusses the proposed methodology, and Section 4 gives the experimental results and discussion, and finally concludes with Section 5.

3.Proposed Methodology:

This section depicts the methodology used to detect the new vessels in the optic disc region of the retinal image. Figure.1 shows the process in a form of the block diagram of the proposed technique.

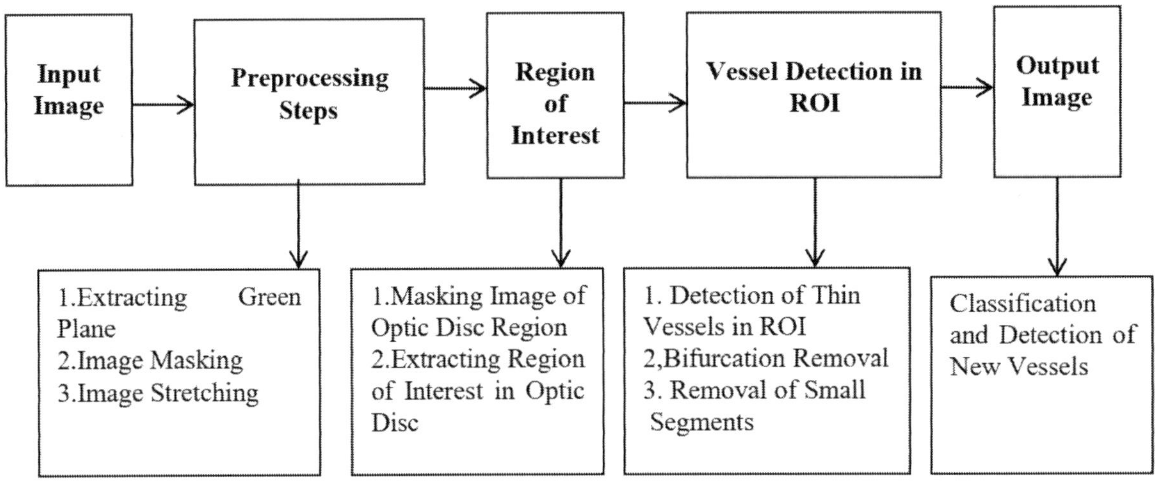

Figure 1. Block Diagram of the Proposed Method

3.1. Image Acquisition

Initially, the technique starts by collecting the retinal images to pre-process, segment, extract features, train, and test the classifier. The features extracted will help has to classify the images based on the presence of new vessels in the optic disc region of the fundus image. For the images, the commonly available dataset is used for the detection of diabetic retinopathy. From the Kaggle dataset, only 65 images are used for this system. Among 65 images 33 normal and 32 abnormal i.e eyes with symptoms of diabetic retinopathy are collected.

3.2 Pre-processing

In the image, the pre-processing technique is carried to improve the features of the image to extract the desired features from the image.

3.2.1. Green Plane Extraction

The 3-order matrix is used to represent the colored image. The rows are expressed in first-order, columns expressed in second-order and pixel color expresses the third order. In this study, the RGB color image format is used. Hence, the third order of the image takes the value of 3 colors that include Red, Green, and Blue respectively. The rows and columns values are assigned based on the size of the image. Among the other colors, the green color plane is used in the DR analysis has it shows the better contrast difference between the blood vasculatures in the optic disc region, and the background retina, and secondly it would reduce computational overhead if a single green channel is used for vessel detection instead of all the three red, green, and blue channels.

IOP Conf. Series: Materials Science and Engineering 1187 (2021) 012030 doi:10.1088/1757-899X/1187/1/012030

The Green channel produces high contrast and vessels are differentiable from the background while in the red and blue channels vessels are not much different and showing similar shades as the background.

3.2.2. Contrast Adjustment

The grey levels of the image were stabilized by stretching the contrast of the image to cover the whole dynamic range of the pixels, the surrounding pixels with dark borders and with any image labels are excluded. The contrast of any fundus image is normally very low due to many reasons, like insufficient or varying lighting conditions while capturing retinal images, low dynamic or nonlinear range of the imaging sensor such that non-uniform distribution of illumination within the image. So, it is important to make the contrast high to make the edges of vessels more visible and separable from the background. For this reason, many contrast enhancement methods are applied, i.e., methods based on spatial domain and histogram-based techniques like histogram equalization technique, Contrast limited adaptive histogram equalization technique, Adaptive histogram equalization technique, and local and global contrast stretching techniques, etc. However, the study is made use of the sigmoid function-based contrast enhancement technique which is a key and distinguishing part of this method. A sigmoid function is a mathematical function that has an "S" like shape, it is the logistic function's special case and having the formula:

$$f(x) = \frac{1}{1+e^{gx}} \quad (1)$$

Where "g" is the gain and it controls the actual contrast and "x" is the cutoff which denotes the normalized greyness factor and its value is varied around it. Its normal initial value is 0.5 which is mid of the greyscale. Different images may require a different value of g to be enhanced. This function has the property of the continuous function. Its output ranges between 0 and 1. It is differentiable and a real-valued function, bearing a first derivative that is either non-positive or non-negative.

3.3. Region of Interest

A Region of Interest is a particular sample from the dataset that is recognized for desired purposes. This concept can be applied to any area. For example, in medical image processing, the tumor boundaries are identified in the image or in a volume, to measure its size. Here, the Optic disc is considered as a region of interest. In this method, 15 and 0.5 values are used for gain and cutoff respectively. All the images for study are resized before locating the optic disc, as its diameter ranges up to 300 pixels approximately. The region of the optic disc region is then masked and extracted by using morphological operations.

3.4. Vessel Detection

3.4.1 Detection of Thin Vessels

This process is carried out to segment thin vessels that are normal vasculatures present in the optic disc region. In the comparison of normal vessels with the abnormal ones, the abnormal vasculatures are found to be small and are more twisted than the normal vasculature. The lamina cribrosa provides a mesh-like structure that acts as a bright background for the detection of vessels with greater contrast. Hence the vessels forming dark ridges are identified by using the strength of ridges k is given by

$$k = \frac{L_x^2 L_{yy} + L_y^2 L_{xx} - 2L_x L_y L_{xy}}{(L_x^2 + L_y^2)^{3/2}} \quad (2)$$

Where L provides the Gaussian filtered image that is used to determine the scale of the ridges. L_x, the subscript indicates the partial derivative and L_{xx} is the second-order partial derivative concerning x. the k value becomes positive for vessel ridges and negative for the valley in between the vasculature. K value is unspecified when the slope reaches zero in both directions.

3.4.2 Removal of Bifurcation and Small segments

On Detecting the thin vessels from the ROI of the Optic disc. The filter is used to remove the bifurcated vessels. The undesired pixels in the image that includes dots or thin lines forming noises are identified by applying thresholding. Then the undesired pixels are removed by undergoing further processes that are to be performed to improve the image and maintain the undesired blood vessel pixels as a post pre-processing step. To accomplish this step, an image opening (morphological operation) is used to exclude the desired pixels that are limited to or less than 90 pixels. Now due to image opening, little gaps between vessels have appeared which is removed to some extent with image dilation.

3.5 Detection of New vessels

An SVM classifier shows a discriminative and better performance in classifying DR in comparison with other classifiers. It is a supervised learning technique that is used to map the input features in high-dimensional space. The classes are separated by constructing a margin of the hyperplane. The mapped points that lie closest to the decision of separation affect the position of the support vector. The probability of the classification error of the mapped points is minimized when it is non-separatable. To separate two different classes linearly by using Optimal hyperplane (HO) is given as follows,

$$y_i\,(w.x_i\ + b) \geq 1 \qquad i = 1 \ldots m \tag{3}$$

The extracted features are normalized by using the following formula,

$$f' = \frac{f - m}{s} \tag{4}$$

Where f represents the feature and f' represents the normalized value, respectively. m and n values are the mean and standard deviation for the feature. The values are calculated to train the classifier by using the features. The abnormality in the vessels is detected by taking the probability of the highest abnormalities and comparing it with the threshold values. Different points are taken into consideration by varying the threshold values of the abnormal. The classifier is trained by using the images except for the test image. The feature values are normalized before the classification and it is calculated each time excluding the test image.

4.Experimental Results and Discussion:

In this paper, the results are discussed based on problem recognition. The SVM classifier technique is applied to extract the features from the input fundus images for the early prediction of diabetic retinopathy. The results are processed by using MATLAB software and the processed images are displayed by using Graphical User Interface (GUI). The below images in figure 2. show the output of each processing technique.

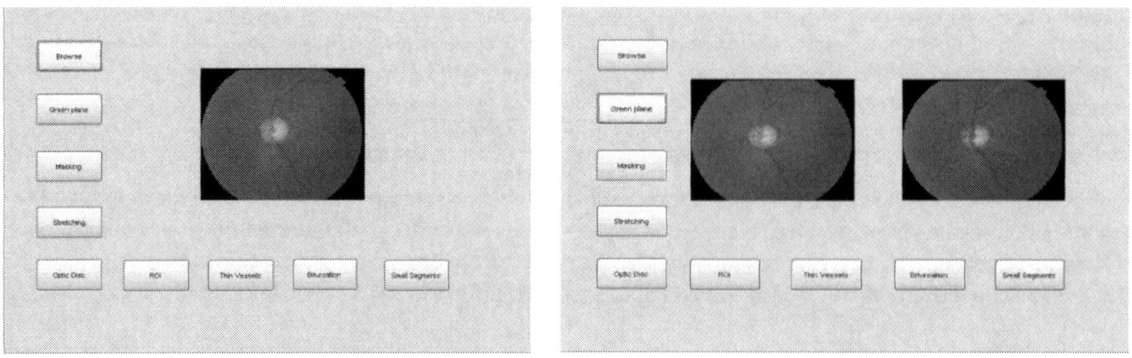

(a) Input Image (b) Extracting Green Plane

IOP Conf. Series: Materials Science and Engineering 1187 (2021) 012030 doi:10.1088/1757-899X/1187/1/012030

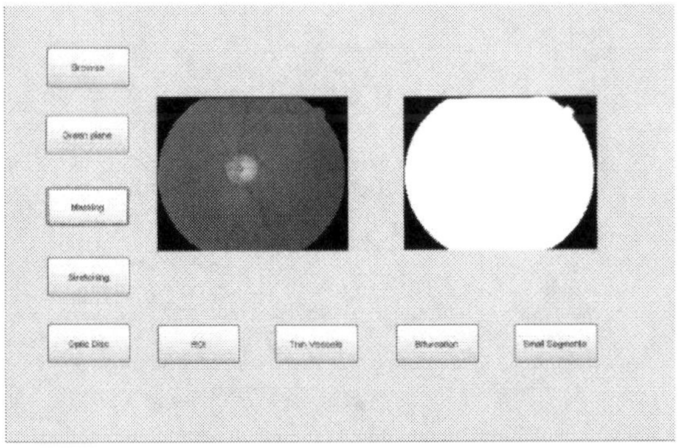

(c) Image Masking

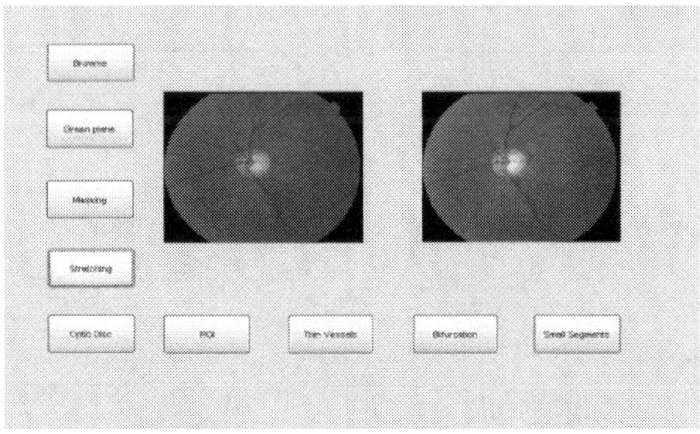

(d) Contrast Stretching

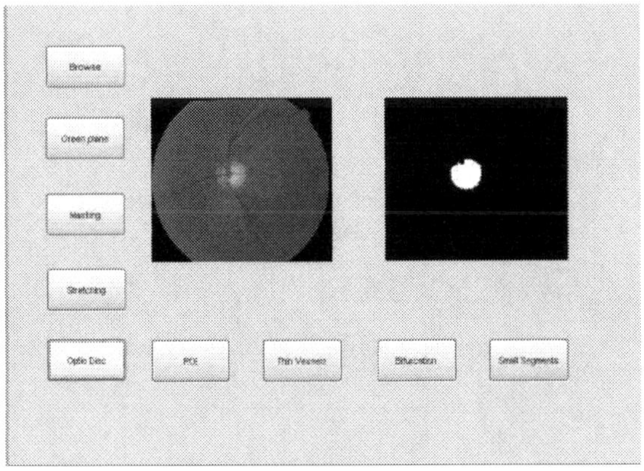

(e) Optic Disc Region

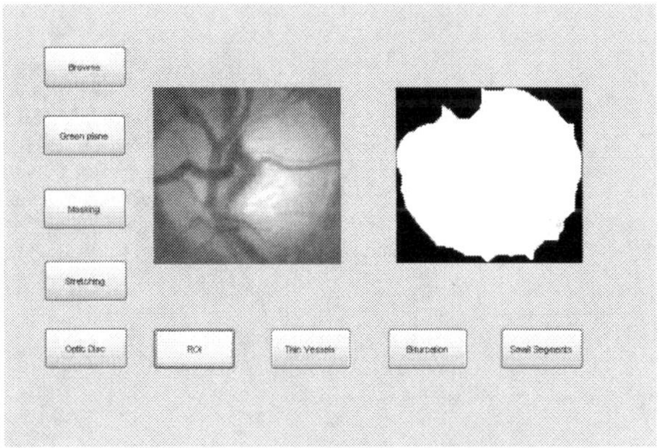

(f) Region of Interest

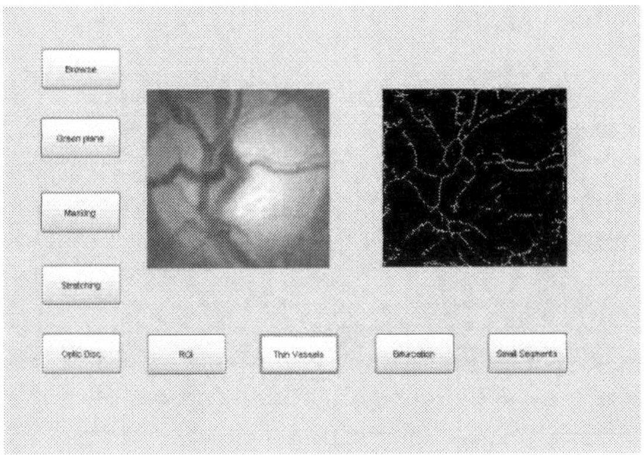

(g) Thin Vessels Detection

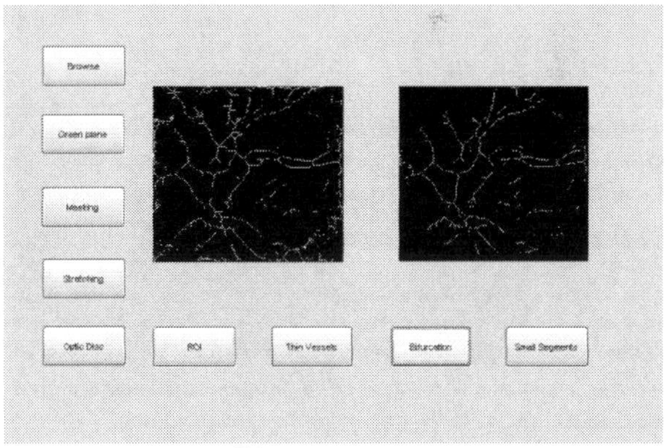

(h) Bifurcation Removed

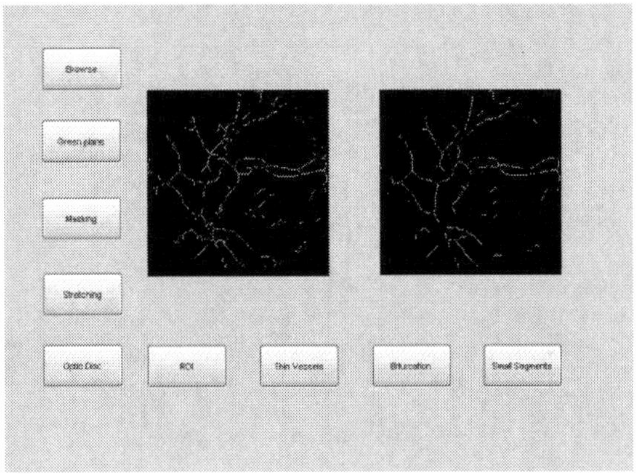

(i) Removal of Small Segments

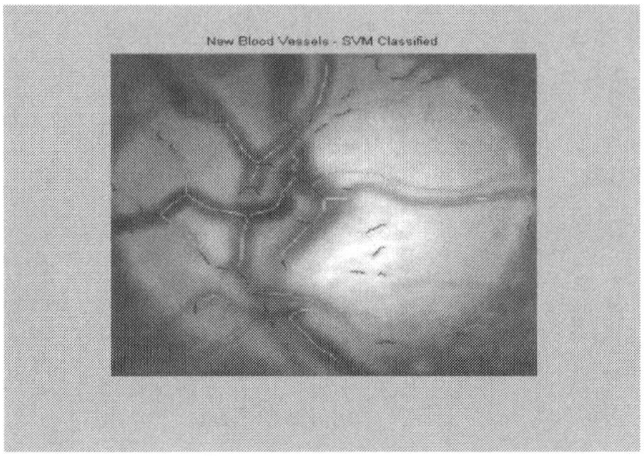

(j) Detection of New Vessels

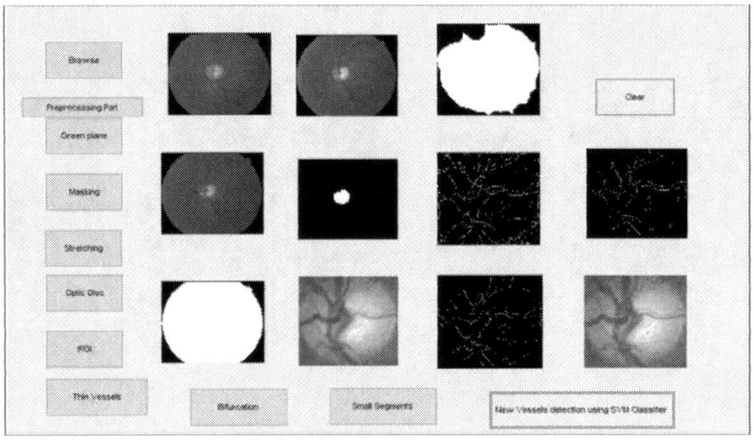

(k) Overall Output Window

Figure 2. New Blood Vessels Classification using SVM classifier

The performance of the SVM Classifier is analyzed by using False Positive (FP), False Negative (FN), True Negative (TN), and True Positive (TP) are given as follows:

- True Positive (TP): Normal Eye is correctly identified as a normal image.
- False Negative (FN): Normal Eye is incorrectly identified as Diabetic Retinopathy.
- True Negative (TN): Diabetic Retinopathy is correctly identified as Diabetic Retinopathy.
- False Positive (FP): Diabetic Retinopathy is incorrectly identified as normal.

From the above classes, the accuracy, sensitivity, and specificity of the classifier values are calculated by using the following formulae.

$$Accuracy = \frac{TP+TN}{TP+TN+FP+FN} \quad (5)$$

$$Sensitivity = \frac{TP}{TP+FN} \quad (6)$$

$$Specificity = \frac{TN}{TN+FP} \quad (7)$$

Table 1. shows the confusion matrix to evaluate the performance of the classifier. The correct and incorrect predictions are classified into each class. This shows the way by which the classifier predicts the different labels with confusion. It is plotted along with Normal and abnormal blood vessels in the optic disc of the retinal image with respect to true or false labels.

Table 1. Confusion Matrix for SVM Classifier

Actual/Prediction	Positive	Negative
Positive	31(TP)	2 (FP)
Negative	1 (FN)	31 (TN)

The Accuracy of the SVM classifier is calculated as 95.38% respectively. Then the sensitivity and specificity of the classifier are measured as 96.8% and 93.9% respectively. The comparison of the accuracy of the proposed system with existing works is analyzed and represented in Figure 3.

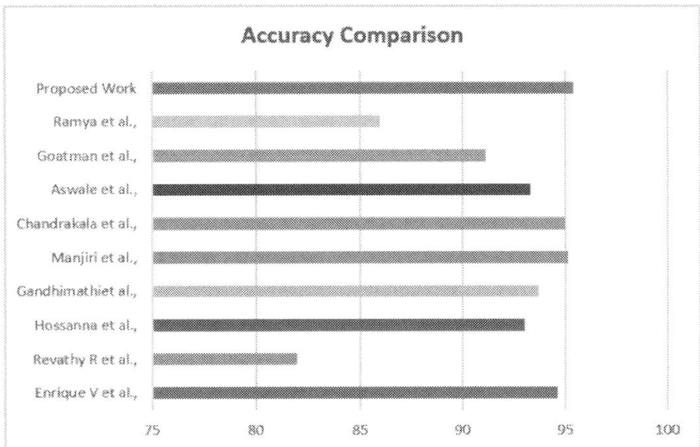

Figure 3. Comparison of Classification Accuracy

The proposed system showed a better performance in the classification of diabetic retinopathy and hence it can be used by the ophthalmologists to predict diabetic retinopathy in the patient at the earlier stage in an efficient way.

5.Conclusion and Future Scope:

The project deals with the development of an automated system for the detection of new vessels from the optic disc of retinal eye image by using a machine learning algorithm. The SVM Classification model gives an accuracy of 95.4%, specificity of 96.8%, and sensitivity of 93.94%. The SVM classifier shows better performance in the classification of the disease. To develop a user-friendly interface, this project has been designed with a Graphical User Interface. Image processing operation returns parameters that are necessary for the machine learning model. The output consists of the main window where the user input image is inserted in the main window that undergoes different phases of image processing operation output. This trained model delivers an efficient output. The accuracy of the machine learning model can be improvised by training the model with different machine learning algorithms. Training with various algorithms with different datasets will yield greater accuracy.

6. References

[1] Hariton Costin, "Recent trends in Medical Image Processing", *Computer Science Journal of Moldova*, 2014; 2(65).

[2] Akram T Kharroubi and Hisham M Darwish, "The Diabetic mellitus: The epidemic of the centuary", *World J Diabetes*. 2015 jun 25; 6(6): 850-867.

[3] RafalMuc, Agnieszka Saracen, Iwona Grabska-Liberek, "Association of diabetic retinopathy with retinal neurodegeneration on the background of diabetes mellitus. Overview of recent medical studies with an assessment of the impacts on healthcare systems", *Open Med.*2018; 13:130-136.

[4] Manjiri B. Patwari, Ramesh R. Manza, Yogesh M. Rajput, Manoj Saswade, Neha Deshande, "Review on detection and classification of Diabetic Retinopathy Leisons using Image Processing Techniques", *IJERT*, 2013:2278-0181.

[5] V. Ramya, "SVM Based Detection for Diabetic Retinopathy", *IJRSI*. January 2018, 2321–2705.

[6] Keith A. Goatman, Alan D. Fleming, Sam Philip, Graeme J. Williams, John A. Olson, and Peter F. Sharp, "Detection of New Vessels on the Optic Disc Using Retinal Photographs", *IEEE*, 30(4), 2011.

[7] V.A.Aswale, 2 J. A. Shaikh," Detection of microaneurysm in fundus retinal images using SVM classifier*", IJEDR*. 5(4). 2017.

[8] Chandrakala B, Sindhu Madhuri G, "Analysis of Eye Retina for Diabetic Detection using PCA & SVM Methods", *IRJET*. 077(06), June 2020.

[9] Gandhimathi. K, Ponmathi. M, Arulaalan. M and Samundeeswari. P, "Analysis of Diabetic Retinopathy from the Features of Color Fundus Images using Classifiers", *IJERT*. Vol. 4 Issue 02, February-2015, 2278-0181.

[10] P Hossanna Princye, V Vijayakumar, "Retinal disease diagnosis by morphological feature extraction and SVM classification of retinal blood vessels", *biomedical research*. 2018, s22-s30.

[11] Revathy R, Nithya B S, Reshma J J, Ragendhu S S, 5 Sumithra M D, "Diabetic Retinopathy Detection using Machine Learning", *IJERT*. Vol. 9 Issue 06, June-2020, 2278-0181.

[12] Enrique V. Carrera, Andres Gonzalez, Ricardo carrera, "Automated Detection of Diabetic retinopathy using SVM", 2017.

[13] T. Manikandan, V. Nandalal, S. Joshua Kumaresan, J.L. Mazher Iqbal and A.Muruganandham, "Synthesis and characterisation of magnetic nanoparticles for lung cancer detection and therapy", International journal of psychosocial rehabilitation, vol. 24(5), pp. 2730-2740, 2020.

[14] N.Malligeswari ,C.Rajani , Dr.G.Kavya," A Novel Approach for Lung Pattern Analysis using Neural Networks and Fuzzy Interface System", Advances in Engineering Research (AER), volume 142,pp231-235

IOP Conf. Series: Materials Science and Engineering 1187 (2021) 012031 doi:10.1088/1757-899X/1187/1/012031

Fruit Classification Comparison Based on CNN and YOLO

Riyanshu Raj [1], S S Nagaraj [2], Saurav Ritesh [3], Thushar T A [4] and V M Aparanji [5]

[1,2,3,4] Siddaganga Institute of Technology, Visvesvaraya Technological University, Tumkur, Karnataka
[5] Electronics and Communication Engg., Siddaganga Institute of Technology, Tumkur Karnataka

Abstract Today, agriculture has become more of science than art. The demand for fresh produce is growing, while availability of field workers is diminishing. Also, Farmers keep investing in land, irrigation, and labor, and much of the fruit is left hanging on the trees because there are not enough people to pick it when it is ripe. Automation in the agricultural field using modern information and Communication Technology can increase the efficiency, reliability, and precision; and reduces the need of human intervention. Thus, to meet solution to this problem is by reducing the costs for farmers and increasing the yield of crops which can be achieved by classifying of fruit by using modern technique in the field of Deep Neural Networking that involves CNN, YOLO etc. which provides a helping hand to farmers to optimize their farming patterns and boost yields by allowing them to grow taller fruit trees that can be harvested by classifying based on the condition of fruit and saving significant labour cost. In conclusion, the proposed methods are highly acceptable and recognizing the necessity, and unresolved classifications.

Keywords: Automation, Deep Neural networking, CNN, YOLO and classification.

1. Introduction

Automation in the agricultural field using modern information and Communication Technology increased the efficiency, reliability, and precision; and reduced the need of human intervention. Robots used for agriculture are known as Ag-bots. Recognizing fruit that are ready for harvesting are becoming most difficult task in agricultural field. Which causes loss for agriculturalist in labour-less area, where the fruit gets over ripe and destroy. However constructing online fruit classification system can enhance the growth of fruit which requires high skilled labours to check the condition of the fruit and labour accidents and conflicts in deciding the condition of fruit can be resolved with high classification accurate device that gives good prediction of fruit, so the main aim of this study is to provide method for fruit classification on different network models that improve the efficiency in terms of accuracy in identifying the fruit condition and position and enhance the idea of user to extend the production.
The paper is organised into following sections, section 2 provides a review on object detection and classification with both pros and cons in it. Section 3 provides the overall block diagram of different

IOP Conf. Series: Materials Science and Engineering 1187 (2021) 012031 doi:10.1088/1757-899X/1187/1/012031

algorithm that need to perform and section 4 deals with software that planned to implement on different algorithm to enhance the result. And regarding results of the research algorithm with respective results are exhibited and section 5 provides conclusion and future scope for the study.

2. Related work

A traditional method which uses texture feature and colour for fruit classification which is done by Support Vector Machine (SVM) which is based on statistical and Co-Occurrence features from wavelet transform. It has three processes like pre-processing, feature extraction and texture feature the proposed model has an accuracy of about 95.3% [1]. The proposed model uses six convolutional layers with pooling layer and fully connected layers, which is better than three state of the art approaches- voting based, vector machine, wavelet entropy and genetic algorithm. It has an accuracy of 91.44% [2]. Among the fruit there are several changes because of similarity in shape, edge feature and colour. The proposed model gives a Pure Convolutional Neural Network (PCNN) with minimal number of parameters. The PCNN consists of seven layers (convolutional) also the proposed model involves Global Average Pooling (GAP) layer which is every effective. The proposed approach using PCNN with GAP provides superior performance and overcome overfitting problem and has the accuracy of 98.88% [3]. Dataset has been prepared for training, for which the image is acquired using Charged Coupled Device (CCD). K-means algorithm is used to segment an image and polygon fitting method is implemented to define boundaries in case of overlying fruits [4]. A classification of three green vegetables of different size, which involves software like Tensorflow, YOLO algorithm from darkflow is presented in [4], which is a version of Tensorflow in Opencv and preprocessing is done with drawing bounding boxes for vegetables manually, YOLO is used to detect and classify and achieved 50% accuracy in image detection and 70% in video.

Double track method which involves classification Neural Network using different matrix. 1st classification of fruit with background and 2nd classification based on image with ROI (Region of Interest), both results are aggregated and predict the class member with certainty factor, and here CNN classify fruit in whole image, YOLO for detects object in whole image and again CNN classification with single object [5]. Real time object classification has gained more popularity and innovations, a simple approach YOLO takes at object detection and paper gives background of CNN and also gives types of object detection algorithm like CNN- classification based algorithm and YOLO- regression based algorithm [6].

IOP Conf. Series: Materials Science and Engineering 1187 (2021) 012031 doi:10.1088/1757-899X/1187/1/012031

3. Block diagram of proposed system

Figure 1 shows the system architecture of fruit classification, which includes dataset, and training based on classifying the fruit on their morphologic content and Colour, to which module that contains train, test and validation is built followed by processing unit, where the input is divided or framed on processing convenient upon which processing unit detects the object or fruit by localizing the position and display it out with classification specification.

.

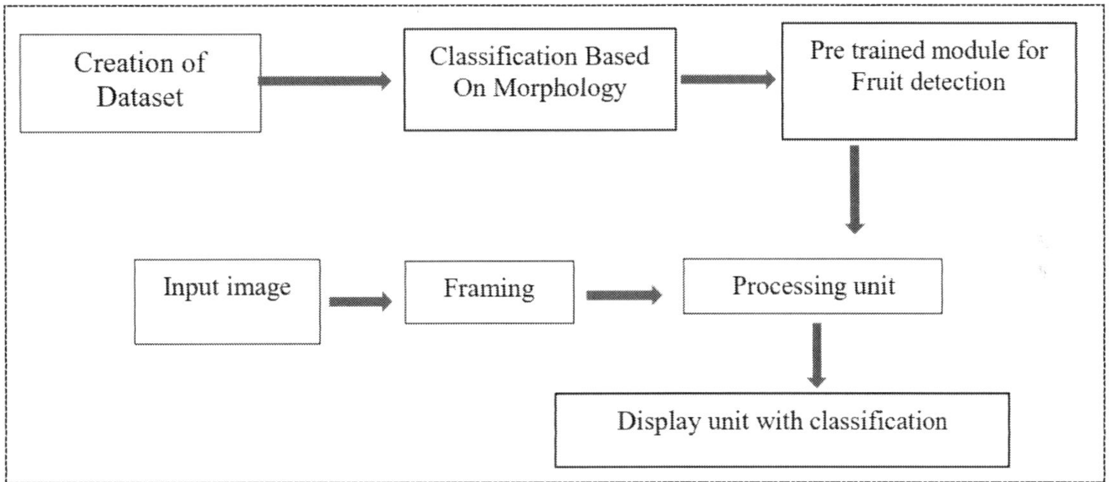

Figure 1. Block Diagram of fruit classification

4. Software description

4.1. Convolution Neural Network

Convolutional Neural Network (CNN) is a feed-forward artificial neural network. It is inspired by the biological visual cortex having small regions of cells. And this architecture shown below in Figure.2 deals with multiple convolution with filter specified, pooling followed by Artificial Neural Network (ANN). In the beginning, when the input is flattened to 1D, neurons formed due to weight with respective image number (grey number) and bais is of bigger size, which is difficult for computation, so using filter operation makes process lighter, where original input is taken and convolved using the kernel of user choice, which then followed by Relu to reduce the nonzero elements and make the process linear therefore speed up for classification. Then to reduce overfitting in ANN, further convolution is carried out with kernel and Relu as used in previous convolution and before feeding into ANN, max pooling is done.

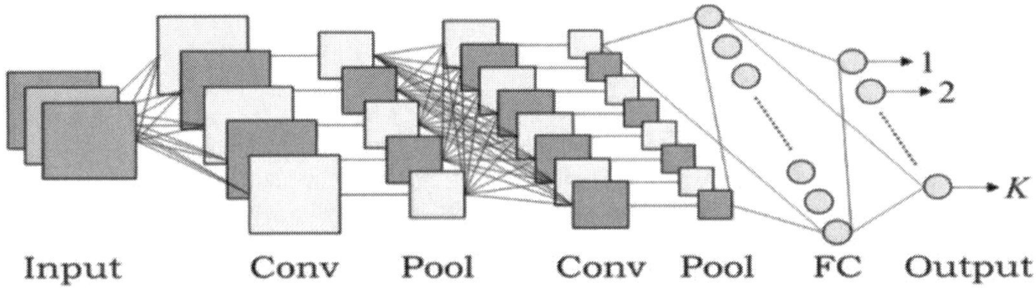

Figure 2. : Architecture of Convolution Neural Network

Here an image of fruit represented as a 100x100 matrix of values, among which 3x3 matrix is taken and slide that 3x3 window around the image for convolution. Then Relu function $R(x) = 0, forall\ x < 0\ and\ R(x) = x\ for\ x > 0$ is carried out to expell all negative values from the matrix of After which, matrix multiply window by 3x3 kernal followed by max pooling of layer 2 is done for 2nd time. As a result, 2D is flattened to 1D value in block and fed to ANN for detection. The fruits detected using this model is shown in Figure 3.

Figure 3. : Fruits Detected using CNN

4.2. You Only Look Once (YOLO)

You Only Look Once (YOLO) algorithm basically works in three steps or techniques, Residual blocks, Bounding box regression and intersection over union. First the image is divided into grids which has dimension SxS next the bounding box is an outline that highlights the object in an image, the bounding box consists of attributes width, height, class and bounding box centre it represents the probability of an object appearing in the bounding box. Intersection over union provides an output box that surrounds the object perfectly. The grid cell is responsible for prediction of outer boxes. The intersection over union output is 1if the predicted output is equal to the real box. The above three techniques are applied to the input image which results in the detection of the fruit. In Figure 4 represents the output of multiple fruit detected, which is represented using vector consist of class, probability and bounded in box with location specified using x-y coordinate and size of the box. And all other fruit in the figure due to loss produced with trained dataset.

Figure 4. : Fruits Detected using YOLO

4.3. MATLAB

MATLAB is a high-performance language for technical computing. It integrates computation, A color detection algorithm finds pixels in an image that are the same color or color range as a specified color. To differentiate them from the rest of the image the color of detected pixels can then be adjusted. The Figure 5 shows the basic color detection model in Matlab.

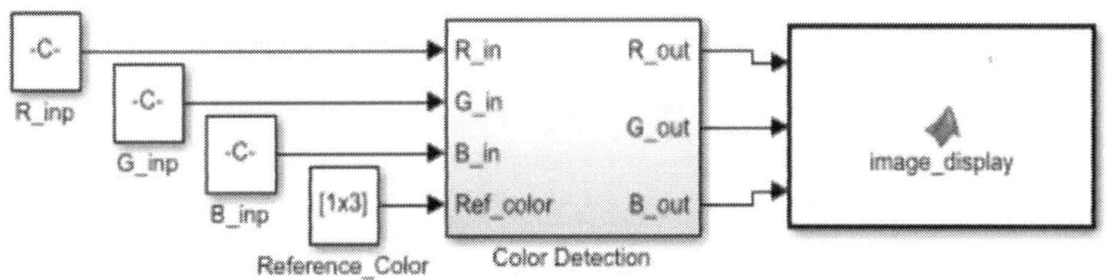

Figure 5. Color detection model in Matlab.

The fruit image is given as source file, which is read using imread() function and stored in the given cluster. Then, Gaussian filter has been created using the fspecial() function which returns a Gaussian filter of the given size let's say, hsize=9 with standard deviation Sigma(sigma=3), that is rotationally symmetric. Use im filter function and apply local smearing using the Gaussian filter which is thereafter, converted from RGB to YIQ values of NTSC image as shown in Figure 6. Using the function imfindcircle, the detected fruits are circled according to their provided radii range and as red tomatoes color range with circle shape are ripen, and green color are unripen, as it can be in Figure 7.

IOP Conf. Series: Materials Science and Engineering 1187 (2021) 012031 doi:10.1088/1757-899X/1187/1/012031

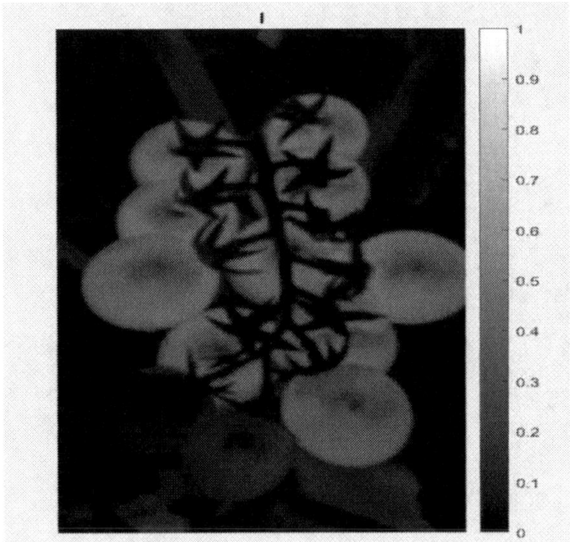

Figure 6. YIQ value image for ripen tomato

Figure 7. Ripe fruit detection.

5. Result

Fruit detection algorithms are simulated YOLO, CNN and in matlab plathform.

Table 1. Comparison of image processing models.

Models	Accuracy
YOLO	>= 85%
Colour Detection (Matlab)	>≡ 74%
CNN	>= 63%

The Table1 shows the comparison of You Only Look at Once (YOLO) in python, convolution neural networks (CNN) in Python and the colour detection in MATLAB. In YOLO algorithm accuracy achived is 85% and in matlab color detection algorithm using gaussian filter has shown the accuracy of 74 % and in CNN biological visual cortex, accuracy of 63 % is achived.

6. Conclusion

From the above results on the implemented algorithms, YOLO achive maximum accuracy, when compared to other algorithm like CNN and colour detection using mathlab and mathlab implementation is for circle shape fruits only and it is able to recognize all the ripen fruit in YIQ value used. Thereby YOLO hold advantages in classification of objects than any other algorithm and futher image classification can be made using increment in demensions like 3D by keeping much more obective or parameters. Once the simulation classifying the fruit is completed, a camera can be placed on a drone to capture the image of the fruit, classify the fruit and pluck the fruit by using gripper connected to the drone. This work can lead to atonomous drone which can help in harvesting process.

7. References

[1] R.Shantha Selva Kumari and V.Gomathy *Fruit Classification using Statistical Features in SVM Classifier* 2018 *Int. Conf. on Electrical Energy Systems (Chennai)* pp 526 – 529

[2] Siyuan Lu, Zhihai Lu, Soriya Aok and Logan Graham *Fruit Classification Based on Six Layer Convolutional Network* 2018 *Int. Conf. on Digital Signal Processing (Shanghai)*

[3] Asia Kausar, Mohsin Sharif, JinHyuck Park and Dong Ryeol Shin *Pure-CNN: A Framework for Fruit Images Classification* 2018 *Int. Conf. on Computational Science and Computational Intelligence (Las Vegas)* pp 404 – 408

[4] Upulie H.D.I and Lakshini Kuganandamurthy *Real-Time Object Detection using YOLO: A review* 2021 *Research gate publication (Srilanka)* 351411017

[5] Rudnik Katarzyna and Michalski Paweł *A Vision-Based Method Utilizing Deep Convolutional Neural Networks for Fruit Variety Classification in Uncertainty Conditions of Retail Sales* 2019 *Appl. Sci.* 9(19), 3971

[6] Sachin C, N Manasa, Vicky Sharma and Nippun Kumaar A A *Vegetable Classification Using You Only Look Once Algorithm* 2019 *Int. Conf. on Cutting Edge Technologies in Engineering (Barabanki)* pp 101 - 107

Development of data driven adaptive edge detectors for image processing

Likhith H G[1], Nishanth H L[2],Lohit E[3], Rudramurthy M S[4], Sushma S A[5] and Keerthan Kumar T G[6]

Department of Information Science, Siddaganga Institute of Technology, Tumakuru, Karnataka, India.

Abstract: In most vision processing activities, the early stage involves identifying the features in an image that provide cues to structure and properties of the object in the scene. Most common features in an image or in a scene are edges. Edges are significant local changes in intensity within an image. Most important goal of edge detection is to produce a line drawing from an image representing the scene. The significant features of an image such as line, curve and corners can be extracted from edges. During the stage of discovering and exploring the information from an image of that scene, edge detection is the most important and early-stage activity and as such it is prominent active area in image processing. Most popular edge detection algorithm such as Robert, Sobel, Canny, Prewitt and Laplacian of Gaussian (LoG), etc. are currently in use. This paper emphasis on an experimental study of limitations of conventional edge detectors and to devise a novel approach to resolve the conflicting issues i.e., limitations of these edge detectors in adaptive space utilizing novel methods such as Bi-dimensional Empirical Mode Decomposition (BEMD), Image Empirical Mode Decomposition (IEMD), Complete Ensemble Empirical Mode Decomposition (CEEMD) and Multivariate Decomposition techniques. Further, to study the performance of these modified edge detectors on the images of complex scenes which are of societal and agricultural importance.

Keywords: *Bi-dimensional Empirical Mode Decomposition (BEMD), Edge Detection, Image Processing.*

1. INTRODUCTION

An image is characterized as a 2D work G(x,y) where 1(x,y) are spatial coordinates, and the sufficiency of G anytime pair of directions (x,y) is called the power of a picture. At the point when G and (x , y) are in constrained range then it is called as computerized picture. A picture can likewise be depicted as 2-dimensional cluster which is resolved in terms of sections and lines. By considering an image, it comprises of set of components where each component is called as pixel. Each pixel has esteem at specified areas. There are numerous kinds of picture specifically black and white image, binary image, 8-bit and 16-bit shading position. Each picture is comprised of pixel matrix.

1.1 Image Processing

Digital Image Processing is a technique of processing a digital image using digital computer. We use image processing algorithms, to either enhance the quality of the image or to retrieve valuable information from the image. Many image processing techniques have been developed in last couple of years. Many of the image processing technique is used to extract useful information from an image. Image processing technique is becoming more popular since it is easily available for personal computers and may other software's. Digital image processing is a technique of performing some set of operations on an image in order to obtain some useful information from it or for the purpose of image enhancement. The applications of Image Processing are Medical Imaging, Finger-Print Recognition, Face Recognition, Automated Traffic Controlling Systems etc. As of now many techniques have been developed using edge detection for extracting edges from an image.

1.2 Edge Detection

Edge location is a kind of Image Segmentation procedures which decides the nearness of an edge or line in an image and diagrams! them in a suitable manner. The principle motivation behind edge identification is to rearrange the picture information so as to limit the measure of information to be handled. For the most part, an edge1can be characterized as the pixel limit that associate 2 separate districts with adequacy traits, for example), unique consistent luminance! and improvement esteems in the picture. In this work, we present strategies for edge division of pictures; we utilized five procedures for this class; Sobel administrator method, Prewitt system, Laplace strategy, Canny strategy, Roberts system, and they are contrasted with each other so as with pick the best procedure for edge recognition portion picture. These methods applied on one picture to pick base speculations for division or edge identification picture. In this project an endeavour is made to contemplate the exhibition of most normally utilized edge discovery procedures for picture division and furthermore the correlation of these methods is completed with an analysis by utilizing MAT LAB programming. We will utilize the edges to discover coinciding between objects. Edge location is a procedure engaged with picture handling which assists with finding the edge or a line in a predefined picture and gives them in a suitable manner. The key significant motivation behind edge discovery is to portion the picture with the goal that information included will be limited.

1.2.1 Bi-dimensional Empirical Mode Decomposition

The Empirical Mode Decomposition was recently presented with a new time-frequency analysis tool to non-stationary and non-linear signals . As Empirical Mode Decomposition is self-adaptive and is able to find the intrinsic modes of the signal . It does not have any implication on optimality. Some instance may arise where certain optimality will be considered hence, we need a decomposition of signal and reconstruction scheme. We will give a modified version of empirical mode decomposition algorithm. For this purpose, two formulations are proposed. Firstly, signal which utilizes a linear weighting for the Intrinsic Mode Functions (IMF). Secondly, algorithm adopts a bidirectional weighting. Bi-directional weighting mainly will not use weights for IMF modes. The two empirical mode decomposition methods proposed will extend the capability of traditional empirical mode decomposition which will be well matched for optimal Signal Recovery MATLAB simulation studies have been done to showcase the application of the proposed optimal empirical mode decomposition algorithms to denouncing problem.

The developments going on in the analysis of the non-linear and non-stationary data have received large attention by the Image Analyst. As we know that Huang in 1998 introduced the Empirical Mode Decomposition in Signal Processing this is related to only seismic and biomedical signals. The main intention of our method is to apply the Empirical Mode Decomposition to texture extraction and image filtering this is widely recognized as a difficult and challenging computer vision problem. The development of an algorithm based on Bi-Dimensional Empirical Mode Decomposition to extract features at different scalesfrequencies . Hence these features are called Intrinsic*Mode Functions these functions are extracted by a shifting process. The Bi-Dimensional sifting process is realized using classical operators to detect local maxima and it's all because of radial basis function for surface interpolation.

$$s(x) = p_m(x) + \sum_{i=1}^{N} \beta_i \varphi(|x - x_i|), \quad x \epsilon R^d, \beta_i \epsilon R \; \ldots\ldots\ldots\ldots\ldots\ldots\ldots\ldots (1.2.1)$$

As shown in the equation (1.2.1) the performance of the texture extraction algorithms is demonstrated using the Bi-dimensional Empirical mode decomposition method in the experiment with both synthetic and natural images.

1.2.2 Ensemble Empirical Mode Decomposition

A new Ensemble Empirical Mode Decomposition (EEMD) was introduced. This new technology involves filtering a gathering of background noise signal (information) and combines the mean as the last obvious outcome. Limited not minuscule, plentifulness repetitive sound important to compel the outfit to deplete every single imaginable modification in the filtering procedure, therefore making the distinctive scale signs to gather in the correct Intrinsic Mode Functions (IMF) directed by the dyadic channel banks. As EEMD is a period space examination technique, the additional background noise arrived at the midpoint of out with adequate number of preliminaries; the main persevering part that endures the averaging procedure is the segment of the sign (unique information), which is then treated as the valid and increasingly physical important answer. The impact of the additional background noise; in this way, the additional commotion examines the segment of the sign of practically identical scale in one IMF. With this troupe mean, one can segregate scales ordinarily with no from the prior passionate premise decision as in the abnormality test for the first EEMD estimation. This new system utilizes the full ideal situation of the quantifiable characteristics of tedious sound disturb the sign in its real course of action neighbourhood, and to check itself ensuing to filling its need; thusly, it addresses an impressive improvement over the first EMD and is a truly upheaval helped data examination methodology.

2. LITERATURESURVEY

It gives the brief explanation about the various traditional edge detection techniques and the existing work related to the edge detection in image processing domain. The purpose of edge detection is to reduce amount of data to be processed at the same time preserving the structural properties for processing an image. In a Grey scale image the edge is a basic segregating feature that it is used to separate the regions which the grey level is low or high with in different values of the edge. If an image consists of noise then it is very difficult to detect an edge, we know that if noise is high frequency component and because of this noise true edges will not be detected. If we process a noisy image we end up with distorted or blurred image.

2.1 Steps in Edge Detection

The 3 steps of detecting an edge is as follows,

• **Smoothing an image**: In the first step noise is reduced to maximum extent and image filtering is performed in order to improve the performance and to obtain the best possible result

• **Detection of an edge**: In this step edges are identified and extracted based on the specified threshold value.

• **Localization of an edge**: This step is used to select candidate edge point and candidate which are true are considered as proper edges for future reference.[1][2]

2.1.1 Performance Criteria. The following parameters are used for determining the performance criteria of edge detectors and they include,

1. Good Edge Detection: The probability of failing to mark the true edges must be low and as well as probability of marking non edge points falsely should be low. [2]
2. Good Edge Localization: The points which are marked as the edges must be more accurate for finding true edges. [2]
3. Single response for single edge: If more than 2 edge response is detected for same edge, then only edge should be considered as true edge.

In the below shown Fig 2.1.1(a) gives the Noise and the step edge is shown in Fig. 2.1.1(b) and Fig. 2.1.1(d) and the outputs of the convolutions are in the Fig. 2.1.1(c) and Fig. 2.1.1(e). All these figures mark the centre of the edge at the local maximum in the output of the convolutions.

Thus, filter in Fig. 2.1.1(d) performs much better than that in Fig. 2.1.1(b) because of the threshold value of the latter chosen for the detection of the edges. [1][3]

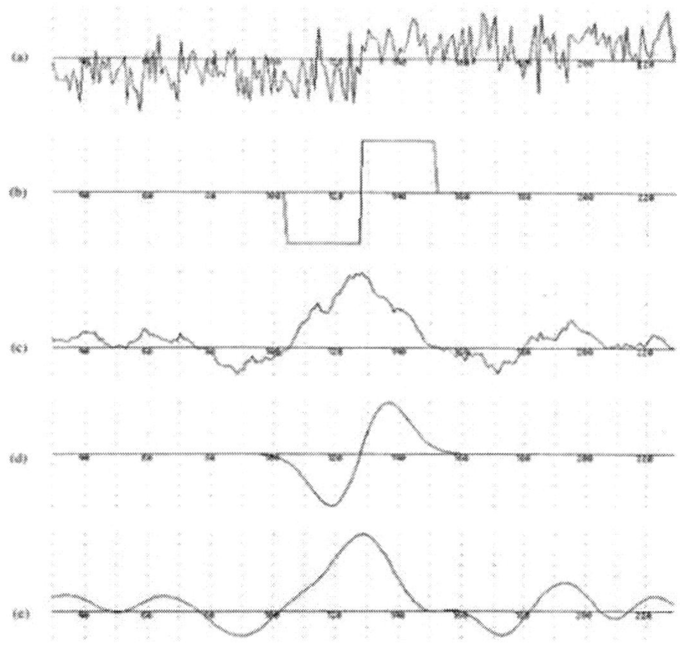

2.2 Classification of Edge Detectors

The edge detectors are broadly classified as Gradient based edge detector and Laplacian based edge detectors. Gradient based detectors are also called as first-order edge detectors and Laplacian based edge detectors are also said to be second-order edge detectors. Fig 2.2.1 depicts the classification of edge detectors.

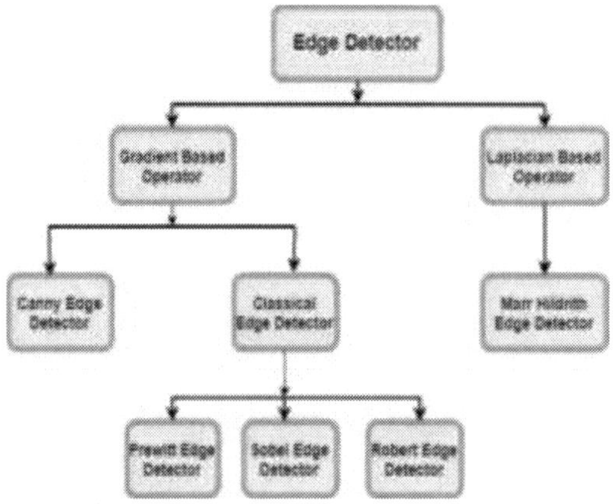

Fig 2.2.1 Classification of edge detectors

2.2.1 Gradient based Edge Detector

Most edge discovery techniques deal with the supposition that the edge happens where there is an irregularity in the force work or a precarious power slope in the picture. Utilizing this presumption, if one takes the subordinate of the force an incentive overthe picture and discovers focuses where the subordinate is most extreme then the edge could be found. The slope is a vector, whose parts measure how quick pixel esteem is changing with separation in the x and y bearing. [4]

2.2.2 Laplacian based Edge Detector

The Laplacian is a 2-D proportion of the second subordinate of a picture. The Laplacian of a picture features areas of fast force change and is in this manner frequently utilized for edge discovery zero intersection edge finders. The Laplacian is regularly applied to a picture that has first been smoothed with something approximating a Gaussian smoothing channel to decrease its affectability to commotion. The administrator ordinarily takes a solitary dark level picture as info and produces another paired picture as yield. The zero intersection locator searches for places in the Laplacian of a picture where the estimation of the Laplacian goes through zero for example focuses where the Laplacian changes sign. Such focuses frequently happen at edges in pictures for example focuses where the power of the picture changes quickly, yet they likewise happen at places that are not as simple to connect with edges. It is ideal to think about the zero-intersection identifier as an element finder as opposed to as a particular edge locator. Zero intersections consistently lie on shut shapes, thus the yield from the zero intersection identifiers is typically a double picture with single pixel thickness lines indicating the places of the zero intersection focuses. [5]

2.3 Classical Operators

Some of the classical edge detectors also called the traditional edge detectors namely Canny, Sobel and Prewitt edge detectors are explained in detail below.

2.3.1 Canny Edge Detection

The traditional canny edge detection method is commonly used in grey scale image processing. But this classical edge detection algorithm was unable to deal with colour images and the parameters in the algorithms were difficult to determine. [2]

Edges are used to distinguish boundaries in an image and hence problem of basic importance in image processing. Edges which is present in an image are location with high intensity contrasts or a change in intensity from one pixel to other. Edge detection of an image reduces amount of data to be processed, while retaining the important structural properties in an image. The Canny edge detector is one among the famous optimal edge detector. He was very much successful in achieving his ideas and goals and methods which can be found in his own paper named "A Computational Approach to Edge Detection". In this paper he followed list of methods in order to improve edge detection. The basic thing is error rate should be very low as possible. The basic thing is image should not be missed out and hence there should not be any response to non-edges. The second thing is that the Edge points should be localized in the best way. In other words, the distance between the edge pixels as found by the edge detector and the actual edge should be at minimum. A third criterion is that only one response should be given for one edge i.e., one edge provides one response. Based on these three criteria, the canny edge detector initially smooths the image to reduce the noise. Smoothing helps to reduce noise. It is used to find image gradients in order to showcase region with very high spatial derivative. The region found through algorithm suppresses any pixel that is not at maximum. [1]

2.3.2 Steps of Canny Edge Detection

Canny edge detection method which has received much attention during the recent years due to its many applications in different fields. Edge detection is one of those challenging problems and up to date, there is no technique that provides a robust solution to all situations and different applications that edge detection method may encounter. The steps followed by canny method are as follows. [1][2][3]

Step 1:Noise Reduction by Smoothing: Image is smoothed by reducing the intensity of the noise present in the image. A Gaussian Filter is used to reduce the noise in the image.

Step 2: Finding Gradients: Detection of the edges where the grey scale intensity is maximum. Most of the times the sobel operator is used to determine the gradient of each pixel in an image to detect the edges. Gradients in both x(Gx) and y(Gy) directions are determined and the magnitude of the gradient at each pixel is calculated.

Step 3: Non-Maximum Suppression: It is carried out to preserve the local maxima of the gradient of the image.

Step 4: Hysteresis Thresholding: The output of the non-maximum suppressions contains local maxima created by the noise. For avoiding the problem of streaking two thresholds are used.[1]

2.3.3 Sobel Edge Detector

Sobel operator is an edge detection technique that is used to calculate gradient approximation of the image for detecting an edge. For each pixel of an image this operator provides gradient vector or normal vector. It takes image as an input and calculate magnitude and direction. But compared to Robert computation speed is very less but kernel size is large so less sensible to noise as compared. Having bigger mask, errors which are produced due to noise effect is minimized by local averaging method. It uses set of 3*3 convolution kernels or masks. Maximum edges are identified with respect to the perpendicular angle. The sobel operator is shown below. [4][5] The Gradients Gx, Gy is used to determine gradient with x and y direction respectively.

-1	0	+1
-2	0	+2
-1	0	+1

Gx

+1	+2	+1
0	0	0
-1	-2	-1

Gy

2.3.4 Prewitt Edge Detector

The function of Prewitt edge detector is a type of edge detector which is similar to that of sobel Edge detector but will be having different kernels, performance of Prewitt is better compared to that of Sobel edge detector. The Gradients Gx, Gy is used to determine gradient with x and y direction respectively. [4][5]

-1	0	+1
-1	0	+1
-1	0	+1

G_x

+1	+1	+1
0	0	0
-1	-1	-1

G_y

2.4 Bi-dimensional Empirical Mode Decomposition

We, propelled imaging is persistently created in various applications, for example object affirmation, satellite imaging, biomedical, Internet, etc. The image quality spoils as a result of the tarnishing of a couple of sorts of fuss. This clatter pollutes the image during getting, transmission and limit. Thus, the upheaval decline is a critical development in picture assessment and the essential most noteworthy development to take before the photos are readied. The Bi-dimensional Empirical Mode Decomposition (BEMD) has made better approaches to manage taking care of filtering and denoising pictures. Better nature of results is performed than existing breaking down procedures, for instance, Fourier Transform and Wavelets Transform strategies. [6]

The BEMD is particularly sensible for picture taking care of, for instance, surface examination, picture filtering and denoising. It is a nonlinear and self-flexible channel and it is a profitable system appeared differently in relation to the next philosophy reliant on wavelet root or Gabor Transforms. The BEMD separates an image into various leveled fragments called a Bi-dimensional Intrinsic Mode Functions (BIMFs) and a development, this rot relies upon the close by spatial assortments or sizes of the image. This paper presents the BEMD approach arranged in DWT system for picture denoising and its assessment with the most used methodologies in the denoising space as the center channel, DWT change strategy and the ordinary procedure BEMD reliant on histogram one. It shows the effect of the number of IMFs isolated on the visual idea of a denoised picture in terms of Peak Signal to Noise Ratio (PSNR) and Mean Squared Error (MSE). The accompanying fragment presents the most gigantic of the Bi-dimensional Empirical Mode Decomposition (BEMD) and the nuances execution of the sifting process. [7]

3. SYSTEM DESIGN

An edge is the easily visible and is easily perceivable and are the significant boundaries between the different regions of the image showing different intensities. The intensity may be due to the pixel value of the image or due to the light gradient. But we focus on the edges that are due to the different pixel intensities.

Initially, in the beginning of the process an image is given as the input to different classical edge detection algorithms. The classical edge detection techniques involve Canny, Sobeland Prewitt edge detection techniques. These techniques have their own merits and the demerits.

The advantage of these classical edge detection algorithms is that these techniques are easy in implementation and are very easy to understand because these algorithms are very simple. The limitation of the techniques is that these techniques are not able to remove the noise from the image. Noise include white noise and the partial gaussian noise. Hence most of the edges that are detected by these techniques contain noise as very major component. Hence, to eliminate noise from the image the methodology in this work is based on utilizing the potential ability of Bi-dimensional Empirical Mode Decomposition (BEMD).Another method called Image Empirical Mode Decomposition (IEMD). Both BEMD and the IEMD methods decompose a given image into various Intrinsic Mode Functions (IMF's) of different spatial frequency using the image input itself i.e. that these methods are Data Driven. They help in reducing the white noise as well as the partial gaussian noise.

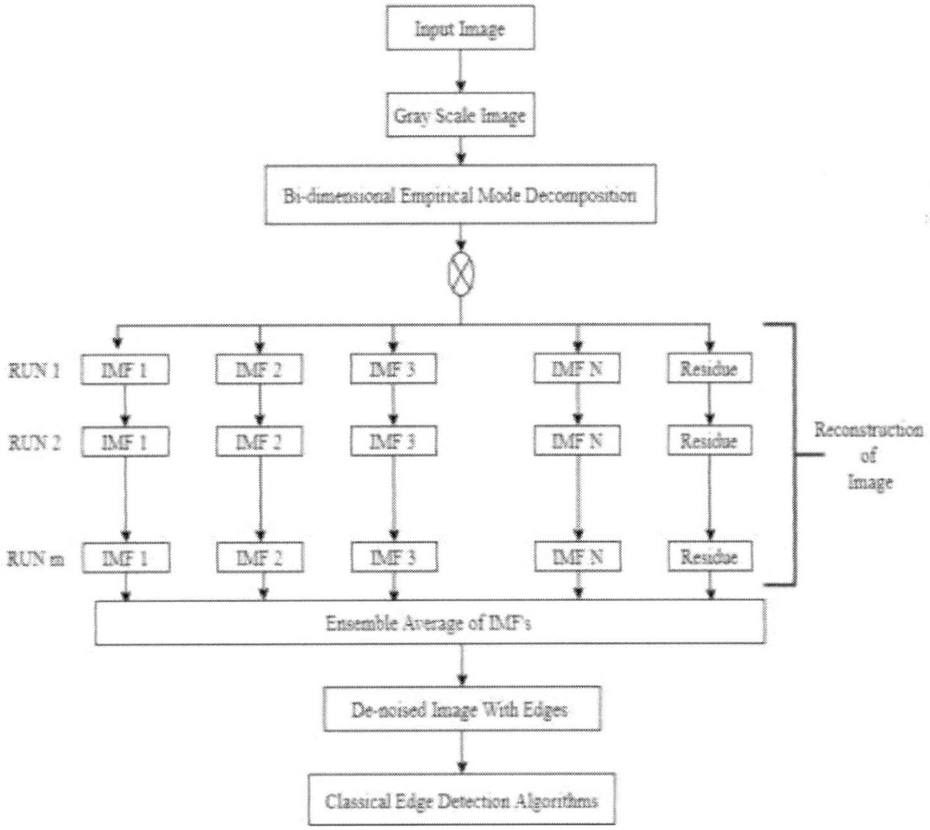

Fig 3.1.1 Flow Diagram of the Edge Detection Algorithm

3.1 Flow of Traditional Edge Detectors

Traditional edge detectors like the Canny, Sobel and the Prewitt are the most important edge detection algorithms. It is observed from the literature survey that Canny Edge detector is one of the most optimal edge detection algorithms compared with all other traditional edge detection algorithms. The flow diagram of the canny edge detector is shown in fig 3.2.1. It is because the canny edge detection algorithm has a very error rate compared to the other traditional edge detection techniques.

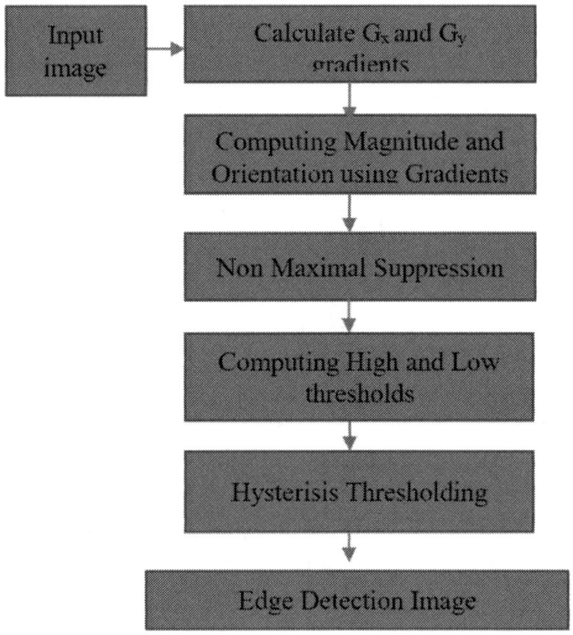

Fig 3.1.1 Flow Diagram of the canny edge detection

The other traditional edge detection techniques face the problem of detecting the false edges as against to detecting the true edges and even produces a thin or a thick line. These problems are mainly due to the interpolation of the noise with the pixels. To avoid interpolation of noise with the image, the image is primarily provided as the input to the Bidimensional Empirical Mode Decomposition (BEMD) algorithm which results in a set of Intrinsic Mode Functions (IMF's) and here we consider only those IMF's in which noise has been eliminated and gives fine edges including all the micro edges.

If the output of the BEMD algorithm is provided as the input to the traditional edge detectors namely Canny, Sobel and Prewitt methods, then these methods are expected to provide better performance than earlier as the noise present in the image has already been eliminated by BEMD algorithm. Hence, the edges will be detected more accurately and thus image segmentation will be more accurate and provides a way to improve the performance of theof the edge detection techniques adaptively.

Another limitation of the classical edge detector is that there may be the possibility of malfunctioning at the corners curves, and where the grey level intensity function varies. The algorithms may not find the correct orientation of an edge in a direction due to the usage of the

Laplacian filter. Some filters like the Gaussian filter may use the complex computations and may detect the false or the zero crossings which may lead to detection of the false edges.

Sometimes there may be problem of the edges missing due to the selection of the wrong and irrelevant threshold values in thetraditional edge detection algorithm. The threshold value is the value of the gradient above which a pixel is detected as the edge. There is another parameter called the Sigma value. The Sigma value indicates the amount of blurring that is needed to be done for the noise. Both the threshold and the sigma values help in determining the correct edge. But due to the wrong selection of these values there may be an edge missing resulting in false edge detection.

BEMD helps in detecting the true edges and there are no missing edges as the noise will already been eliminated in the process of decomposing the image into various IMFs' and hence this new way of edge detection technique will be helpful accurately identifying the edges without any error. This can be used in practical applications where it is very important to get the correct and accurate edges without any false edges.

The Primary objective of the work is to detect and discover the most common features of an image using the edges. During the discovering and exploring the information from an image, edge detection is most important and early-stage activity and the most prominent active area in the image processing. BEMD and IEMD are those techniques that play a very crucial role in this process of detecting the accurate edges of an image in the process. The adaptive and the data driven method, improves the performance Of the image processing by eliminating the noise and providing accurate edges for image segmentation. Edge identification is troublesome in up roarious pictures, as both the edge and, the commotion contain high recurrence content. Hence it is very difficult to eliminate the noise without disturbing the pixel value of the edges. BEMD succeeds in differentiating the noise and the edges and provides a adaptive way called the data driven way to detect and analyze the edges of the image. If the output of this BEMD is provided as the input to the traditional algorithms it gives high performance and better-quality edges.

4. SYSTEM IMPLEMENTATION

4.1 Data Collection

The novel approach of edge detection takes input from different domain. The input image is taken from the various domains of interest and they include the images from the satellite, the images captured by the drones for agricultural importance and the military purpose, the images that are taken to do research on the wildlife using the drones in the forest and the wildlife sanctuary. All these sources of the images are for the societal and environmental importance. It is because they have a wide range of application in the society and helpful in the development and the sustainable usage of the natural resources present on the globe.

4.2 Data Processing

The image that we provide to the edge detection algorithm is considered as the data. Image that is needed to be processed and analyzed for various applications has a very large size. It is because the image is a high dimensional matrix of pixels. Each pixel of the image contains a large information about the image.

In Digital image processing a pixel is smallest element or the smallest unit of the image. Any point in the image corresponds. to a pixel, the value of a pixel at that point is the intensity of the light at that point. Each pixel stores a value proportional to the light intensity at that location. As each pixel stores the information at a single point location, it has a very huge amount of information stored in it. So, the matrix of the pixel values (image) has a very large amount of information regarding the scene. Hence it becomes potentially difficult to process the entire image for any application. It is also a time-consuming process.

To avoid processing of the entire pixels of the image, edge detection plays a very crucial role. Edge detection is the early stage in most of the vision processing activities. The early stage involves identifying the features of an image. The edge detection process is the vital process in the image processing. It helps in identifying the significant and the most important features of the image. Hence, it reduces the amount of the data needed to be processed, as the entire image is not being processed only the significant features of the image will be processed. It reduces the time required to process the entire image.

Images that are received by the satellite, the images received by the drones for the military purpose will be of very huge size. Even the images for the agricultural importance will be very large in size. Hence edge detection of these images leads to the identification of the significant and important features and helps in reducing the amount of data that is to be processed for the purpose of analyzing. Large data will be compressed into a small useful data named as the edge of the image.

4.3 Data analysis

The intrinsic mode components obtained from each method are carefully analyzed and the IMF which is physically meaningful is subjected to edge detection process using traditional edge detectors such as Canny, Sobel, Prewitt. In this method of combining BEMD with the traditional edge detectors expected to provide superior edge detection performance and some of the weaknesses of traditional edge detectors expected to be masked or eliminated.

4.4 Steps to Decompose Image into IMFs'

Steps to decompose an image into IMF are as follows,
Step1: To first IMF is gotten with the picture itself as info signal

h10(m, n)=x(m, m).--------------------------------------- (4.3.1)

The main record shows the IMF number, l=1,. .... ,L, and the subsequent list demonstrates the emphasis number, k = 1,. ,K, in the filtering procedure; m and n speak to the two spatial measurements thus the name Bi-dimensional Empirical Mode Decomposition or 2D - Empirical Mode Decomposition as shown in equation (4.3.1). To locate the following IMF, utilize the buildup relating to the recently discovered IMF as information signal given by equation (4.3.2) below.

h20(m, n) =r1(m,m) --------------------------------(4.3.2)

Step 2: The positions along with the amplitudes of all local maxima and the local minima of the input signal is to be found.

Step 3:Make the upper envelope by spline addition of the nearby maxima and the lower envelope by spline introduction of the neighborhood minima. Denote the envelopes as e+(m, n) and e−(m, n), respectively.

Step 4: For each position (m, n), figure the mean of the upper envelope and the lower envelope shown below in equation (4.3.3).

e lk(m, n) = (e+(m, n) +e−(m,n))/2 -- (4.3.3)

The sign elk(m, n) is added to as the envelope mean.

Step 5:Take away the envelope mean sign from the past sign given by equation (4.3.4):

hlk(m, n) = hl(k−1)(m, n) +elk(m, n)... .... (4.3.4)

This is one emphasis of the filtering procedure. The subsequent stage is to check if the sign hlk(m, n) from stage 4 is an IMF or not. The procedure stops when the envelope mean sign is sufficiently close to zero given below by equation (4.3.5).

e lk(m, n)| < ε∀(m, n).-- (4.3.5)

The estimation of ε in the stop basis influences the EMD so that if it isn't sufficiently little, at that point there won't be an adequate number of IMFs to isolate every inborn mode in the sign. Then again, if the number ε were excessively little, the cycles will take long time

Step 6: Check if the mean sign is sufficiently close to zero, in light of the stop measure. If not, rehash the procedure from stage 1 with the subsequent sign from stage 4 as the info signal an adequate number of times. At the point when the stop rule is met, the IMF cl(m, n) is characterized as the last aftereffect of (4) by equation (4.3.6).

cl(m, n) =hlk(m, n). -- (4.3.6)

After the IMF is found, characterize the buildup as below equation (4.3.7).

Rl(m, n) = hl0(m, n) −cl(m, n). ------------- (4.3.7)

Step 7:The following IMF is found by beginning once again from stage 1, presently with the buildup as the info signal as below equation (4.3.8).

h(l+1)0(m, n) =rl(m,n). -- (4.3.8)

Steps (1) to (7) can be rehashed for all the resulting rj. The EMD is finished when the buildup, preferably, doesn't contain any extrema focuses. The sign can be communicated as the total of IMFs and the last buildup.

5. EXPERIMENTAL RESULTS

In this paper, we have experimented various edge detection techniques such as Sobel, Prewitt, and canny. All of these are traditional edge detection techniques. But providing original image shown in fig 5.1(a) as input directly to the traditional edge detection techniques, output will not have more detailed edges and original edges.

IOP Conf. Series: Materials Science and Engineering 1187 (2021) 012032 doi:10.1088/1757-899X/1187/1/012032

Fig 5.1(a) original image

So, we came up with a new adaptive traditional edge detector which consists of various processes as follows. As shown in fig 5.1(a) we use the original image as the input to the BEMD (Bi- dimensional empirical mode decomposition) which is of the size 480x480x3,where the original image is a RGB coloured image which consist of 3 layers i.e. Red layer component of 480x480 pixel, Blue layer component of 480x480 pixel and a Green layer component 480x480 pixel. But we cannot have this 3-layer image to be given as a input to BEMD algorithm, so we convert that original image into a grayscale image which is of the size 480x480 pixel size as shown in Fig5.1(b).

Fig 5.1(b) Grayscale image

We add up some Gaussian noise to the grayscale image of value between 0.1 and 0.2 to beat already present noise in the image and finally fed that noisy grayscale image into BEMD function for producing the IMF'S (Intrinsic Mode Function) as shown below.

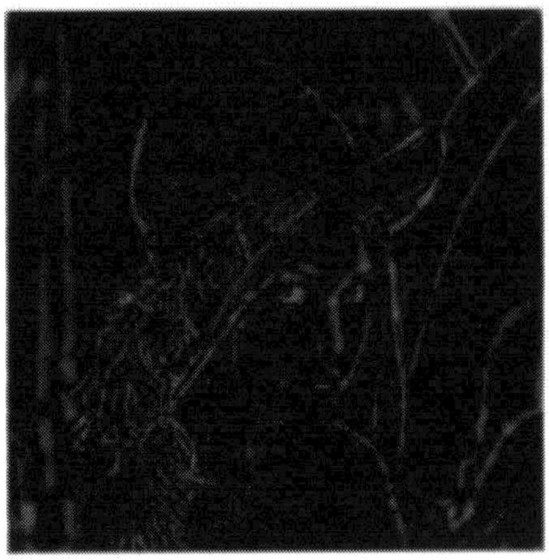

Fig 5.1(c) RUN1BEMD IMF1

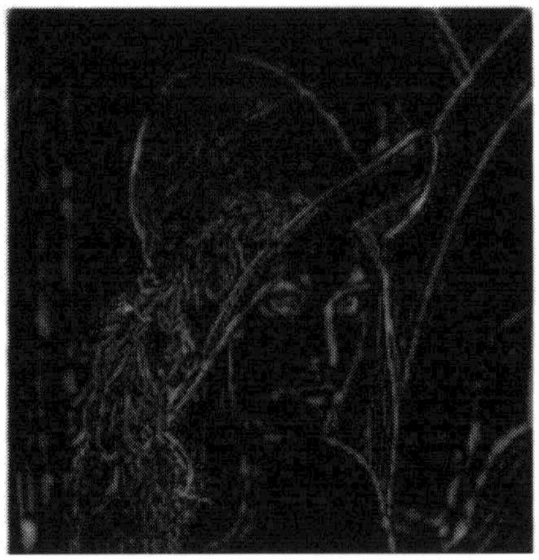

Fig 5.1(d) RUN1 BEMDIMF2

IOP Conf. Series: Materials Science and Engineering 1187 (2021) 012032 doi:10.1088/1757-899X/1187/1/012032

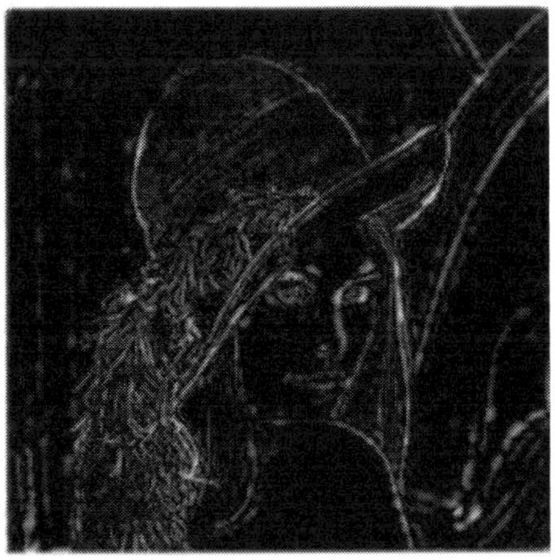

Fig 5.1(e) RUN1BEMDIMF3

Fig 5.1(f) RUN1 BEMDIMF4

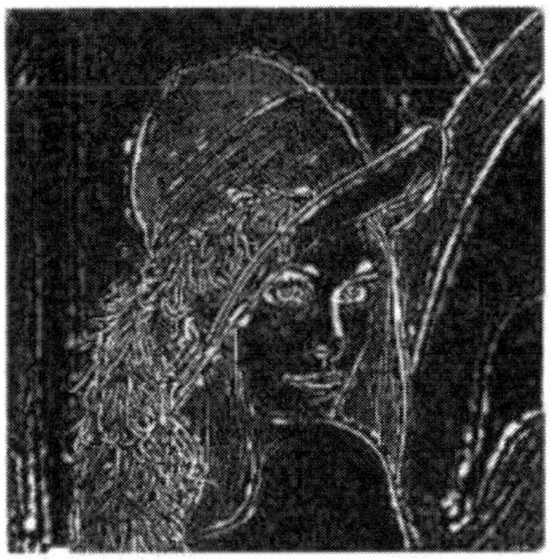

Fig 5.1(g) RUN1 BEMD IMF5

The fig 5.1(c), 5.1(d), 5.1(e), 5.1(f) and 5.1(g) shown above are the output of the BEMD function from RUN1. Fig 5.1(c) shown above is of the size 480x480 pixel is the IMF1 of RUN1 which as some noise and much of the edges are not visible. So, fig 5.1(d), 5.1(e), 5.1(f) and 5.1(g) are of the size 480x480 pixels and the edge are visible as IMF number increases in this case IMF5 as visible edges as compared to other 4 IMF's. Also, some of the micro edges are not visible in IMF 1, 2, 3, but in IMF 4 and 5 micro edges are visible and it can be improved by sharpen those IMF's.

In obtained IMF's as shown above some of them are having low and high frequency. Mean IMF1 contains having high frequency components and we know that even noise is of high frequency. So Mean IMF1 will be having high frequency noise interpolated along with it and thus, will not be having better quality edges or the true edges, since it mostly consists of high scale noise. Then we have obtained subsequent IMF's whose frequency values goes on decreasing as the IMF number increases. In this case Mean IMF5 is the highest IMF obtained and will be having lower frequency components where noise will be eliminated completely, hence it contains high quality edges and accurate edges as compared to another Mean IMF's. In the subsequent Fig 5.7 shown below gives comparison of canny edge detector with different threshold values for different IMF's. Different threshold value having high threshold value 0.20 and low threshold value of 0.17. also, high threshold value of 0.30 and low threshold value of 0.20, but in the threshold value between 0.22 and 0.27 which as obtained a best output image with that threshold value and is considered as the final Canny output image. But Canny has a best edge detection property as compared to other traditional edge detector like Sobel and Prewitt. So canny edge detector with BEMD consider to be best way for Data Driven Adaptive Edge Detectors for Image Processing.

LT=0.17 HT=0.20 **LT=0.22 HT=0.27** **LT=0.20 HT=0.30**

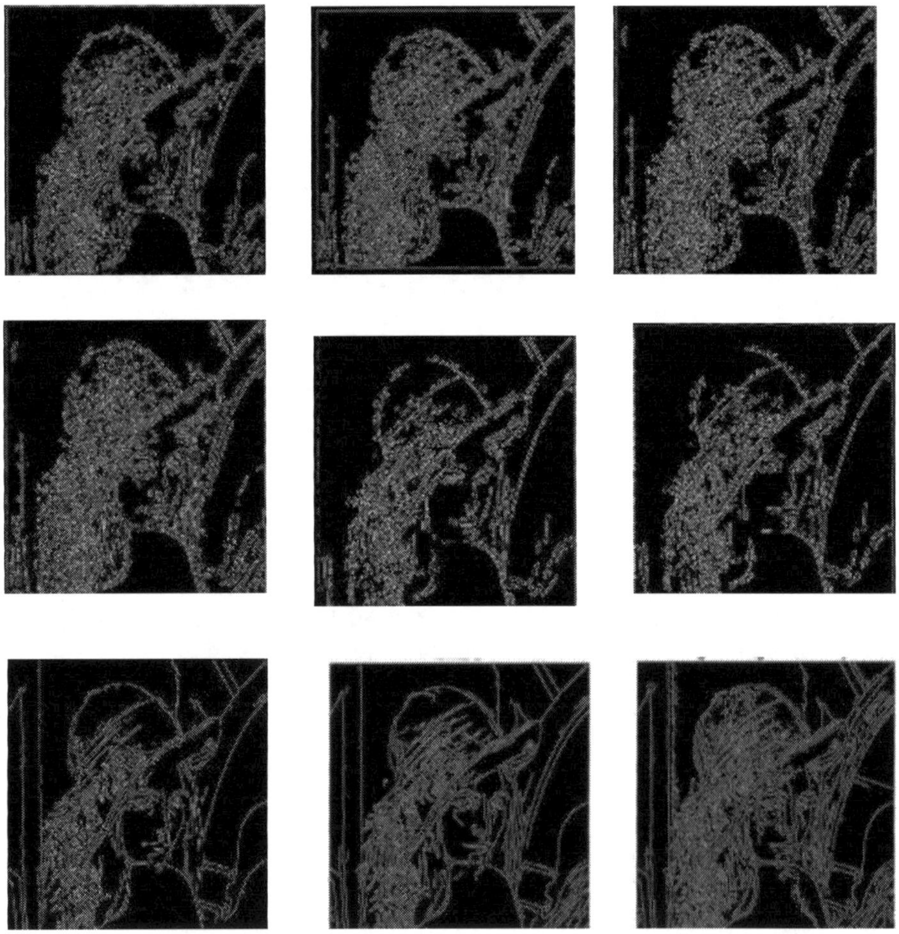

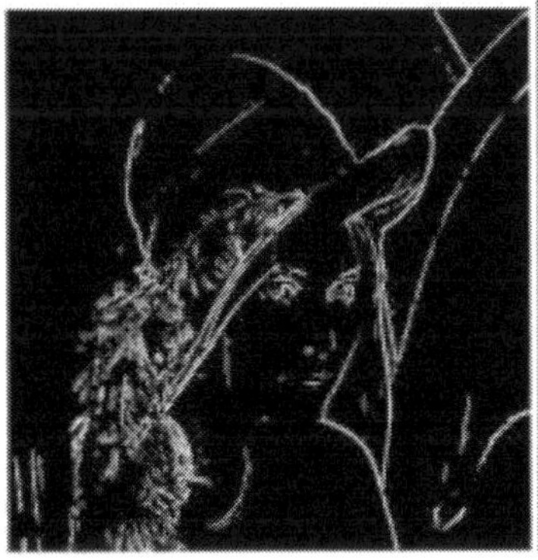

Fig 5.8(a)

Mean IMF 5 given as input to Prewitt operator.

Fig 5.8(b)

Grayscale given as input to Prewitt operator.

Fig 5.9(a)

Mean IMF 5 given as input to Sobel operator.

Fig 5.9(b)

Grayscale given as input to Sobeloperator.

In above shown Fig 5.8 is the output image of Prewitt edge detector with a threshold value of 0.02 and Prewitt edge detector doesn't have much micro edges visible in the output image as shown above. But as compared to Sobel in some better in terms of edge detection of macro and micro edges.Fig5.9shownaboveistheoutputofSobel edge detector with the threshold value of 20. Sobel doesn't have that much edge detection as compared to canny and Prewitt.

From the experimental results we can conclude that canny has a best edge detection property as compared to other traditional edge detector like Sobel and Prewitt. So canny edge detector with BEMD consider to be best way for Data Driven Adaptive Edge Detectors for Image Processing. Canny gives the best edge detection when given the threshold between 0.22 and 0.27 for ensembled IMF 5 where all the noise will be eliminated by the process of noise beating.

6. CONCLUSION AND FUTURE ENHANCEMENTS

The edge detector performance criterion and methods of evaluation provide us a good understanding of possible ways of finding out the effectiveness of each edge detection technique. From the result obtained, Canny, Sobel and Prewitt prove to have more sensitive and time-consuming. But in cases where simplicity and speed are not dominant factors, Canny, Sobel, and Prewitt could be very suitable and robust. Sobel operator will spot out noise found in real-world image as edges though and the detected edges could be thick. The Canny edge detector does the same algorithm to solve these problems by first blur those real world image slightly and then applying an algorithm that effectively thins the edges. In Prewitt presence of noise false edges get detected. The first-order derivative is more affected by noise as compared to other operators. The Laplacian technique of detecting the edge by using zero-crossing techniques. The LoG can reduce noise due to the presence of the Gaussian filter. As compare to all operator Canny Edge detector is the better in noise suppression and micro edge detection.

The results shows that traditional edge detection performance can be improved when combined with BEMD on grayscale images. However, efficient is needed in terms of speed and recognition of the edges in images rate to determine the choice of the best algorithm needed for Development of Data Driven Edge Detectors for Image Processing. Future work will focus on finding and further improving the edge detector to achieve higher and accurate detection in real-time. With speed in mind, fast processors will also be investigated. The different modern edge detectors like empirical mode decomposition and many other techniques will be studied and the results of these techniques will be analyzed to get the best adaptive edge detection techniques for various images like a flame and the nest of gazelle.

We have used the BEMD function to extract the IMF from the image to give as an input to the traditional edge detection technique. In future work, we may complete the ensemble empirical mode decomposition technique, where it has much accurate output and less noise image output.

REFERENCES

[1] John Canny, "A Computational Approach to Edge Detection", IEEE transactions on pattern analysis and machine intelligence, November1986.
[2] Amruta L Kabade, "Canny edge detection algorithm", International Journal of Advanced Research in Electronics and Communication Engineering (IJARECE), May2016.
[3] Dhanalakshmi Renganathan, Riyas K S, Anu George, "Implementation of Canny Edge Detection Algorithm in FPGA Using HDL", March2017.
[4] KM Spoorthy, Suchetha G Hegde, N Vijetha, M S Rudramurthy, Keerthan Kumar TG, SA Sushma," "Performance Analysis of Bird Counting Techniques Using Digital Photograph", 5th International Conference on Intelligent Computing and Control Systems (ICICCS),2021
[5] Gunasheela B.G, Sushma S.A, Kavana G, Nagateju T. V, Pavanaraju,"Optimized Fire Detection on Surveillance System using Image Processing and Convolutional Neural Networks", International Journal of Advanced Science and Technology,Vol-9,Issue-10,SERSC.
[6] AnnaLinderhed, "Image Empirical Mode Decomposition:A new Tool for Image Processing", Advances in Adaptive Data Analysis Vol. 1, No. 22009.
[7] Keerthan Kumar T G, Harika E, N Vani, AishwaryaPatted "A Novel Creative Photo grid Android mobile Application for Photo Editing", International Journal of Advanced Research in Computer and Communication Engineering. Vol. 5, Issue 2, February 2016.
[8] Kajol, R; Akshay, Kashyap K; Keerthan, Kumar T, G. International Journal of Information Engineering and Electronic Business; Hong Kong Vol. 10, Iss. 2, (Mar 2018): 17. DOI:10.5815/ijieeb.2018.02.03

Assessment of Asphalt Road Pavement Rehabilitation and Maintenance Practices: The Case of West Showa Zone in Oromia, Ethiopia

Lensa Merga Kenea[a], Leevesh Kumar[b], Sitesh Kumar Singh[c], Geleta Chala[d], Jifara Chimdi[e]

[a,b,d,e] School of Civil and Environmental Engineering, Ambo University, Ambo - 19, Ethiopia.

[c] Department of Civil Engineering, Wollega University, Nekemte – 395, Ethiopia

Abstract. This study mainly aimed at assessing asphalt road pavement rehabilitation and maintenance practices in West Showa Zone with particular reference to the roads from Ambo to Gedo which covers over 64kms. The study aimed at identifying pavement distresses and causes of pavement distresses. According to data collection and analysis, it was found that the material properties and composition (28.1%), construction quality (18.8%), and road maintenance standards (15.6%) were perceived to be the major causes of pavement distresses on the roadway under study. On the other hand, the result of survey condition data obtained from ERA was analyzed and major distresses were identified including longitudinal joint crack, bleeding and tripping, potholes, generalized marked fretting, deformation, subsidence, rutting, patch, wide block crack, isolated scabbing, low shoulder and wheel path rutting. Besides, limited skilled man power ($\bar{x}$=3.78), lack of appropriate pavement management system ($\bar{x}$ =3.93), budget constraint ($\bar{x}$=4.43), absence of appropriate machineries ($\bar{x}$=3.53), and lack of monitory and quality control ($\bar{x}$=3.75) were among the major challenges related to effective pavement work, road maintenance and rehabilitation practices on the roadway under study.

Keywords: *Pavement, Distresses, Rehabilitation, Maintenance, Evaluation.*

1. INTRODUCTION

1.1 Background

The road network is a largest infrastructure, the rehabilitation and maintenance of which needs special necessity (Fakhri and Dezfoulian, 2017). As the nationwide highway connection demands keep on to grow, hence ensures the necessity for maintenance and construction practices that will support cover the lifespan of pavements at all levels. Apparently, no pavement is constructed that does not need rehabilitation and maintenance. To understand the which maintenance activity is taken for proper rehabilitation and maintenance of bitumen pavements, it is essential to use details evaluation of pavement to identifying and understanding the causes of failures of pavement.

In a number of developing countries road degrade due to lack of care has turn out to be a growing issue. Due to lack of maintenance the problem was discussed and the outcomes have been well-defined and calculated. Nevertheless, the solutions are still not regularly understood and the expanse of the problem is not fully appreciated. Similarly, the actions required to correct the deficiencies are under-estimated. These contain the scale of care and capacity growth required, and the time-scale essential for formation an effective road managing system. Such a system would pause road network weakening and safeguard that economic, resources investments are made in a way which preserves the quality and value of the assets and, in calculation, expands the network in relation to the demands and importance of the users (Addis Ababa Project Administration Service, 1984).

Rapid data collection based on subjective decision has been used, because the collection of detailed pavement condition data is enormously costly and time-consuming, furthermore, preservative seal, restorative seal, crack seal, patching, surface milling, base repair, seal coat, full depth reclamation, resurfacing, widening, reconstruction, curb removal and replacement, pavement marking, and replacement, sidewalk removal and grading, mulched seeding, tree trimming (removal), striping and shrill overlay have been considered as activities for rehabilitation and maintenance (Fakhri and Dezfoulian, 2017). After the assessment of pavement surface physical characteristics, GIS software, as a practical tool, was used to display pavement section distress and the result of arrangement.
According to Ebeling (2007), for better tracking and choice making when it comes to the users' needs GIS helps answer questions and solve problems by looking at data in a way that is quickly understood and easily shared the collected data can be presented on a map. In the case of road data, GIS would have extensive pavement data. Additional information may include traffic volume, weather data, political districts, as-built information, population, sign retro reflectivity, and any other data that may have an effect on the road functioning at a high level. The present asphalt road surface condition needed some sophisticated tool to systematically identify pavement distresses which enable stakeholders take remedial action. As a result, the application of Arc GIS software was used, severe pavement distresses were identified and the results were diagrammatically supported.

The following could be mentioned among the local studies conducted which were partially related to asphalt road rehabilitation and maintenance practices. To this end, Temesgen Girmay (2016) studied asphalt road rehabilitation and maintenance management, (Kalid Ali, 2016) undertaken a study on assessing Ethiopia Roads Authority's (ERA) pavement management system, and finally (Fikir Alebachew, 2005) studied at capital city Addis Ababa arterial roads pavement distresses, maintenance options and causes. All of the studies did not address the application of Arc GIS software in studying asphalt road pavement maintenance and rehabilitation practices. On the other hand, among the international studies on asphalt road pavement maintenance using GIS include researches conducted by Robert J.(2011), Acquah and Fosu (2017) and Fakhri and Dezfoulian (2017). These studies addressed the application of GIS on road maintenance both as a system of prioritization and management approach. Therefore, in this study, effort was made to fill this gap by using Arc GIS as a tool to study asphalt road pavement rehabilitation and maintenance practices in West Showa Zone in Oromia National Regional State.

2. RESEARCH METHODOLOGY

The research methodology section of the study highlights on research design employed to guide the study, description of the study area, study period, sample size, and sampling technique, tools for data collection, data collection procedures, data processing and analysis, data quality assurance and ethical issues.

2.1. Study area

West Showa Zone is situated in the west portion of Addis Ababa, the Ethiopia capital city. It is one of the twenty zones in Oromia National Regional state and runs from Holeta to Bako which covers a distance of about 254 kms including the distance between Holeta and Mugher asphalt road. The study area can be depicted in Fig. 1. The map is adapted and taken from ECDSWC: Transport Design and Supervision Works Sector (2016).

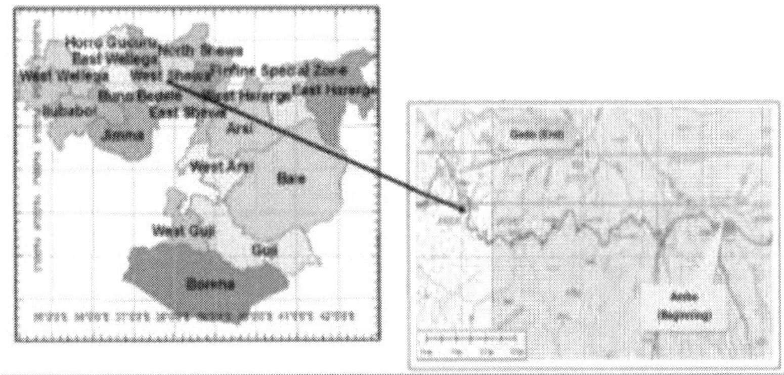

Fig. 1. Study area

2.2. Research design

This research employed descriptive survey design. So, this study was guided by mixed methods research which involves both quantitative and qualitative data to attain the specific aims of this study. Quantitative data was collected by using questionnaire (close-ended and open-ended), whereas qualitative data was collected through field observation.

2.3. Study period

This study took almost one year starting from the inception up to the completion of the final research report. Accordingly, it was begun in the late October, 2019 and completed in the early of November, 2020.

2.4. Sampling technique and sample size

Since the total study population was small in number, the researcher took all the available participants (i.e., the total population) as a sample. The participants were composed of professionals who were working as clients, in contracting companies and consultant offices to fill the questionnaires. Hence, 6 clients (Regional and Federal Roads Authorities), 18 contractors, and 8 consultants were taken as participants in the study.

3. DATA COLLECTION AND ANALYSIS

In this research Primary data has been collected from clients, contracting companies and consultant offices by using questionnaire. Similarly, field observation was used to collect visual condition data on the construction area which would help to make proper evaluation of the pavement surface. Finally, secondary data source (i.e., ERA document) was used to obtain information archived during the maintenance processes in the roadway under study. In order to collect pertinent data for the study, three key instruments were used.

3.1. Questionnaire

In current study Questionnaire was the primary data collecting instrument employed. With the help of questionnaire collection for survey data, providing organized, often mathematical data, being able to be managed without the presence of the investigator, and frequently being moderately straightforward to investigate (Wilson and McLean 1994, cited in Cohen, Manion and Morrison, 2007:317). The first section was supposed to collect demographic information about respondents. The second section was meant to obtain data about the causes of distresses on the roadway, the third was questions related to pavement rehabilitation and maintenance practices and associated challenges. The final section was supposed to gather data about pavement evaluation and challenges affecting pavement activities. The questionnaire was completed by respondents whose designations were office engineers, project managers, site engineers and surveyors who worked in five different organizations. In connection to this, primarily 36 respondents were planned to complete the questionnaire, but later it was confirmed that 4 members resigned from their respective organizations. So, the final questionnaire (both the close-ended and open-ended questionnaire) data was collected from 32 respondents.

3.2. Field observation

Field observation was used to collect visual condition data on the respective roadway which would help to make proper evaluation of the surface of the pavement. Thus, checklist was prepared (PASER rating scale, University of Wisconsin, 2002) to effectively conduct pavement evaluation. The checklist contained surface rating, visible distresses, general condition and treatment measures taken on the road surface. The ratings ranged from excellent (10) to failed (1). Twenty locations were randomly taken and observation was conducted by the researcher. In doing this, station numbers for each location were recorded, then visible distresses were evaluated and backed by sample photos which were taken on the actual road to illustrate the pavement surface condition.

3.3. ERA document

Since the road connecting Ambo to Gedo was under maintenance by ERA (Ethiopian Roads Authority), the researcher officially consulted ERA authorities to obtain pertinent data. This secondary data consisted of documents such as condition survey data, environment condition and cross section (i.e., both vertical and horizontal alignment) data. More specifically, the document related to condition survey data helped to properly identify and show the pavement distresses/failures and major causes of distresses/failures by using Arc GIS software.

3.4. Data processing and analysis

Quantitative data collected from contactors, consultants and clients through questionnaire was analyzed by using descriptive statistics such as frequency distribution, mean, mode, range, variance and standard deviation. On the other hand, the data collected through field observation was qualitatively analyzed and discussed along with the result obtained from questionnaire analysis as shown in Table 1.

Table 1. Data Evaluation by PASER rating scale (University of Wisconsin, 2002)

Surface rating	Visible Distress	General Condition/Treatment Measures
10-Excellent	Not Found	Newly constructed
9-Excellent	Not Found	Newly overlay. Like new
8-very good	No longitudinal cracks except reflection of paving joints. Rare transverse cracks, frequently spaced (40' or greater)	Newly sealcoat or new cold mix. slight maintenance required.
7-Good	Surface shows nearly traffic wear, very slight or no raveling. Due to reflection or paving intersections Longitudinal cracks (open ¼"). Transverse cracks (open ¼") spaced 10' or more apart, little or slight crack raveling. No repairing or very few repairs in excellent condition.	First symbols of aging. Maintain with routine crack filling. Structural
6-Good	Longitudinal cracks (open ¼"-1/2") due to reflective cracking Slight raveling (loss of fines) and traffic wear. Transverse cracks (open ¼") spaced 10' or more apart, little or slight crack raveling. Minor to moderate polishing. Occasional repairing in good condition.	Shows signs of aging. Sound structural condition. Could extend life with sealcoat.
5-Fair	Moderate to severe raveling (loss of fine and coarse aggregate). Longitudinal and transverse cracks (open ½") Shows first signs of slight raveling and secondary cracks. First signs of longitudinal cracks near pavement edge. Transverse cracking and first signs of block cracking. Slight crack raveling (open ½"). Extensive to severe flushing or repairing. Some patching or edge wedging in good condition.	External aging. Sound structural condition. Needs sealcoat or thin non-structural overlay (less than 2")
4-Fair	Severe surface raveling. Multiple longitudinal and transverse cracking with slight raveling. Block cracking (over 25-50% of surface). repairing in fair condition. Slight rutting or distortions (1" deep or less).	Significant aging and first signs of need for strengthening. Would benefit from a structured overlay (2" or more).
3-Poor	Closely spaced longitudinal and transverse cracks often showing raveling and crack erosion. Block cracking over 50% of surface Some alligator cracking (less than 25% of surface). Patches in fair to poor condition. Moderate rutting or distortion (1" or 2" deep). Occasional potholes.	Needs patching and repair to major overlay. Milling and removal of deterioration extends the life of overlay.
2-Very poor	Alligator cracking (over 25% of surface). Severe distortions (over 2" deep). Extensive repairing in poor condition. Potholes.	Severe distortion. Needs reconstruction. Needs reconstruction with extensive base repair. Pulverization of old pavement is effective.
1-Failed	Severe distress with extensive loss of surface integrity.	Failed. Needs total reconstruction.

Similarly, the data obtained from ERA document was used both to consolidate the data gathered through questionnaire and field observation. The survey condition data that was obtained from ERA document was fed to Arc GIS application to help show road pavement distresses/failures and indicate the major causes of pavement distresses/failures.

3.5. Data quality assurance

The questionnaire was pre-tested on (accounts 10% from the total sample size) before conducting the definite data collection. In the study pre-test results were not included, but used for reliability of the questionnaire items. The researcher checked the completeness of each questionnaire at the end of every response by the respondents. Finally, with the help of colleague's completeness of the questionnaire was rechecked.

4. CONCLUSIONS AND RECOMMENDATIONS

4.1. Conclusions

The main aim of this study was to assess asphalt road pavement rehabilitation and maintenance practices in West Showa Zone particularly on the roadway connecting Ambo to Gedo. It further investigated the maintenance activities and made use of Arc GIS software to clearly elucidate the visible pavement distresses. Following are the conclusions:

1. Some of the major distresses identified by Arc GIS application include, longitudinal joint crack, bleeding and tripping, potholes, generalized marked fretting, deformation, subsidence, rutting, patch, wide block crack, isolated scabbing, low shoulder and wheel path rutting.
2. Material properties and composition, construction quality, and road maintenance standards were perceived to be the major causes of pavement distress on the roadway under study.
3. ERA applied quality maintenance and rehabilitation work, used alternative treatment, and good overlay and patching strategy which are adequate to preserve pavement in a better condition and reduce expensive maintenance work.
4. Limited skilled human resource, lack of suitable pavement managing system, budget restriction, non-availability of suitable machines, and lack of quality control and supervision were found to be the most serious challenges related to pavement work, road maintenance and rehabilitation work on the roadway from Ambo to Gedo.
5. The present pavement work and practice by ERA on the roadway between Ambo and Gedo was found to be at a better condition.
6. The pavement surfaces were evaluated with longitudinal and transverse cracks moderate to severe raveling, extensive to severe flushing and polishing with slight rutting or distortions.

4.2. Recommendations

From the conclusions of this study so far, the recommendations have been forwarded for the improvement of the current road maintenance and rehabilitation practice by ERA particularly on the roadways in the West Showa Zone are as follows.

a) In the system of pavement management surveys of pavement condition are very important. Condition surveys inform the management about the actions to be taken to ensure the effective and efficient use of resources. Moreover, it enables the Federal Roads Authorities and other stakeholders to make early intervention which would help to prolong the life and service duration of the asphalt roads.
b) ERA has to give suitable attention to road maintenance works and seek for several fund options to fill the rehabilitation and maintenance practices. The authority has to emphasis on leading

preventive maintenance for pavements before they weaken into poor conditions; this can save a significant amount of pavement maintenance budget.

c) ERA should asses and adopt alternative maintenance strategies and treatment selection techniques to resist scarcity of resources required machine, labor, fund and has to develop its own organizational maintenance approaches for efficient use of available resources.

d) ERA and other stakeholders should hire adequate and skilled man power to carry out the rehabilitation and maintenance work periodically. Moreover, based on the findings, appropriate machineries should be fulfilled to completely carry out the rehabilitation and maintenance activities. On the other hand, the pavement management should take in to account the nature and type of the soil, environmental condition and traffic volume and construct quality pavement that suits the standard of the roadway under study. This reduces early deterioration.

REFERENCES

[1]. Australian Asphalt Pavement Association (2010). Flexible Pavement Maintenance and Rehabilitation. https://www.aapaq.org>notes.

[2]. Bhaumik, D., Faust, N.L., Estrada, D., Linares, J., (1997). "Three Dimensional Urban GIS for Atlanta," Proceedings of SPIE, The International Society for Optical Engineering, Vol. 3085, 115-124.

[3]. Burningham, S. and Stankevich, N., (2005)."Why Road Maintenance is Important and How to get it done." World Bank Technical Note TRN- 4.

[4]. Chamorro, A., and Tighe, S. L. (2009). "Development of a management framework for rural roads in developing countries: Integrating socioeconomic impacts." *Transportation Research Record*, 2093, 99–107.

[5]. Cohen L., Manion L., and Morrison K., (2007)."Research Methods in Education."6th ed. London and New York: Routledge.

[6]. David R. G., (2005). "Pavement Preservation Definitions Memorandum," Director Office of Asset Management, USA Department of Transportation.

[7]. De Solminihac, H. (2001). *Gestion de infraestructura vial.*Pontificia Universidad Católica De Chile, Santiago, Chile.

[8]. Ebeling, M R., (2007). "Managing Low Volume Roads with Wisconsin Information System for Local Roads." Transportation Research Record, pp. 277-283.

[9]. ECDSWC. (2016). "Ethiopian Construction Design and Supervision Works Corporation: Transport Design and Supervision Works Sector." Addis Ababa.

[10].ERA (2001). Geometric Design Manual. Addis Ababa.

[11].ERA (2014). Standard Technical Specifications and Methods of Measurement for Road Works. Addis Ababa.

[12].FHWA.(2003b). "Distress Identification Manual for the Long-Term Pavement Performance Program. Publication N° FHWA-RD-03-031. "Federal Highway Administration, Department of Transportation.

[13].Fiker Alebachew (2005). "Pavement Distresses on Addis Ababa City Arterial Roads, Causes and Maintenance Options." (M.SC Thesis), Addis Ababa University, Addis Ababa.

[14].Gumsar S. N., Guite T., & Martina N. (2014). Pavement Distress: A Case Study. National Conference on Recent Advances in Civil Engineering. *International Journal of Innovative Research in Science, Engineering and Technology, Vol. 3, Issue 4.*

[15].Helgeson, T J., (2003). "Using Innovation to Respond to Infrastructure Needs."American Water Works Association Journal, 95(5), 42-42.

[16].Haynes F. Donald., Haehnel R., & Zabilansky L., (1993). Repair, Evaluation, Maintenance, and Rehabilitation Research Program: Icing Problems at Corps Projects. *U.S. Army Cold Regions Research and Engineering Laboratory.*

[17].https://www.researchgate.net/publication/23506086

[18].Kalid A. (2016). Assessing Ethiopian Roads Authority's (ERA) Pavement Management System. (M.SC Thesis), Addis Ababa University, Addis Ababa.

[19].Kmetz, R. J., (2011). "GIS Based Pavement Maintenance: A Systematic Approach," College of Technology Directed Projects. http://docs.lib.purdue.edu/techdirproj/36

[20].Mansour F. and Reza D. (2017). "Application of GIS for Pavement Maintenance Prioritization Based on a New Approach at a Network Level GSTF," Journal of Engineering Technology (JET) Vol.4 No.2.

[21].Osorio A., (2015). "Development of Performance Models and Maintenance Standards of Urban Pavements for Network Management." (M.SC thesis), University of Waterloo and the Pontificia Universidad Católica de Chile. Waterloo, Ontario, Canada / Santiago, Chile.

[22].Pavement Surface Evaluation and Rating (PASER) Manuals (2002): Transport Information Center Publications, University of Wisconsin-Madison.

[23].P.C.Acquah and C. Fosu (2017). "Implementation of Geographic Information System Application in the Maintenance Management of Roads in Ghana: A Case Study of Roads in Kumasi Metropolis." American Journal of Geographic Information System, 6(3): 90-102.

[24].Petzold, R. (1994)."The American City & County." Pittsfield, Vol. 109, Issue 3, 56-64.

[25].Project Administration Service (1984). Operation and Management Improvement Study, City Of Addis Ababa.

[26].Sadek, S., (2005)."A GIS-Based Landslide Hazard Framework for Road Repair and Maintenance." The Electronic Journal of Geotechnical Engineering, 10-15.

[27].TAC.(2013). *Pavement Asset Design and Management Guide*. Transportation Association of Canada, Ontario, Canada.

[28].Temesgen Girmay (2016). "Asphalt Road Pavement Rehabilitation and Maintenance: A Case Study in Addis Ababa City Roads Authority," M.SC Thesis, Department of Civil Engineering, Addis Ababa University, Addis Ababa

[29].Tighe S., Huen K., and Haas R. (2007)."Environmental and Traffic Deterioration with Mechanistic-Empirical Pavement Design Model."*Journal of the Transportation Research Board*, 1989(2), 336–343.

[30].Wasike, W. S., (2001). Road infrastructure policies in Kenya: historical trends and current challenges (No. 1). Kenya Institute for Public Policy Research and Analysis.

[31].Workie Tegen, (2017). "Life Cycle Cost Analysis for Road Maintenance Interventions." (M.SC thesis), Addis Ababa University, Addis Ababa.

[32].Understanding Asphalt Pavement Distresses. Asphalt: The Magazine of the Asphalt Institute. http://asphaltmagazine.com/understanding-asphalt-pavement-distresses

IOP Conf. Series: Materials Science and Engineering 1187 (2021) 012034 doi:10.1088/1757-899X/1187/1/012034

Effect of Thermal Barrier Coating in Sustainable Power production of Gas Turbines

S Ghosh, M Saha, S Bakshi and S Mondal
Mechanical Engineering Department, JIS college of Engineering, Kalyani, West Bengal 741235, India

Abstract. Gas turbine play an important role in producing power for aviation industries. A sustainable operating life with high efficiency of a gas turbine requires very high temperature operation. The increase in thermal conductivity with temperature results in high thermal stress which leads to permanent deformation reducing blade life and efficiency. A thermal barrier coating is crucial to remove such detrimental effects. In the present work, an assessment of gas turbine blade materials has been carried out to study the effect of various coating techniques to improve longevity of blades. Coating methods have been found to play an important role in determining the sustainability of gas turbine performance. It was revealed that such coatings considerably reduce the heat penetration rate and thus increase the service life of gas turbine blades.

Keywords: TBC, Gas Turbine, Coating, Sustainability, Power

1. Introduction

Gas turbine performance is improved when operating life is prolonged, operating efficiency becomes high and emissions are reduced. Design and development of the leading gas turbine blades and other engine components which are exposed to various harsh operating conditions, has been an interesting topic of research for many years. Conventionally the gas turbine blades have been manufactured using nickel, chromium, cobalt and titanium-based superalloys. However, the most suitable material has been the nickel based super alloys in the design of turbine components [1]. To improve the thermodynamic efficiency and sustainability of gas turbine power generation, increasing turbine entry temperature is highly effective. At earlier stages gas turbine engine components were manufactured from conventionally cast (CC) alloys which lacked required heat resistance. In course of time nickel based super alloys became more popular for better heat resistance and more sustainable operation [2].

Table 1 Nickel based alloys for Gas Turbine blades [3]

Type	Alloy	Components (weight %)												
		Ni	Cr	Co	Mo	w	Al	Ti	Ta	Nb	Hf	C	B	Zr
CC	IN-738LC	61.5	16	8.5	1.75	2.6	3.4	3.4	1.75	0.9	-	0.11	0.01	0.04
	Rene 80	57	14	9	5	5	4	5	-	-	0.8	0.17	0.016	0.01
DS	CM247LC	61.7	8.1	9.2	0.5	9.5	5.6	0.7	3.2	-	1.4	0.07	0.015	0.007
	CM186LC	72.5	6	9.3	0.5	.4	5.7	0.7	3.4	-	1.4	0.07	0.015	0.005
SC	René N4	61.6	9.79	7.39	2	6	3.7	4.2	4.73	0.5	-	-	-	-
	CMSX-4	71.6	6.4	6.4	0.1	5.6	2.9	0.6	6.4	-	-	-	-	-
	CMSX-10	74.4	2.4	3.3	0.4	5.3	5.7	0.2	8.2	0.08	0.03	-	-	-

To withstand the increasing operating temperature of gas turbines, directional solidification (DS) was introduced by Bridgman method where solidification slowly takes place in one direction. It removes grain boundaries which are aligned perpendicular to the stress axis. The eliminate all grain boundaries,

Pratt and Whitney used Single-crystal (SC) blade technology for JT9D-7R4 engine in 1982. This technology was useful to manufacture blades ready to operate at even high temperature compared to DS type blades. Table 1 summarizes the chemical compositions of some major alloys applicable in this perspective. At highly enhanced temperatures and harsh environments there is a tendency of core elements to get damaged if directly exposed to flame without any thermal barrier. A Thermal Barrier Coating (TBC) can be applied over those parts which get direct exposure to very high temperature in the range of 1370°C-1590°C. If a coating of low thermal conductivity is applied on gas turbine blades, there can be a possible reduction of 100°C-300°C in the associated components. On the other hand with this technology, 250°C higher combustion gas temperature can be utilized in comparison to the melting point of super alloys demonstrated earlier [2]. TBCs also increase the service life because they reduce the temperature of metal components. Fig. 1 shows a gas turbine blade before and after TBC coating. The cooling holes have enormous importance to control the blade temperature and make the whole operation sustainable.

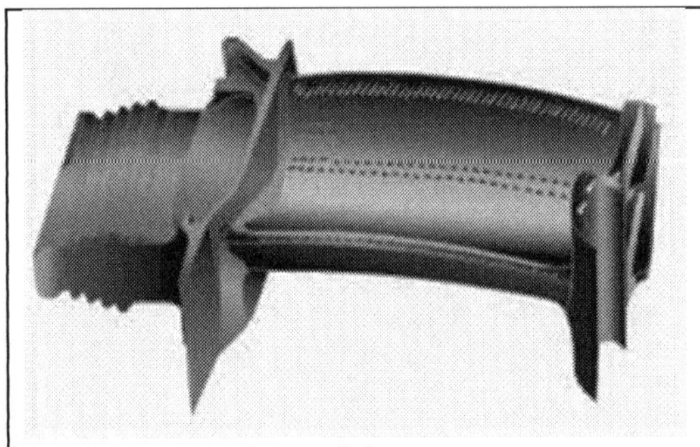
(a)

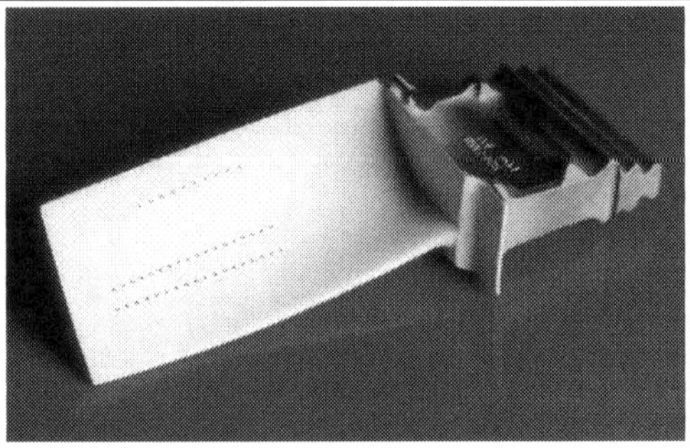
(b)

Figure 1. Gas Turbine Blades with cooling slots (a) without Coating (b) with TBC Coating

2. Blade Life Prediction and influence of operating temperature

2.1 Analysis of Gas Turbine Blade Life

When a material is exposed to stress along with high temperature for a long time, it undergoes permanent deformation, such as creep. Creep occurs at stress less than yield stress and below melting temperature of the material. Creep deformations are responsible for mechanical degradation and eventual breakage of the materials. Thus, the estimation of creep is primary in the perspective of the reliability and sustainability of blades. Generally, Creep life prediction models uses time-temperature parameters to predict long-term creep life [4]. It uses lesser experimental data to derive the creep life prediction curve. The power law model [5,6] for steady-state creep rate ϵ_{ss}, as a function of temperature T and stress has been demonstrated in Eq. 1.

$$\dot{\epsilon}_{ss} \propto A\sigma^n exp\left\{\frac{-Q_c}{RT}\right\} \quad (1)$$

Where, σ is applied stress, A is proportionality constant, Q_c is creep activation energy, R is ideal gas constant and n is the creep exponent. Larson-Miller parameter (LMP) method is widely used for predicting long-term creep life from short-term creep tests. The equation 2 depicts an empirical prediction of creep life [7] where stresses are assumed constant.

$$P_{LM} = f(\sigma) = t(C_{LM} + logt_r) \quad (2)$$

where, P_{LM} is the Larson-Miller parameter, C_{LM} is the Larson-Miller constant and t_r is the rupture time for creep test. However, the model doesn't properly explain how the deformation changes with temperature and applied stress. Newly developed models like Wilshire model and Normalized power law [8,9] shows improved predictability of creep life where the stress is normalized with tensile strength instead of depending on yield strength.

2.2 Effect of thermal conductivity on blade life
To cope up with improving gas turbine inlet temperature towards a more sustainable power generation, manufacturing process of superalloys have experienced major developments. The operating temperature has been influenced by the melting point of nickel and cobalt in the range of 1500°C. Generally, the free electrons in a metal dictates the thermal conductivity of alloys while the same is influenced by lattice vibration in nonmetals. Easier heat dissipation and cooling of the blades need higher thermal conductivity. If the temperature distribution inside the blade is uniform, the thermal stress is reduced during operation. Various chemical compositions of superalloys greatly improve the mechanical properties. But they can cause point defects in lattice which further reduces the thermal conductivity. Figure 2 shows the trends of increasing conductivity for all type of casting samples with increasing range of temperature [10]. Such higher conductivity may result in permanent deformation in the blade profile due to higher thermal stress which an affect the life and efficiency of turbine blades. Thus, a thermal barrier coating can be fruitful to have a control over conductivity for sustainable power generation.

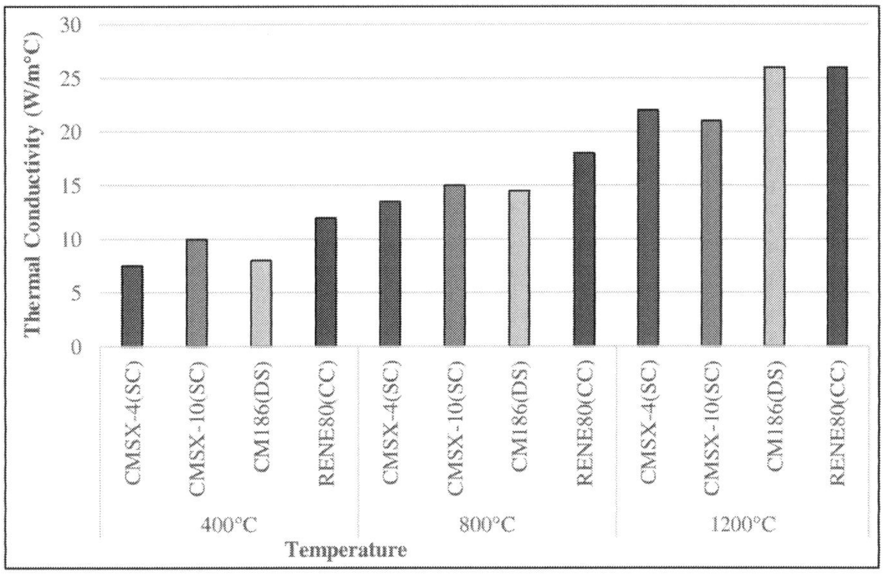

Figure 2. Thermal conductivity as a function of temperature for nickel-based superalloys [10]

3. Thermal Barrier Coating techniques

For protecting the gas turbine blades and associated components, a ceramic layer is plasma-sprayed over a metallic bond-coat layer which acts as a thermal barrier. TBCs primarily consists of a top coating of ceramic material that helps to block the flames to reduce heat transfer. Under the top coating there is a metal matrix called bond coating which works as a binder between top coating and the base material (Fig. 3). Primary component of the top coating is yttria-stabilized zirconia (YSZ), that is useful for protecting and insulating metallic components of gas-turbines especially in aircrafts. Bond coating consists of a layer of MCrAlY alloy that helps to prevent penetration of flame towards base material of superalloy. It is also relieves the thermal stress developed and improves corrosion resistance. Thermally grown oxide (TGO) forms when oxygen penetrates from the top coating and get mixed with Al component of the bond coating [11]. So, the choice of coating deposition methods must consider the high temperature properties like the heat resistance and formation of TGO.

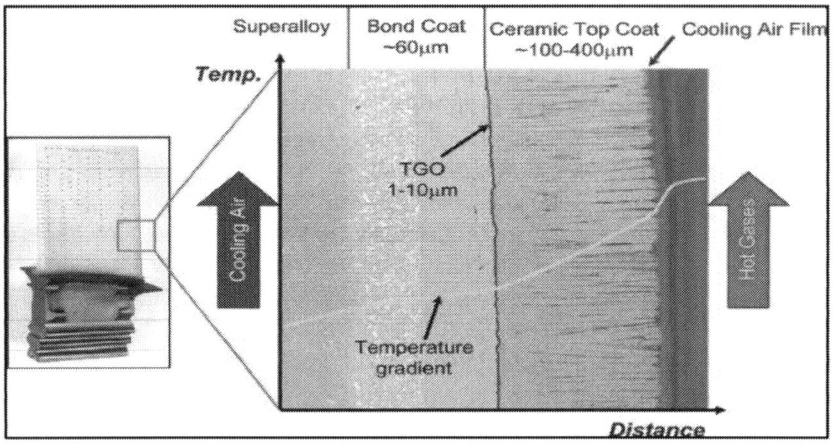

Figure 3. Thermal barrier coating under electron microscope [11]

The mechanism and behaviour of failure for a TBC Coatings in hot and harsh environment is still not completely explored in literature. The microstructure of TBC affects oxidation resistance and coating durability which are some important performances parameters with respect to the sustainable performance of turbine blades. However, microstructure is dependent on materials used along with the methods adopted. It is crucial to analyse various deposition techniques of TBCs depending on the material and temperature used. There are three types of TBC methods: Chemical vapor deposition (CVD), Atmospheric plasma spray (APS) and Electron beam physical vapor deposition (EB-PVD). Moreover, the choice of deposition technique depends on the parameters like surface roughness requirements, coating composition and thickness as well as orientation of cooling holes. APS technique uses high-speed plasma jet consisting of He, Air and N_2 at ultra-high temperature for depositing the coating powder. It prevents the regeneration and early delamination of the TBC [12]. CVD is used for coating thin films of bond layer with a slow rate of deposition [13]. On the other hand, the EB-PVD method employs an electron beam in vacuum chamber to evaporate YSZ ingot, which is deposited on a preheated substrate with a thin layer of Thermally Grown Oxide (TGO). Generally, the top coatings are made using EB-PVD and APS techniques while CVD method improves the coting performance in a substantial manner. The use of TBCs can reduce the temperature as much as 300°C at the metallic surface, which improves the durability of the engine components and enhance performance. The development in aviation industries always demand higher efficiency and durability towards better sustainability in power production necessitating betterment in coating technologies.

4. Parametric effect of TBC on sustainability of gas turbine blades

4.1 Effect of change in thermal conductivity

The most important advantages of TBC coated superalloys are the lesser thermal conductivity and mechanical stability at high temperature. Presence of pores inside the material is another useful attribute towards improved blade performance. Depending on the temperature, the thermal conductivity of polycrystalline materials used as TBCs has a range of 1-30 W/m-K [13]. Among the oxides, zirconia shows very small thermal conductivity in the range of 2 W/m-K. For the YSZ with increase of yttria in zirconia, the thermal conductivity decreases by more than 50% [12]. In this way Zirconia and similar rare earths elements increase the durability of TBCs. This in turn increases the acceptable temperature of gas turbine entry for more improvement in efficiency. A 250 μm thick coating with 4% rare earth metals like Gadolinia, Ytterbia etc. possess an approximate thermal conductivity of 0.88-1.02 W/m-K at 500°C [14]. As discussed earlier, such conductivity range acts as the biggest thermal barrier.

4.2 Effect of Coating Thickness

During the processing of TBC an adjustment must be made between the relative thickness of the top and bond Coating. The interfacial stress behaviour with the change of top coating thickness has been shown in Figure 4. It has been reported in literature that performance of thermal barrier coating improves by increasing the thickness of coating [15]. Stress generated between top coat and bond coat is also reduced with the increasing thickness. However, thick coatings cannot be applied in case of high-pressure turbine blades and nozzles due to strong centrifugal forces [16].

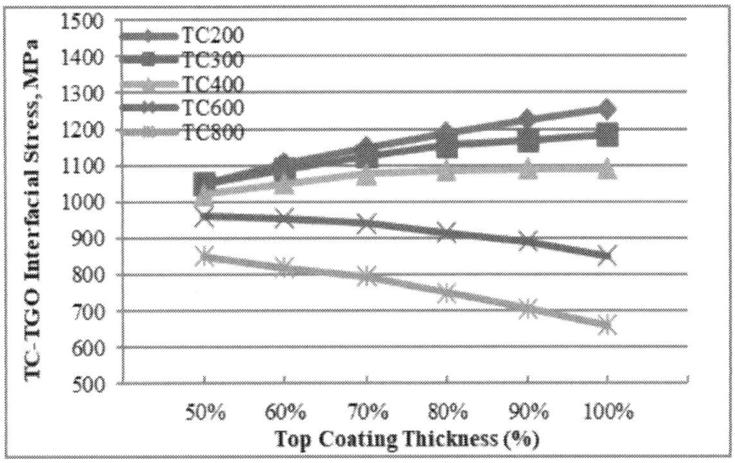

Figure 4. Stress variation with TBC coating thickness

4.3 Effect of Cracks and Pores present in the coatings

Different types and orientations of cracks have varying effects on the performance of TBC. The presence of cracks in the top coating can lead to the failure of TBC coating. However, durability can be sometimes increased by artificially inserting vertical cracks during the coating process. Presence of these cracks reduce the chances of thermal fatigue. As the vertical crack approaches TGO, the stress intensity factor also decreases as it reduces stress concentration. Higher number of pores reduces the heat transfer to the base material by decreasing thermal conductivity of top coating. It also decreases interfacial thermal stress which is useful to improve fatigue life of TBC. But excessive pores can fail the whole coating as it can reduce the adhesion between the top and bond coating.

IOP Conf. Series: Materials Science and Engineering 1187 (2021) 012034 doi:10.1088/1757-899X/1187/1/012034

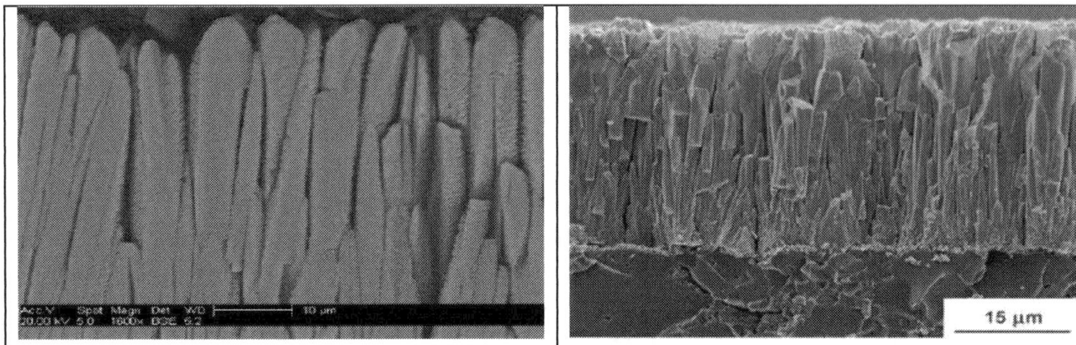

(a) (b)

Figure 5. Microstructure of TBC deposited by (a) EB-PVD method [16], (b) CVD method [17]

4.4 Effect of coating process on mechanical properties of TBC

Figure 5 shows a microstructure comparison from Scanning Electron Microscopy (SEM) of YSZ coating cross-section, deposited using EB-PVD and laser CVD at 230 µm/h [16]. Columnar microstructures are formed in the EB-PVD process which reduces the planar modulus. Strain generation is common in this method due to the difference of thermal expansion coefficients between the base material and the ceramic top coating. CVD method is ideal for coating thin films due to very slow deposition rates. Process parameters can be optimized for non-oxides easily but for oxides appropriate selection of CVD chamber is necessary to control the rate of deposition [17, 18].

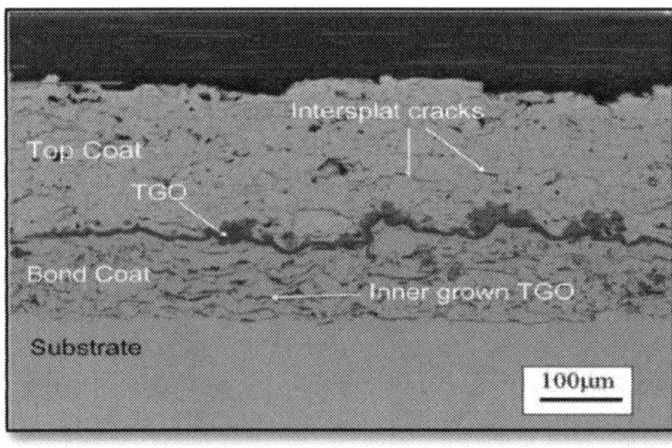

Figure 6. SEM image of a heat-treated APS TBC showing the irregular grain structure [20]

For coatings of ceramic and metals of high melting point, APS based thermal spraying technique is used. In this process sprayed materials at a high speed enters the substrate resulting in high density and adhesion strength of the coating. In this process Al_2O_3 is formed during preheating, deposition and subsequent heat treatments. It plays an important role to prevent the premature delamination of TBCs. In APS process the powder forms a layered structure stacked over the superalloys. Such coatings are characterized by several pores which can reduce the thermal conductivity with temperature to as low as 0.8-1.1 W/m-K. During deposition various parameters like nozzle diameter, plasma gas power, grain size, base material temperature and roughness play an important role over final output. In Figure 6, a representative microstructure of a heat-treated APS sample shows that the topcoat is uniform with some small cracks which are not large enough to cause failure. The TGO is also observed at the interface which has two layers as discussed earlier.

5. Conclusion

A review on the performance of thermal barrier coating has been presented in the perspective of sustainable operating life. Several methods of thermal barrier coating have been analysed to assess their application areas. The study shows that TBC coating is required for protecting gas turbine blades from prolonged high-temperature operation, necessary for efficient power generation. The thermal conductivity also reduces due to the presence of pores generated in the coating process. At the same time, excessive pores deteriorate other mechanical properties. When applying on existing materials or preparing a novel coating, it is essential to select the appropriate coating process. Therefore, the development of thermal barrier coating is crucial in improving the durability of gas turbine blades leading to sustainable power generation.

References

[1] King D, Inderwildi O and Carey C 2009 Advanced aerospace materials: Past present and future *Aviat. Space. Environ.* **3** pp 22–27

[2] Yoo Y S 2014 Nickel base superalloys *Korean J. Chem. Eng.* **17** pp 1–9

[3] INCO Alloy IN-738 Technical Data The International Nickel Company Inc: New York NY USA Available online: https://www.nickelinstitute.org/media/1709/in_738alloy_preliminarydata_497_.pdf

[4] Wang Q, Yang M, Song X L, Jia J and Xiang Z 2016 Rationalization of creep data of creep-resistant steels on the basis of the new power law creep equation *Met. Mater. Trans. A* **47** pp 3479–87

[5] Kim W G, Yin S N, Kim S H, Ryu W S, Lee C B and Kim S J 2009 A numerical approach to determine creep constants for Time-Temperature Parametric methods *Met. Mater. Int.* **15** pp 559–564

[6] Abdallah Z, Perkins K and Arnold C 2018 Creep lifting models and techniques *Creep* p 115 IntechOpen Available from: https://www.intechopen.com/chapters/58367.

[7] Abdallah Z, Gray V, Whittaker M and Perkins K A 2014 A Critical analysis of the conventionally employed creep lifing methods *Materials* **7(5)** pp 3371–98

[8] Kim S C, Shim J H, Jung W S and Choi Y S 2018 Short-term creep data based long-term creep life predictability for grade 92 steels and its microstructural basis *Met. Mater. Int.* **25** pp 713–722

[9] Whittaker M T and Harrison W 2014 Evolution of Wilshire equations for creep life prediction *Mater. High Temp.* **31(3)** pp 233–238

[10] Zielińska M, Yavorska M, Porêba M and Sieniawski J 2010 Thermal properties of cast nickel-based superalloys *Arch. Mater. Sci. Eng.* **44** pp 35–38

[11] Terada Y, Ohkubo K, Miura S, Sanchez J M and Mohri T 2003 Thermal conductivity and thermal expansion of Ir3X (X = Ti Zr Hf V Nb Ta) compounds for high-temperature applications *Mater. Chem. Phys.* **80** pp 385–390

[12] Wang L, Yang J, Ni J, Liu C, Zhong X, Shao F, Zhao H, Tao S and Wang Y 2016 Influence of cracks in APS-TBCs on stress around TGO during thermal cycling: A numerical simulation study *Surf. Coat. Technol.* **285** pp 98–112

[13] Klemens P and Gell M 1998 Thermal conductivity of thermal barrier coatings *Mater Sci. Eng. A* **245** pp 143–149

[14] Craig M, Ndamka N, Wellman R and Nicholls J 2015 CMAS degradation of EB-PVD TBCs: The effect of basicity *Surf. Coat. Technol.* **270** pp 145-153

[15] Padture N P, Gell M and Jordan E H. 2002 Thermal barrier coatings for gas-turbine engine applications *Science* **296** pp 280–284

[16] Nicholls J, Lawson K, Johnstone A and Rickerby D 2002 Methods to reduce the thermal conductivity of EB-PVD TBCs *Surf. Coat. Technol.* **151** pp 383–391

[17] Goto T 2005 Thermal barrier coatings deposited by laser CVD *Surf. Coat. Technol.* **198** pp 367–371

[18] Clarke D 2003 Materials selection guidelines for low thermal conductivity thermal barrier coatings *Surf. Coat. Technol.* **163** pp 67–74

[19] Abedi H Salehi M Shafyei A 2018 Microstructural, mechanical and thermal shock properties of triple-layer TBCs with different thicknesses of bond coat and ceramic top coat deposited onto polyimide matrix composite *Ceram. Int.* **44** pp 6212–22
[20] Li C, Jacques S D M, Chen Y, Xiao P, Beale A M, Michiel M. di, Markossan N, Nylen P and Cernik R.J. 2016 Precise strain profile measurement as a function of depth in thermal barrier coatings using high energy synchrotron X-rays *Scripta Materialia* **113** pp 122-126

Comparison of D-Vine and R-Vine Techniques for Virtual Network Embedding Problem

Preksha Satish [1], Deeksha Lingraj [2], Anjan Kumar S[3], Keerthan Kumar T G [4]

[1][2][3] UG Scholar, [4] Assistant Professor
Department of Information Science and Engineering, Siddaganga Institute of Technology, Tumakuru
{[1]prekshas573, [2]deekshalingraj10, [3]anjansa35, keerthanswamy[4]}@gmail.com

Abstract. In Network virtualization, Virtual Network Embedding(VNE) is the process of mapping virtual nodes and links of a virtual network request(VNR) on a substrate network to fulfill the demands of the request. Embedding virtual network requests helps in achieving network virtualization efficiently. This paper presents Vineyard, a set of VN embedding algorithms namely D-Vine and R-Vine, to introduce a finer correlation between the node mapping and link mapping phases. Deterministic Virtual Network Embedding(D-Vine) algorithm is used to embed virtual nodes on to substrate network based on the capacity constraint when it is not satisfied the system takes the Randomised Virtual Network Embedding(R-Vine) algorithm to map the nodes based on location constraint. Subsequently, after the nodes are embedded the link mapping phase is done based on the distance constraint. A window-based virtual embedding algorithm(W-Vine) is also introduced to evaluate the effect of lookahead in Virtual Network Embedding. After the mapping of multiple virtual network requests, analysis is done to compare the node and CPU utilization in both the algorithms and the variations are conserved.

1. Introduction

In Cloud Computing, Network virtualization is a trending technology for the internet of the future. This situation is dealt with by using internet ossification where necessary and innovative solutions are used for sharing the physical network resources with proper accountability[1][2][13]. During Network Virtualization, the first object that is studied is the virtual network. The active and passive network components are mostly distributed over the substrate network. A virtual topology is formed by connecting all virtual nodes through virtual links[2][5]. It is the responsibility of the InP to operate and maintain the physical infrastructure whereas the SP is responsible for aggregating the resources from INPs to construct virtual networks and end-to-end services are provided to the users.[9] Introducing network virtualization decouples the management and service roles of SP by recognizing three distinct components: A Virtual Network Provider (VnP) gathers virtual resources from an InP or several InPs, Virtual Network Operators[10] (VNOs) implement, control, and operate the VNs based on the requirements from the SP, Service Providers are freed from management and can focus on business by using the VNs for providing custom services[2][3][4].

In the environment of network virtualization, multiple service providers can form different virtual networks to provide end-to-end services to end-users by subletting shared resources from one or more infrastructure providers without committing to physical infrastructure[1][4]. Substrate nodes and Virtual nodes exist within the NVE[7]. The virtual nodes are embedded on the different substrate nodes and are mutually connected by the lane in the substrate networks that correspond to the virtual links, this is called the Virtual Network embedding problem[2][6]. There is a possibility that more virtual nodes might share the same substrate node resources, ensuring efficient request mapping will increase substrate node utilization. For a successful and efficient mapping, the embedding problem for Virtual Network Embedding must be considered an important challenge[5][6][11][12]. After the successful mapping of virtual nodes, it is necessary to check whether virtual nodes correspond to substrate nodes based on the location and capacity constraints[2][6]. The overall system aims to develop and allocate the resources to the VN that satisfies the conditions of the substrate network and perhaps increases the substrate node utilization and also scheduling of requests is done so that upcoming requests are set in a queue for the allocation purpose.

2. Literature Survey

Virtual network embedding is the procedure of embedding virtual network requests onto substrate networks to satisfy the virtual network's requirements. It is common for virtual network embedding algorithms to lack coordination between the node and link mapping and delay for packet transfer. In this context, various research papers focused on existing virtual network embedding algorithms and how to address the shortcomings of existing algorithms. During the study phase, however, it was found that a wide range of embedding algorithms was based on a variety of factors, including resource demands, cost considerations, and the provider's revenue.

In the paper[2], the author describes how the mapping of virtual nodes on substrate nodes is done using exact VNE algorithms. When used in small-scale scenarios, it provides near-optimal embedding of each Virtual Network request. As a consequence, these algorithms cannot be applied in continuous-time scenarios due to their huge computational complexity.

Paper[4] Author describes the mapping of virtual nodes using the Subgraph isomerism detection problem. An isomeric subgraph is found that accomplishes the request within the substrate network and using a hop-limit constraint applies to substrate paths used to map the request. As a result, this leads to getting stuck in local optimum positions far from the actual optimum.

In paper[5], the author describes that the embedding of virtual network requests is done only when the virtual network requests are static. Virtual network demands and resources are matched with the substrate networks and embedding is done. Virtual requests change dynamically; efficient allocation of resources cannot be done when the system is static.

In paper[6], Node mapping and link mapping are done using the uncoordinated VNE technique MCMCF provides a multipath routing solution for each virtual link based on optimal linear programming. The problem with this technique is that it drives up the price of single or multiple paths in the virtual link mapping phase, which leads to a low acceptance rate and low long-term revenue.

In paper[8], the Novel node ranking approach is used to rank all the substrate nodes and virtual nodes before embedding each virtual network request. Virtual link propagation delay is considered a link constraint for embedding. This algorithm uses an iterative approach to rank all the nodes that can be further improvised.

In paper[12], Linear programming derived from Mixed Integer programming is used to select the virtual nodes for embedding. The Markov Chain approach is used to select the candidate substrate nodes and complete the link configuration. This approach is only leveraged to splittable link mapping.

IOP Conf. Series: Materials Science and Engineering 1187 (2021) 012035 doi:10.1088/1757-899X/1187/1/012035

3. Experimental

A Virtual Network Embedding problem involves mapping virtual network requests, which have different conditions on virtual nodes and links, onto particular physical nodes and links within a substrate network that has countable resources. Figure 1 shows the flowchart of the proposed system. The vineyard algorithms, which contain Deterministic Virtual Network Embedding (D-Vine) and Randomized Virtual Network Embedding (R-Vine) are proposed, to contribute a better correlation between the node and link mapping. Virtual nodes are mapped against substrate nodes in a form that promotes the mapping of virtual links to physical paths during the succeeding phase. Augmented substrate graph construction is done where the cluster is formed for each virtual node, finding out the possibility of the number of substrate nodes for the embedding. Once the location constraints are satisfied for the substrate and the virtual nodes, only those nodes are moved ahead in the embedding process; the rest of the substrate nodes are considered for the next request. Deterministic and Randomized Rounding techniques are used for node and link mapping purposes where the capacity constraint is the foremost constraint that has to be satisfied to embed the node, deterministic algorithm mainly concentrates on mapping of virtual nodes to substrate nodes based on the capacity constraints those nodes which fail in the D-vine follows R-vine where location constraint is considered for embedding the node. Once the node embedding is done, the next part is to link the physical paths to the virtual paths. With the help of Euclidean distance, the distance is calculated between the two nodes in the substrate network and the virtual network. Based on the distance constraint, the link mapping is done. After the embedding process, the analysis is done for the virtual requests over the time obtained by plotting the graph and the observations are made.

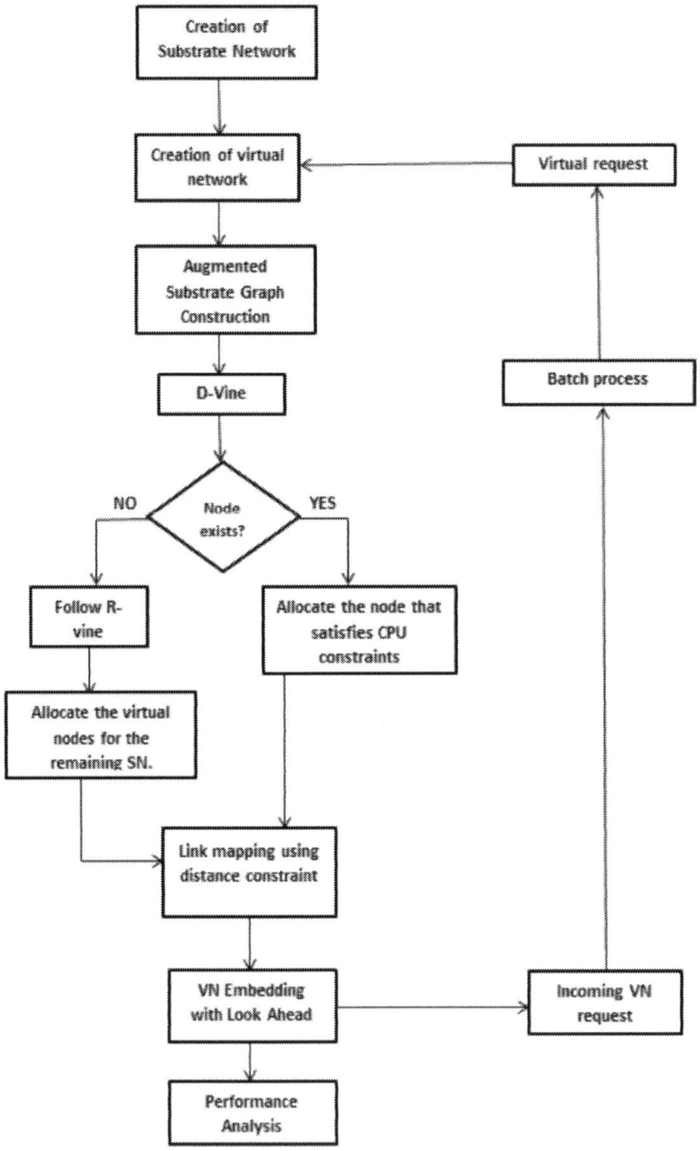

Figure 1. Flowchart of the proposed system

3.1 Construction of Substrate Network: The substrate network is implemented as a weighted bidirectional graph, $G^s = (N^s, E^s)$, N^s indicates substrate nodes set, and E^s indicates a substrate link set. On a graph, each and every substrate node $n^s \in N^s$ is allocated with CPU capacity weight values c(n^s) as well as location loc(n^s) on a general coordinate system. Every substrate link $e^s(i,j) \in E^S$ between two substrate nodes i and j are allocated with the distance d(x,y) value obtained by using the euclidean distance.

IOP Conf. Series: Materials Science and Engineering 1187 (2021) 012035 doi:10.1088/1757-899X/1187/1/012035

3.2 Construction of Virtual Network Request: The Virtual Network Requests implemented as a weighted bidirectional graph with each virtual network request having the form $G^v = (N^v, E^v)$, N^v indicates virtual nodes set, and E^v indicates a virtual link set. Each Virtual network request(VNR) is having a positive value D^v that indicates the distance of virtual node $n^v \in N^v$ will be embedded according to its favored location $loc(n^v)$. A landmark-based approach is used in our model to request virtual nodes so Infrastructure providers do not have to expose their network structures.

3.3 Augmented Substrate Graph Construction: As soon as the virtual network request arrives, each virtual node must choose the substrate nodes on which embedding can be done. With augment substrate graph construction, clustered nodes are created based on virtual node location constraints. The cluster is formed if the substrate node location x is less than the location of the virtual node, otherwise, the particular substrate node is rejected. Prior to actual embedding, augmented substrate graph construction focuses on forming the cluster that includes substrate nodes for each virtual node.

$$x\_loc(N^S) < x\_loc(n^v) \in N^V \qquad (1)$$

According to *equation(1),* the x coordinate location of the substrate node should be lesser than the x coordinate location of the virtual node.

3.4 D-Vine and R-Vine for embedding: The D-Vine algorithm is implemented once the augmented substrate graph construction is done. As inputs, the algorithms take the augmented substrate graph and the virtual network request. Using the D-Vine algorithm, we check if the virtual node is possible to embed on the substrate node based on capacity constraints if the capacity of the substrate node is lesser than the virtual node, the embedding is not possible, or else if the capacity of the substrate node is greater than that of the virtual node, the difference is calculated and the node with the minimal residual capacity is allocated for the virtual node. Those nodes which are not embedded still undergo the R-Vine technique and allocation is done based on the location constraints. After the residual capacity is calculated, there are chances that a single substrate node can be embedded on the virtual node, in that case, D-Vine follows to consider the substrate node which is having minimum residual capacity, so that unique embedding of the virtual node on to the substrate node property is conserved. Once the node embedding is completed, the system follows in embedding the links. The distance between the two nodes is calculated by using the euclidean distance. Once the distance is calculated between all the nodes of substrate network and virtual network, then link mapping is done based on the distance constraint, i.e if the distance between the 2 nodes of the virtual network request(VNR) is lesser than the substrate network then the splittable mapping is done, also the virtual link mapping is started in the descending order of the distance value because of the condition that link with the highest distance is difficult to map. If the distance of the virtual link is more than that of the distance of the substrate node, then the link mapping fails.

$$R_N(n^S) = c\,(n^S) - \sum_{\forall n^V \uparrow n^S} c\,(n^V) \qquad (2)$$

Where $x \uparrow y$ indicates that the virtual node name y is introduced on substrate node x

$$R_E\,(c) = min\ R_N(n^S) \qquad (3)$$

Equation (2) indicates that residual capacity is obtained by calculating the difference between the substrate and the virtual node.
Equation (3) shows that the node with minimum residual capacity is selected.

$$\sum_{w \in \Omega(m)} x_{mw} = 1,\ m\ N^{S'} \setminus N^{S} \quad (4)$$

$$\sum x_{mw} \leq 1,\ N^{S} \quad (5)$$

Equation (4) states that each virtual node should be embedded on a single substrate node *Equation (5*) indicates that a unique substrate node should be selected for embedding.

$$d(x,y) = \sum_{i=1}^{n} (y_i - x_i)^2 \quad (6)$$

$$distance(x,y) = \sqrt{d(x,y)} \quad (7)$$

Equation (6) and equation (7) are used to calculate the euclidean distance.

$$distance(x,y)(E^{v}) <= distance(x,y)(E^{S}) \quad (8)$$

Distance *Equation (8)* indicates that the distance of the virtual link should be less than or equal to the distance between 2 nodes of the substrate network.

3.5 VNE Embedding with lookahead: Virtual requests don't arrive in time and there may be a clash in the incoming multiple virtual requests. The proposed system handles the multiple virtual requests in an efficient way, once the first virtual request starts processing and during the execution of the request, if other requests start arriving then automatically the system calculates the maximum waiting period for the particular request and is stored in the batch process, In order to select the virtual requests out of the queue, a window period mechanism is used where only those requests which are having the maximum waiting period less than window period are selected for execution, if there are a number of requests satisfying the time constraints, then the requests are sent for execution using first come first serve basis technique, in this way all the virtual requests are handled in an efficient way.

$$G^{v} = (N^{v}, E^{v})(t_w) < (W_P) \quad (9)$$

t_w= waiting period W_P =window_period

Equation (9) indicates that the virtual network request waiting period must be less than the window period.

4. Results

Figure 2 shows the analysis of Node utilization using R-Vine and D-Vine technique, D-vine algorithm mainly concerned with the CPU resource utilization, only those nodes which satisfy the capacity constraint is allocated with the nodes, hence the percentage of allocation of nodes using D-vine is lesser when compared to the R-Vine technique. R-Vine technique follows checking of location constraint, therefore the possibility of allocating the node using R-Vine is more with respect to the D-vine, the above graph shows the variation in the percentage of node allocation using R-Vine and D-Vine with the number of substrate nodes, From the above graph, it is indicating that node utilization in R-vine is more when compared with the D-vine and through the variation, one can note down the observations that R-vine is independent of the capacity constraints, only D-vine considers capacity constraints, based on that allocation is done.

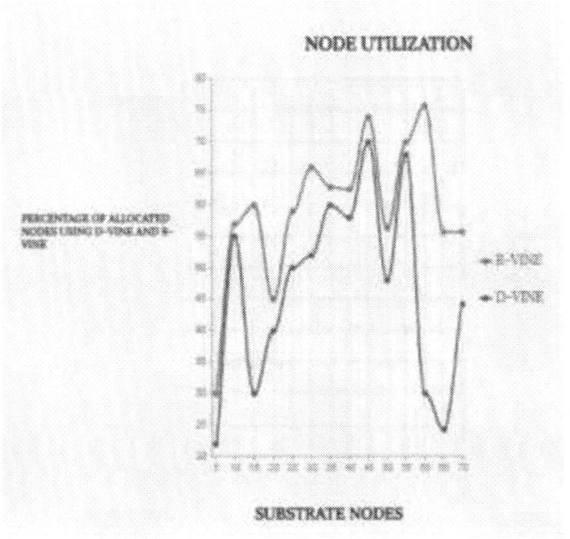

Figure 2. Node utilization analysis

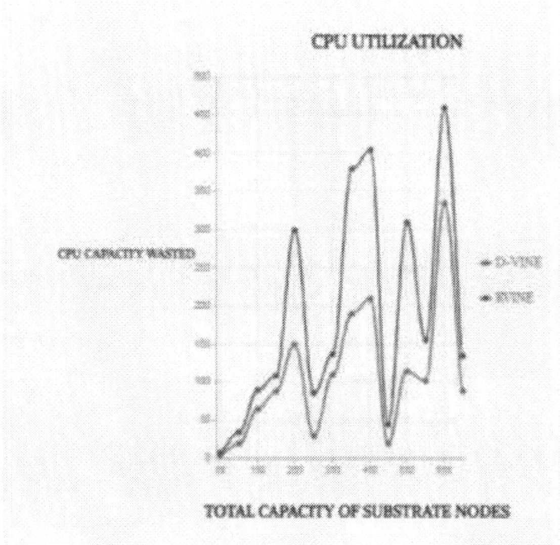

Figure 3. CPU utilization analysis

Figure 3 shows the CPU utilization using R-vine and D-vine. The analysis is done based on finding the CPU capacity wasted after the execution of both D-vine and R-vine. It is observed that D-vine focuses on CPU resource utilization and it allows for the embedding only if the nodes satisfy the capacity constraints. The total substrate node capacity is compared with some of the capacity of the nodes embedded using D-vine and R-Vine, the analysis shows that in D-vine there is lesser wastage of CPU capacity compared with R-vine, Therefore D-vine is more efficient than R-vine in CPU utilization process.

5. Conclusion

In the process of making network virtualization, an intrinsic entity of the upcoming generation of internet design, systematic and pragmatic algorithms for embedding the virtual networks are very much important. In this paper, the algorithms that are proposed perform efficient embedding and establish a step ahead correlation between the node and link mapping juncture. The correlation notably hikes the solution stretch and the endowment of heuristic algorithms. We started with setting up the network and then used the algorithms to accept multiple virtual network requests and embed them on the substrate networks considering the residual capacity, location, and distance constraints. A window-based mechanism is used to store the requests in a batch system and allow the requests which satisfy the time constraints one after the other for execution based on a first come first serve basis. Both algorithms are compared with respect to the node and CPU utilization and the analysis is drawn. Considering the node utilization inspection, it is observed that R-Vine performance is better than D-vine because of the fact that R-vine only focuses on the location constraint, and allocation of the node is done irrespective of the capacity of the nodes. With the help of CPU utilization analysis, it is observed that D-vine is more efficient than R-vine because the CPU capacity wasted is less when compared with the R-vine and it is important that lesser the CPU wastage more is the efficient binding of nodes and that increases resource utilization and the overall performance of the system is increased.

References

[1] Fischer, A.; Botero, J.F.; Till Beck, M.; de Meer, H.; Hesselbach, X., "Virtual Network Embedding: A Survey," in Communications Surveys & Tutorials, IEEE, vol.15, no.4, pp.1888-1906, Fourth Quarter 2013. DOI: 10.1109/SURV.2013.013013.00155.

[2] M. Chowdhury, M. Rahman and R. Boutaba, "Vineyard: Virtual network embedding algorithms with coordinated node and link mapping", IEEE/ACM Trans. Netw., vol. 20, no. 1, pp. 206-219, Feb. 2012.

[3] N. M. M. K. Chowdhury, "Network virtualization: State of the art and research challenges", IEEE Commun. Mag., vol. 47, no. 7, pp. 20-26, Jul. 2009.

[4] T. Anderson, L. Peterson, S. Shenker, and J. Turner, "Overcoming the Internet impasse through virtualization," Computer, vol. 38, no. 4, pp.34–41, 2005.

[5] J. Turner and D. Taylor, "Diversifying the internet," in Proc. IEEE GLOBECOM, 2005, vol. 2, pp. 755-760

[6] N. Feamster, L. Gao, and J. Rexford, "How to lease the Internet in your spare time," ACM SIGCOMM Computer Communication Review, vol. 37, no. 1, pp. 61–64, 2007

[7] J. Lu and J. Turner, "Efficient mapping of virtual networks onto a shared substrate," Washington University, Tech. Rep. WUCSE-2006-35, 2006.

[8] Cheng, X.; Su, S.; Zhang, Z.; Wang, H.; Yang, F.; Luo, Y.; Wang, J.,"Virtual network embedding through topology-aware node ranking," SIGCOMM Computer Communication Review, vol. 41, pp 38–47, April 2011. DOI: 10.1145/19711162.1971168.

[9] Even, S.; Itai, A.; Shamir, A., "On the complexity of time table and multi-commodity flow problems," Foundations of Computer Science, 16th Annual Symposium on, pp.184-193, 13-15, October 1975. DOI: 10.1109/SFCS.1975.21.

[10] K. Tutschku, T. Zinner, A. Nakao, and P. Tran, "Network virtualization: Implementation steps towards the future internet", Proc. Workshop on Overlay and Network Virtualization at KiVS, march 2009.

[11] A. Haider, R. Potter and A. Nakao, "Challenges in resource allocation in network virtualization", 20th ITC Specialist Seminar, vol. 18, pp. 20, 2009.

[12] Lingnan Gao, George N Rouskas, "Virtual Network Reconfiguration with Load Balancing Migration Cost Considerations", IEEE Infocom 2018.

[13] Keerthan Kumar T G, H. K. Virupakshaiah and Nanda K V, "Ensuring an online Chat Mechanism with accountability to sharing the non-downloadable file from the Cloud," 2016 2nd International Conference on Applied and Theoretical Computing and Communication Technology (iCATccT), 2016, pp. 718-721, DOI: 10.1109/ICATCCT.2016.7912093

IOP Conf. Series: Materials Science and Engineering 1187 (2021) 012036 doi:10.1088/1757-899X/1187/1/012036

Power Allocation Scheme for Non-Orthogonal Multiple Access System

R. Seetharaman [1*], M. Tharun[1], S. Gayathri[2], K. Anandan[3]

[1]Department of ECE, CEG Campus, Anna University, Chennai-600025, India

[2]Department of EEE, CEG Campus, Anna University, Chennai-600025, India

[3]PS Division of E & E, TUV SUD South Asia Pvt Ltd, Chennai-600041, India

*Corresponding author email: rseetharamanece@gmail.com, rseetharamanece@annauniv.edu

Abstract. Since there exists a principal power-defer tradeoff in the NOMA framework, power proficiency additionally becomes basic for Ultra-Reliable Low Latency Communication (URLLC), particularly where countless gadgets are battery-fueled. Joining these contemplations, we reproduce Dynamic Power Allocation (DPA) for power area non-symmetrical different access (PD-NOMA) with client versatility. The exhibition of the DPA is contrasted and Static Power Allocation under client versatility. The characterization of clients in (NOMA) depends on power, while it depends on schedule, recurrence and code in Orthogonal Multiple Access (OMA). Since there exists an essential force defer tradeoff in the NOMA framework, power effectiveness likewise becomes basic for Ultra-Reliable Low Latency Communication (URLLC), particularly where countless gadgets are battery-controlled. Consolidating these contemplations, we reenact Dynamic Power Allocation (DPA) for power area non-symmetrical numerous entrance (PD-NOMA) with client portability. Force effectiveness becomes basic particularly for little and cumbersome battery-controlled IoT gadgets. Further, adaptability is additionally critical to speak with different machine-type gadgets just as human clients while meeting an assortment of nature of administration (QoS) prerequisites. The presentation of the DPA is contrasted and Static Power Allocation under client portability.

1. Introduction

The fifth era is the approaching unrest of portable innovation which has an immense exhibit of creative components like accelerate to 10 Gigabits each second, multiple times more gadgets, Faster reaction time, Virtually '0' inertness, Ubiquitous network, bigger information volume per unit region (high framework otherworldly proficiency). One of the basic methods utilized here is Power Domain multiplexing. In remote correspondence data transmission and time are the two characterizing provisions of a channel. Clients can approach channels through numerous entrance procedures. There are two sorts of different access methods. One is OMA and the other one is NOMA. In the OMA procedure, frameworks are symmetrical to one another. A portion of these procedures are TDMA, FDMA, CDMA, OFDMA. Because of high traffic volume, OFDMA couldn't completely fulfill the ghastly productivity and force use. By considering those challenges the NOMA method has been proposed. In this procedure, the individual client is permitted to involve the entire range and multiplexed from each other in the force space. With SIC the most vulnerable sign is extricated by eliminating the most grounded signal. NOMA offers 30% more throughput than OFDMA. It gives higher ghastly productivity because of various clients on a similar recurrence and time. In this venture, the NOMA strategy and dynamic force allotment of clients are clarified unmistakably. The essential standard of NOMA in uplink and downlink is introduced and the proposed NOMA with numerous receiving wires is executed [1]. The NOMA is introduced and power assignment factor is incorporated and the inclusion district for fixed BS is determined [2]. A unique examination of NOMA and the Rayleigh blurring coefficients are determined and another assessment method is proposed with NOMA ghostly productivity [3]. In a different access innovation for subcarrier and a clever utility, capacity is presented, and the non-convexity issue is arranged utilizing a two-venture iterative strategy [4]. The Non-symmetrical various access (NOMA) for 5G organizations is addressed and range proficiency (EE-SE) of NOMA and OFDMA strategies are thought about, the recreated the outcomes concerning energy and range effectiveness [5]. A hypothetical examination of ghastly productivity of shut type of NOMA in Nagakami is proposed and a blurring file is presented [6]. Two distinctive force distribution plans are presented, one depends on channel data, and the other procedure on pre-characterized QoS for every client is presented and the expected addition of NOMA is dissected [7]. In the Non-symmetrical different access (NOMA) that has an ability for a 5G framework is proposed and another helpful NOMA is presented and broke down the blackout likelihood and variety are plotted. Existing NOMA for 5G is contemplated and the current test is recognized as energy utilization and proposed a clever energy-proficient NOMA. Here another sign-based hypothesis is presented in NOMA and a greatly simultaneous NOMA(MC-NOMA) and the channel limit of multi-client MIMO is examined and reproductions are performed.

In power-area multiplexing, various clients are assigned diverse force coefficients as indicated by their divert conditions to accomplish a high framework execution. Specifically, different clients' data signals are superimposed on the transmitter side. On the beneficiary side, progressive obstruction undoing (SIC) is applied for translating the signs individually until the ideal client's sign is acquired, giving a decent compromise between the throughput of the framework and the client's reasonableness. In code-area multiplexing, various clients are assigned various codes and multiplexed throughout a similar time-recurrence asset. Despite the fact that code-space multiplexing can possibly improve unearthly proficiency, it requires a high transmission transfer speed and isn't effectively pertinent to the current frameworks. Then again, power-area multiplexing has a straightforward execution as impressive changes are not needed on the current organizations. Likewise, it doesn't need extra data transmission to work on ghastly productivity.

2. System model

Non-symmetrical different access is arising as a promising innovation for 5G with its prevalent phantom productivity and client decency [8]. We are considering a downlink transmission station (BS) for the two users. User 1 is far/weak as he is far away from the transmitting BS. User 2 is the near/strong users. Let d_1 and d_2 denote their distances from the Base Station. The BS has two distinct messages x_1 to user 1 (far user), and x_2 to user 2 (near user). α_1 and α_2 are the power allocation factors for the far and the near user respectively ($\alpha_1 + \alpha_2 = 1$). In NOMA, to advance client decency, more force is given to the far client and less capacity to close client. The objective is to explain the advantages of NOMA over symmetrical numerous entrance (OMA, for example, OFDMA embraced by Long-Term Evolution (LTE) [9].

To get the outage probability, we have set the same target rate for both near and far users. It very well may be considered that to be limit is expanded, the blackout likelihood likewise increments on the grounds that the higher the edge esteem, the more will be the odds of the rate falling underneath the edge and more will be blackout likelihood. The correlation between blackout likelihood and SNR of dynamic force portion, the blackout likelihood diminished with an increment in SNR.

In the downlink, the base station communicates the sign for the close to client and the far client.

In our proposed model instate the length of the information. The client information for the all over clients are produced utilizing irregular number age. The transmission is done by BPSK adjustment. The SNR is fixed for information length. The superposition coding is executed in the transmitter to discover the capacity achievable rate for both far user and near user [10]. The AGWN is added for both all over clients. The demodulation is applied to recover the all over client. Gotten client information is again tweaked as x_1 and is changed into a BPSK signal. The assessed yield is demodulated to acquire x_2. The outcomes are investigated by looking at got information and bit blunder rate.

3. NOMA system with Rayleigh Fading Channel

The advancement of remote interchanges and the web of things (IoT), non-symmetrical various access (NOMA) has been considered as one of the successful plans to meet the quickly developing client access necessities and the NOMA system is modeled with the Rayleigh fading channel [11]. The near and far users are fixed with the distance value. The power and path loss exponent values are assigned as per the model. Rayleigh fading coefficients are generated as h_1 and h_2. h_1 and h_2 are coefficients of standardized complex Gaussian irregular variable with zero mean and difference. A two-client downlink agreeable non-symmetrical different access (NOMA) short-parcel correspondence network is researched in level Rayleigh blurring channels, where a focal client helpfully transfers the signs expected to a phone edge client [12]. The force of the transmitter in the liners scale is estimated and framework transfer speed is determined. The commotion level is additionally set concerning transfer speed. The SNR esteems for the all over client are determined and the reachable objective rates are additionally registered. The blackout likelihood for the comparing objective rate is determined. The graphical outcomes are plotted for Transmitter power Vs Sum Rate as displayed in Figure 1. Far client has a lower limit (i.e., attainable information rate) than close to client, particularly when the communicate power is high. This is on the grounds that the far client performs direct unraveling without eliminating the close to client's information. However, close to client enjoys the benefit of eliminating far user's information utilizing progressive obstruction retraction.

For dynamic power allocation, our only aim is to maximize the sum rate. For sum-rate maximization (SR maximization) the following constraint must be satisfied. When the sum of rates of users 1 and user 2 is greater than or equal to the sum rate and when the channel coefficient of user 1 is greater than the channel coefficient of user 2 then lower power is allocated to the far user (i.e., user 1) and higher power is allocated to near user (i.e., user 2) and the splitting factor gets the values accordingly based on channel coefficients. Similarly, when the channel coefficient of user 2 is greater than the channel coefficient of user 1 then higher power is allocated to the far user (i.e., user 1) and lower power is allocated to the near user (i.e., user 2) and the splitting factor gets the values accordingly based on channel coefficients. For each iteration, the condition checks if the current sum of the individual rates of user 1 and user 2 is greater than or equal to the previously stored sum rate, then the sum rate will get updated. Therefore, at the end of the iteration, the sum rate variable will store the maximum sum rate for the corresponding transmit power.

For optimal power allocation along with sum-rate maximization, we should also ensure user fairness is maintained. For sum-rate maximization along with user fairness, the following constraints must be satisfied. The first condition is to maximize the sum-rate which implies the sum of rates of users 1 and 2 must be greater than or equal to the sum rate and the second condition mentioned is to ensure user fairness which implies that the individual rates of user 1 and user 2 must be greater than or equal to the minimum rate (1 bps/Hz). When these conditions are satisfied then the channel coefficients are checked. So, when the channel coefficient of user 1

IOP Conf. Series: Materials Science and Engineering 1187 (2021) 012036 doi:10.1088/1757-899X/1187/1/012036

is greater than the channel coefficient of user 2 then lower power is allocated to the far user (i.e., user 1), and higher power is allocated to near user (i.e., user 2) and the splitting factor gets the values accordingly based on channel coefficients. Similarly, when the channel coefficient of user 2 is greater than the channel coefficient of user 1 then higher power is allocated to far user (i.e., user 1) and lower power is allocated to the near user (i.e., user 2) and the splitting factor gets the values accordingly based on channel coefficients. Therefore, at the end of the iteration, the sum rate variable will store the maximum sum rate for the corresponding transmit power.

If any of the above conditions fail, let's say suppose if the sum-rate maximization condition fails, the maximum sum rate will be retained. But if the user fairness condition fails i.e., in some cases the individual rate may be less than 1 bps/Hz. This happens when transmit power is low. So, we will be trying two techniques to maximize the sum-rate only.

1. Dividing the power equally between two users.
2. Dividing the power based on channel condition, i.e., High transmitting power is given to the user with weak channel conditions, and low transmitting power is given to the user with strong channel conditions. Once both the rates fall below 1 bps/Hz, we focus only on maximizing the sum rate as we know that user fairness could not be achieved for very low transmit power. So, using the above two techniques we calculate the sum rate in both cases and adopt the technique that gives the maximum sum rate. It was observed that the maximum sum rate was achieved when power was divided equally between the two users.

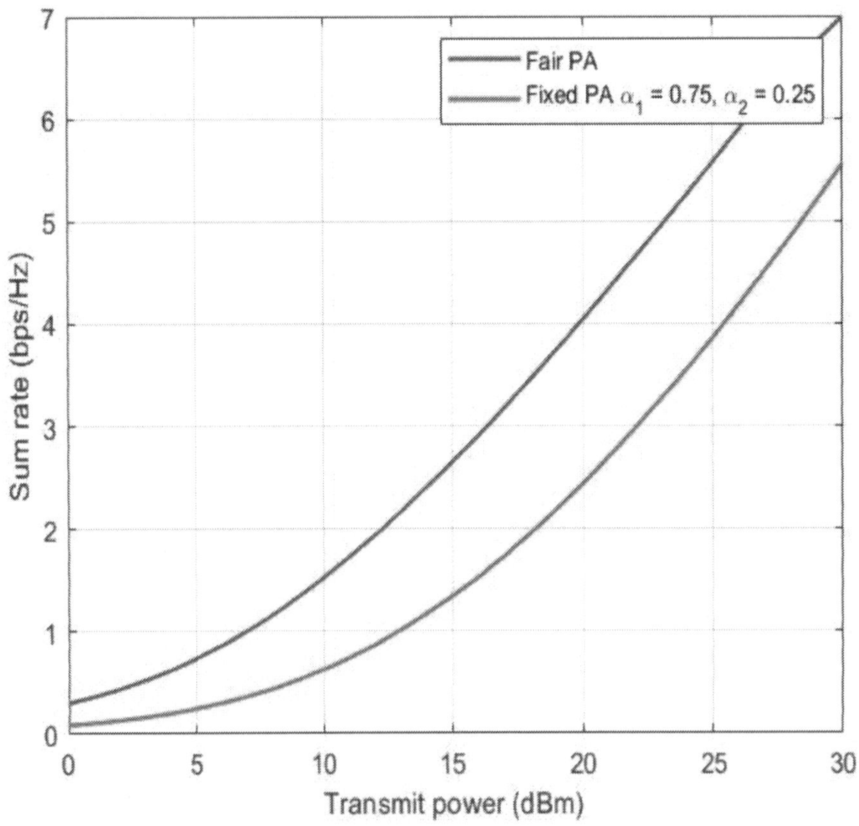

FFigure 1. Transmit Power vs Sum Rate characteristics

The progressive obstruction crossing out is defective at the numerical degree of useful investigation. Consequently, the proposed NOMA is thinking about the flawed and tracking down the debased execution. SIC is great if the close to client precisely translates the far client's information and deducts it from got information by SIC. At the point when it is flawed, the far client's information isn't totally taken out, and still, the buildup of far client's information is available. We can see that flawed SIC detrimentally affects the information pace of the solid client/the client who performs SIC. Attainable information rate corrupts as SIC blunder increments. The presentation examination of the downlink NOMA (DL-NOMA) framework in presence of progressive obstruction wiping out (SIC) blunder over Rayleigh blurring channels is dealt with productively [13].

4. Dynamic Power allocation in NOMA system

SIC is performed at the base station and the received signal at the BS in the uplink. The received signal at BS is represented as Equation (1).

$$y = h_1(P\beta_1 S_1)^{1/2} + h_2(P\beta_2 S_2)^{1/2} \tag{1}$$

Likewise, w is the AWGN with zero mean and fluctuation σ_2. Signal S_1 is decoded and the subsequent client signal S_2 is treated as commotion. Then, at that point the beneficiary takes away the decoded from the got signal. Force levels are distinguished as $P_1= P\beta_1$ and $P_2=P\beta_2$. The force designation coefficient is $\alpha = P_1/P_2$ and reaches. The BER for the close to (client 2) is lower when $\alpha_1<0.1(\alpha_2>0.9)$ than when $\alpha_1\approx0.8$ ($\alpha_2\approx0.2$). Anyway, would we be able to pick $\alpha_1<0.1$ and $\alpha_2>0.9$? The far (client 1) has extremely high BER in this system. The channel gain can be accomplished by multi-transporter transmission and increments with the number of subcarriers [14]. Assuming we need the two clients to reasonably profit from NOMA, the ideal system to work is $\alpha_1\in[0.5,1]$ and $\alpha_2\in[0,0.5]$. This implies an exceptionally huge force is designated to the close to (client 2). Consequently, he has an exceptionally low BER. As $\alpha_1\approx0$ and $\alpha_2\approx1$, the close to (client 2) information is ruling and henceforth, the far (client 1) cannot straightforwardly decipher his sign.

Our reasonable PA offers need to the frail/far client. That is, the force designation coefficients are determined to such an extent that the far client's objective rate is met. The proposed streamlining structure limits the drawn-out time-normal send power use while diminishing the lining delay and ensuring the base time-normal information rates [15]. Solely after gathering the objective pace of far client, all the excess accessible force is apportioned to the close to client. We can quickly see that decent PA is performing ineffectively and its blackout likelihood soaks to constantly when R*>1.5. All in all, the collector is consistently in blackout on the off chance that we utilize fixed PA with R*>1.5. This is on the grounds that, fixed PA neither adventures the immediate CSI, nor considers the objective rate necessities. In any case, our reasonable PA has lower blackout likelihood Bit Error Rate (BER) result for Binary Phase Shift Keying (BPSK) random power allocation technique and dynamic power allocation technique is shown in Figure 2 and Figure 3 we can observe that near user has slightly higher BER than far user. This is because the near user must decode both the far user's data and its own data correctly. Any error in decoding a far user's data or its own data will impact its BER. That is why near user

IOP Conf. Series: Materials Science and Engineering 1187 (2021) 012036 doi:10.1088/1757-899X/1187/1/012036

experiences higher BER than far user. we can observe that the random power allocation technique gave lower error rates. However, user fairness and maximum sum-rate were achieved in optimal power allocation technique and this wasn't present in random power allocation techniques. on the grounds that α_n and α_f and progressively changed dependent on track rate necessity and CSI.

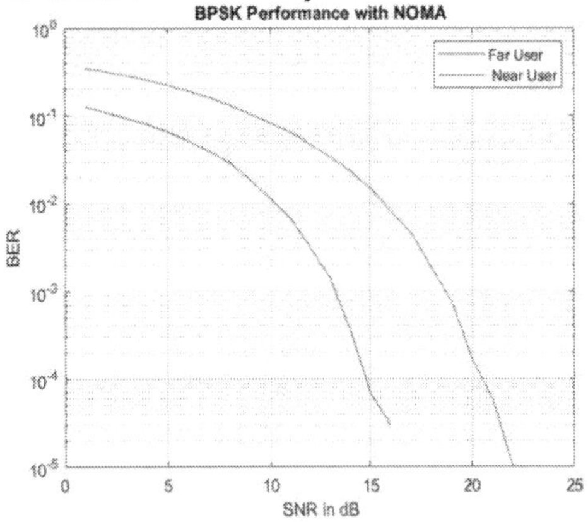

Figure 2. BPSK Random power allocation

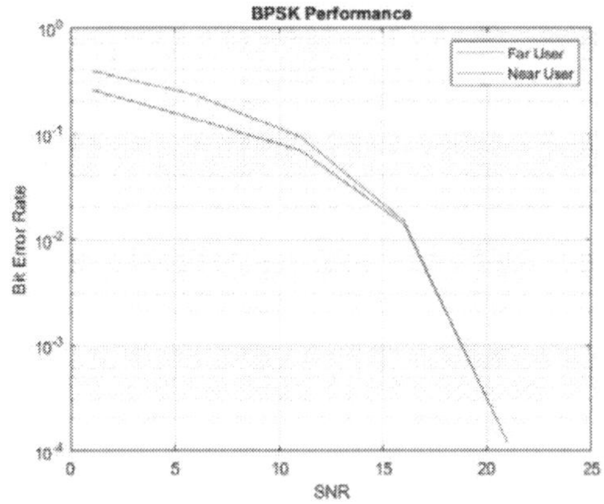

Figure 3. BPSK Dynamic power allocation

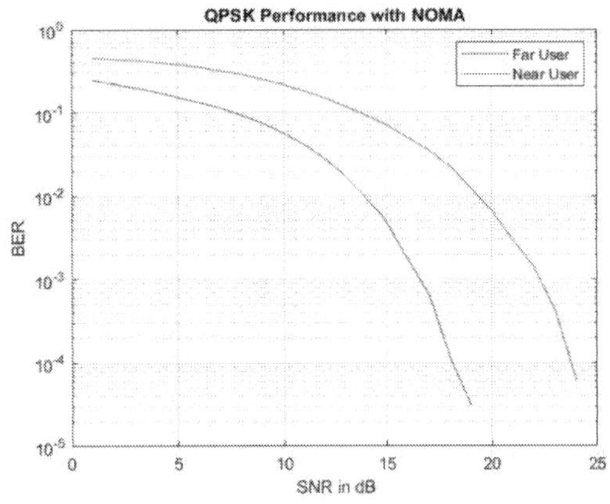

Figure 4. QPSK Random power allocation

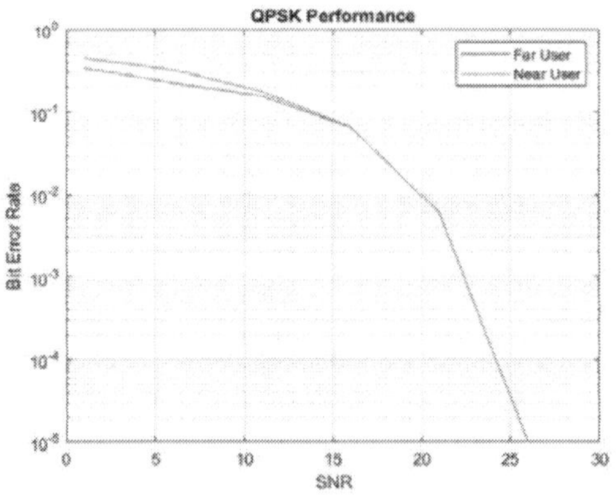

Figure 5. QPSK Dynamic power allocation

Bit Error Rate (BER) result for Quadrature Phase Shift Keying (QPSK) random power allocation technique and dynamic power allocation is shown in Figure 4 and Figure 5. The error rate for QPSK random power allocation technique was similar to that of the BPSK random power allocation technique. But in QPSK random power allocation technique the BER becomes negligible only for higher SNR values compared to BPSK random power allocation. the error rates for random and optimal power allocation techniques were similar but as in the case of BPSK, the optimal power allocation technique split power between the two users in such a way that the sum-rate was maximized and the rate of neither of the users was allowed to fall below a specified threshold, thus ensuring user fairness. Another common feature in both BPSK and QPSK transmission schemes was that data transmitted with high power was received with lower error rates than the data that was transmitted with low power. It could also be observed that BPSK performs better than QPSK in terms of error rate but QPSK is better in terms of the amount of information transmitted. Hence there is a tradeoff between lower error rates and higher information rates and we need to choose accordingly.

In the case of random power allocation, when the transmit power is increased gradually, it can be asserted that outage probability decreases. Moreover, the outage probability of a far user is much less when compared to the outage probability of a near user since more power is allotted to the far user than to a near user. In the case of Dynamic power allocation, when the transmit power is increased gradually, it can be observed that the outage probability of both users decreases and overlaps to a certain extent. The outage probability of the far user is only slightly less than the outage probability of the near user since the power allocation is done dynamically using an optimal power allocation algorithm.

In the case of random power allocation, we can observe that when the threshold rate increases gradually the outage probability starts increasing as the number of times the achievable rate falling below the threshold rate increases with an increase in the threshold rate. Except for low threshold rates, both the users seem to follow the same trend. When higher rates are reached, the outage eventually becomes zero for both the users. In the case of Dynamic power allocation, we can observe that the trend is quite different from that of random power allocation. When the threshold rate increases keeping the transmit power constant, the outage probability also increases. For the far user, distance plays an important role so its outage probability is quite high when compared with that of the near user but when compared with random both users fare quite well as in this case the outage never reaches even when the threshold rate is 10 bps/Hz.

5. Conclusion

In this paper, we clarified the unique force assignment strategy for the Non-Orthogonal various access plot. In this manner, the proposed model dissected various parts of NOMA utilizing separate strategies. Henceforth the framework model considers the Rayleigh blurring channel and changing the blurring coefficients. In this manner, the outcome demonstrated the far client has a lower attainable rate than the close to client due to progressive impedance retraction (SIC). In our proposed work, the close to client plays out the flawed SIC abrogation of the sign by taking away the far client information from the got signal. Our paper centers around power designation for NOMA and presented the force assignment factor and limit of the decent force portion. Henceforth our new proposition powerfully dispensing the force for the far client and the objective rate is acquired. By breaking down the Channel State Information, dynamic force allotment not really settled. Dynamic force portion has a lower blackout likelihood than the static force designation technique. Subsequently the feasible rate is changing powerfully by keeping up with the objective rate at the normal worth. Our proposed work is additionally stretched out by thinking about multiple clients.

5.1 Discussion

The essential justification taking on NOMA in 5G owes to its capacity of serving various clients utilizing similar time and recurrence assets. One of the basic procedures utilized here is Power Domain Multiplexing. NOMA takes advantage of superposition coding at the transmitter and progressive impedance dropping (SIC) at the collector, subsequently multiplexing clients in the force space. NOMA is essentially not quite the same as other different access plans which give symmetrical admittance to the clients either on schedule, recurrence, code or space. In NOMA, every client works in a similar band and simultaneously where they are recognized by their force levels. NOMA utilizes superposition coding at the transmitter to such an extent that the progressive obstruction wiping out (SIC) beneficiary can isolate the clients both in the uplink and in the downlink channels. NOMA can be a reasonable innovation for 5G organizations that can further develop energy proficiency and otherworldly effectiveness beating existing Orthogonal Frequency Multiple entrance strategy (OFDMA) and can oblige a colossal tally of Machine-to-Machine specialized gadgets.

5.2 Future work

A key feature of NOMA is that users with better channel conditions have prior information about the messages of other users. User mobility is also an important factor to be considered for future enhancement of the proposed work. The height is varied and adjusted to different ranges in future work.

References

[1] Khales, FA., Hodtani, GA.: An evaluation of the coverage region for downlink Non-Orthogonal Multiple Access (NOMA) based on Power Allocation Factor. Iran Workshop on Communication and Information Theory,1-5 (2017).

[2] Sedtheetorn, P., Chulajata, T.: Spectral efficiency evaluation for non- orthogonal multiple access in Rayleigh fading. 18th International Conference on Advanced Communication Technology (ICACT), 747-750 (2016).

[3] Baghani, M., Parsaeefard, S., Derakhshani, M., Saad, W.: Dynamic Non- Orthogonal Multiple Access and Orthogonal Multiple Access in 5G Wireless Networks. IEEE Transactions on Communications,67(9),6360-6373 (2019).

[4] Selvam, K., Kumar, K.: Energy and Spectrum Efficiency Trade-off of Non- Orthogonal Multiple Access (NOMA) over OFDMA for Machine-to- Machine Communication. Fifth International Conference on Science Technology Engineering and Mathematics, 523-528 (2019).

[5] Chulajata, T., Sedtheetorn, P.: Theoretical analysis on spectral efficiency of non-orthogonal multiple access in Nakagami fading. IEEE International Conference on Control System, Computing and Engineering, 146-149 (2015).

[6] El-Sayed, MM., Ibrahim, AS., Khairy, MM.: Power allocation strategies for Non-Orthogonal Multiple Access. International Conference on Selected Topics in Mobile & Wireless Networking,1-6 (2016).

[7] Ding, Z., Peng, M., Poor, HV.: Cooperative Non-Orthogonal Multiple Access in 5G Systems.IEEE Communications Letters.19(8) 1462-1465(2015).

[8] Selvam, K., Kumar, K.:Green Point Energy-efficient Analysis of Non- Orthogonal Multiple Access (NOMA) over OFDMA. International Conference on Emerging Trends in Information Technology and Engineering, 1-5 (2020).

[9] Benjebbour, A., Saito, Y., Kishiyama, Y., Li, A., Harada, A., Nakamura, T.: Concept and practical considerations of non-orthogonal multiple access (NOMA) for future radio access. International Symposium on Intelligent Signal Processing and Communication Systems, 770-774 (2013).

[10] Stoica, R., De Abreu, GTF., Hara, T., Ishibashi, K.: Massively Concurrent Non-Orthogonal Multiple Access for 5G Networks and Beyond .IEEE Access, 7. 82080-82100 (2019).

[11] Yin Y, Peng Y, Liu M, Yang J, Gui G. Dynamic user grouping-based NOMA over Rayleigh fading channels. IEEE Access. 2019 Aug 9;7:110964-71.

[12] Lai X, Zhang Q, Qin J. Cooperative NOMA short-packet communications in flat Rayleigh fading channels. IEEE Transactions on Vehicular Technology. 2019 Apr 22;68(6):6182-6.

[13] Jain M, Soni S, Sharma N, Rawal D. Performance analysis at far and near user in NOMA

based system in presence of SIC error. AEU-International Journal of Electronics and Communications. 2020 Feb 1;114:152993.
[14] Liu F, Petrova M. Dynamic power allocation for downlink multi-carrier NOMA systems. IEEE Communications Letters. 2018 Jul 3;22(9):1930-3.
[15] Choi M, Kim J, Moon J. Dynamic power allocation and user scheduling for power-efficient and delay-constrained multiple access networks. IEEE Transactions on Wireless Communications. 2019 Jul 26;18(10):4846-58.

AUTHOR INDEX

Achankunju, Annie 192
Alexander, Libu K 156
Anandan, K. 238, 313
Angeline, A Anita 19, 60
Anushree, N 50
Aparanji, V M 265
Arpitha, H 50
Bakshi, S 297
Balaji, Harish 1
Bali, Jyoti 50
Barla, Prashanth 126
Behera, Soumyajit 90
Bhaaskaran, V S Kanchana 19, 60
Bhagya, B P 24
Bhat, Somashekara 126
Bhat, Vinayambika S 102
Ceylan, Osman 175, 202
Chakkunny, Anna Heba 156
Chala, Geleta 289
Chandran, Banusha 60
Chapala, Shravani 10
Chavda, Khyati 135
Chimdi, Jifara 289
David, Merlin Susan 247
Desai, Chaitra 114
Dikshith, C S 114
Divekar, A.V. 215
Dwivedi, Ved Vyas 84
D'Mello, Demian Antony 223
Garg, Ashwani 90
Garg, Shruti 90
Gayathri, S. 313
Geetanjali, B. 1
Geetha, R S 163
George, Thomas 233
Gharpure, D.C. 215
Ghosh, S 297
Giriyapur, Arunkumar 50
Gohil, Bhumikaba 135
Gohil, NB 84
Gopikrishnan, M. 256
Gupta, Amogh 1
Hakeem, M A Abdul 233
Harshavardhana, M N 184
Jain, V 30
Joseph, Jineeth 192
Joshi, Vinod Kumar 126
Kallur, Arati M 24
Kamath, S. 24
Karibasappa, K G 24
Kenea, Lensa Merga 289
Kodada, Basappa B 223
Kumar, K. Reji 77
Kumar, Leevesh 289
Kumar, S Anjan 305
Kumar, T G Keerthan 272, 305
Kumaraswamy, B G 209
Lakshmi, N 30
Lethakumary, B 233
Likhith, H G 272
Lingraj, Deeksha 305
Lohit, E 272
Mahesh, V. 1
Malladi, Lakshmi Swarupa 10
Megha, R 209
Mondal, S 297
Moorthi, M 256
Mudenagudi, Uma 114
Muduli, Arjuna 42
Murasu, M Mummudi 19
Nagaraj, S S 265
Nair, K R M 192
Nayak, Deepak Kumar 42
Nishanka, T 102
Nishanth, H L 272
Nishmitha, M 102
Pal, Srijani 60
Panda, Dhruba C. 42
Parida, Rajeev K 42
patil, G H Smitha 209
Patil, Ujwala 114
Patro, K Rahul 90
Pooja, V 256
Prajwal, D 102
Prashantkumar, M 209
Priya, M R Ragavi 256
Priya, P. Sasi 19
Priyanka, B B 24
Pushpa, K R 163
Raj, Riyanshu 265
Ravikiran, Y T 209
Reddy, Anvitha 24
Renjith, Shini 247
Ritesh, Saurav 265
Rudramurthy, M S 272
Ruikar, S. D. 144

Saha, M....297
Salimath, Divya S....60
Sarade, Shrenik Suresh....144
Satheesh, E.N....77
Satish, Preksha....305
Seetharaman, R....238, 313
Sheikh, A....30
Singh, Sitesh Kumar....289
Soni, K....30
Sujith, Sanjana....19
Sundareswaran, R....1
Sushma, S A....272
Tarannum, Suraiya....184
Tharun, M....238, 313
Thomas, Bintu....156
Thushar, T A....265
Trivedi, Hitesh....184
Udaykumar, K....209
Van Ngoc, Hoang....70
Varghese, Ancy Sara....192
Vasanthakumar, D K....256
Veluri, Sarada....10
Vernekar, Sameeksha M....114
Vinutha, J N....184
Vyas, Jahanvi H....84
Yilmaz, Ihsan....175, 202